CONTROL AND DYNAMIC SYSTEMS

Advances in Theory and Applications

Volume 21

CONTRIBUTORS TO THIS VOLUME

KERIM DEMIRBAŞ

HOSAM E. EMARA-SHABAIK

J. J. HORSTEN

H. L. JONKERS

RUDOLF KERN

REINHART LUNDERSTAEDT

J. A. MULDER

MICHAEL R. SALAZAR

TRIVENI N. UPADHYAY

JERRY L. WEISS

CONTROL AND DYNAMIC SYSTEMS

ADVANCES IN THEORY
AND APPLICATIONS

Edited by
C. T. LEONDES

School of Engineering and Applied Science
University of California
Los Angeles, California

VOLUME 21: NONLINEAR
AND KALMAN FILTERING
TECHNIQUES
Part 3 of 3

1984

ACADEMIC PRESS, INC.
(Harcourt Brace Jovanovich, Publishers)

Orlando   San Diego   New York   London
Toronto   Montreal   Sydney   Tokyo

ACADEMIC PRESS RAPID MANUSCRIPT REPRODUCTION

ACADEMIC PRESS, INC.
Orlando, Florida 32887

United Kingdom Edition published by
ACADEMIC PRESS, INC. (LONDON) LTD.
24/28 Oval Road, London NW1 7DX

LIBRARY OF CONGRESS CATALOG CARD NUMBER: 64-8027

ISBN 0-12-012721-0

PRINTED IN THE UNITED STATES OF AMERICA

84 85 86 87   9 8 7 6 5 4 3 2 1

CONTENTS

Introduction to State Reconstruction of Dynamic Flight-
Test Maneuvers

H. L. Jonkers, J. A. Mulder, and J. J. Horsten

Optimal Filtering and Control Techniques for Torpedo–Ship
Tracking Systems

Reinhart Lunderstaedt and Rudolf Kern

State Estimation of Ballistic Trajectories with Angle-Only
Measurements

Michael R. Salazar

Information Theoretic Smoothing Algorithms for Dynamic
Systems with or without Interference

Kerim Demirbaş

A New Approach for Developing Practical Filters for Nonlinear
Systems

Hosam E. Emara-Shabaik

Applications of Filtering Techniques to Problems in Communication-Navigation Systems

Triveni N. Upadhyay and Jerry L. Weiss

CONTRIBUTORS

Numbers in parentheses indicate the pages on which the authors' contributions begin.

Kerim Demirbaş[1] (175), *School of Engineering and Applied Science, University of California, Los Angeles, California 90024*

Hosam E. Emara-Shabaik (297), *School of Engineering and Applied Science, University of California, Los Angeles, California 90024*

J. J. Horsten[2] (1), *Department of Aerospace Engineering, Delft University of Technology, 2629 HS Delft, The Netherlands*

H. L. Jonkers[3] (1), *Department of Aerospace Engineering, Delft University of Technology, 2629 HS Delft, The Netherlands*

Rudolf Kern (77), *Department of Mechanical Engineering, University of the German Armed Forces, D-2000 Hamburg 70, Federal Republic of Germany*

Reinhart Lunderstaedt (77), *Department of Mechanical Engineering, University of the German Armed Forces, D-2000 Hamburg 70, Federal Republic of Germany*

J. A. Mulder (1), *Department of Aerospace Engineering, Delft University of Technology, 2629 HS Delft, The Netherlands*

Michael R. Salazar (117), *Nichols Research Corporation, Functional Analysis Directorate, Huntsville, Alabama 35802*

Triveni N. Upadhyay[4] (337), *Radio Communication and Navigation Division, The Charles Stark Draper Laboratory, Inc., Cambridge, Massachusetts 02139*

[1]Present address: *Systems and Research Center, Honeywell, Inc., 2600 Ridgeway Parkway, N.E., Minneapolis, Minnesota 55440.*

[2]Present address: *National Aerospace Laboratory (NLR), Department of Aeroelasticity, Anthony Fokkerweg 2, 1059 CM Amsterdam, The Netherlands.*

[3]Present address: *Department of Industrial Research, Ministry of Education and Science, The Hague, The Netherlands.*

[4]Present address: *The Charles Stark Draper Laboratory, Inc., Cambridge, Massachusetts 02139.*

Jerry L. Weiss (337), *Radio Communication and Navigation Division,
The Charles Stark Draper Laboratory, Inc., Cambridge, Massachusetts
02139*

PREFACE

This is the third and final volume of a trilogy devoted to advances in the techniques and technology of the application of nonlinear filters and Kalman filters. These three volumes comprise a modern treatment of the theoretical techniques from basic to the most advanced, a unique treatment of the computational issues, and now, in this third volume, a selection of substantive examples of the techniques and technology of the application of nonlinear filters and Kalman filters.

There are six chapters in this volume. The first contribution, by Jonkers, Mulder, and Horsten, presents a unique and very much needed treatment of postflight data analysis. While there is much in the literature on linear and nonlinear smoothing techniques, what has been needed is a sufficiently complex and significant treatment of the practical application of these techniques; this chapter accomplishes this. Early applications of Kalman filtering techniques first appeared in the technical report literature in aerospace systems. The chapter by Lunderstaedt and Kern is one of the more recent examples of application to other fields, namely marine systems; it should, therefore, be a much welcome addition to the literature.

Problems of control and dynamic systems begin with the sensors involved, for it is from these that the control processing problem derives its essential information. There is an extremely broad and significant class of problems wherein the sensors involve angle-only information because of the passive nature of these sensors. Salazar's chapter constitutes the first textbook treatment of this significant class of problems. It should serve as an essential reference for many years to come.

Many control and dynamic systems have to function in an environment in which the sensors encounter a very wide variety of interference in their measurement processing. Demirbaş presents a powerful, comprehensive treatment of techniques and issues in this important area of application. Efforts to continually improve systems performance focus on, among other issues, continuing to develop filters that reduce the error covariance matrix, particularly in the transient state. The chapter by Emara-Shabaik presents some of the most powerful results yet developed, along with extensive computer studies that support the powerful utility of his techniques. This volume concludes with a chapter by Upadhyay and Weiss on the major new, broadly significant trends in integrated communication and navigation systems.

CONTENTS OF PREVIOUS VOLUMES

Introduction to State Reconstruction
of Dynamic Flight-Test Maneuvers

H. L. JONKERS
J. A. MULDER
J. J. HORSTEN

Department of Aerospace Engineering
Delft University of Technology
Delft, The Netherlands

SYMBOL DEFINITIONS
AND REFERENCE FRAMES

$\hat{A}$ — Estimated value of A

$A(t)$ — System matrix

A_{x_B}, A_{y_B}, A_{z_B} — Specific forces along the X_B, Y_B, and Z_B axes of the body frame of the aircraft F_B

$B(t)$	Input distribution matrix
$C(t)$	Observation matrix
C_{si}	Sidewash correction factor
C_β	Zero shift of the sideslip wind vane
$D(t)$	Matrix in Eq. (164)
e	Estimation error
$E[A]$	Expected value of A
F	Fisher information matrix
F_B	Aircraft body frame of reference
F_E	Earth-fixed reference frame
F_T	Vehicle-carried vertical frame
g_h	Gravitational acceleration
Δh	Altitude variation relative to the initial altitude
I	Identity matrix
$I_x,\ I_y,\ I_z$	Moments of inertia, respectively, about the X_B, Y_B, and Z_B axes of F_B
I_{xz}	Product of inertia
J	Optimization criterion
k	Discrete-time indication (abbreviated notation of t_k)
K^0	Optimal gain
$K(k)$	Kalman filter gain matrix
$K_s(k)$	Kalman smoother gain matrix
$L_B,\ M_B,\ N_B$	Aerodynamic moments about the axes of F_B
L	Likelihood function
m	Aircraft mass
$m,\ \underline{m}$	Scalar or vector observation residual
$\underline{M}(t)$	System output signal
p	Probability density function
$p_B,\ q_B,\ r_B$	Rotation rates about the X_B, Y_B, and Z_B axes of F_B

$P(k\mid\ell)$	Estimation error covariance matrix at time $t = t_k$ based on ℓ measurements
$\underline{q}(t)$	Vector-valued correction for the random errors in the system output-signal measurements $\underline{M}_m(t)$
$Q(t)$	Covariance matrix of $\underline{q}(t)$
$\underline{r}(k)$	Vector in Eq. (101)
$R(k)$	Matrix in Eqs. (173) and (A.1)
$S(t)$	Matrix in Eq. (202)
t	Time
Δt, T	Sampling-time interval
$\underline{U}(t)$	Vector-valued input signal
$\underline{V}$	Air velocity
V_{x_B}, V_{y_B}, V_{z_B}	Components of $\underline{V}$ in F_B
$\underline{w}(t)$	Vector-valued correction for the random errors in the input-signal measurements $\underline{U}_m(t)$
W	Aircraft weight
$W(t)$	Covariance matrix of $\underline{w}(t)$
$\underline{x}$	Vector-valued state perturbation
x_E, y_E, z_E	Aircraft position relative to F_E
x_{β_B}, y_{β_B}	Position coordinates of the β_v wind vane in the X_B-Y_B plane of F_B
X_B, Y_B, Z_B	Aerodynamic forces along the axes of F_B
$\underline{X}(t)$	System state vector
$\underline{y}(t)$	Discrete-time state perturbation correction vector
$\underline{Y}(t)$	Augmented state or parameter vector
α	Angle of attack
α_v	Wind vane angle of attack
β	Angle of sideslip
β_v	Wind vane angle of sideslip
γ	Flight-path angle
$\Gamma(k, k-1)$	Disturbance distribution matrix

δ_d	Dirac delta function
$\delta_{k,j}$	Kronecker delta
δ_e	Elevator angle
δ_a	Aileron angle
δ_r	Rudder angle
$\epsilon(k)$	Random error in Eq. (126), linear model residual (Appendix)
θ	Angle of pitch
$\underline{\lambda}$	Constant input-signal bias error corrections
σ	Standard deviation
τ	Time
φ	Angle of roll
$\Phi(k, k - 1)$	Transition matrix
ψ	Angle of yaw
$\partial/\partial \underline{x}$	Row vector of partial derivatives

SUBSCRIPT NOTATION

m	Measured magnitude
nom	Nominal
T	Matrix or vector transpose

REFERENCE FRAMES (SEE FIG. 1)

Aircraft Body-Fixed
Reference Frame F_B

The aircraft body-fixed reference frame F_B is a right-handed system $O_B X_B Y_B Z_B$ of orthogonal axes. The origin O_B of F_B lies in the aircraft center of gravity. The $X_B O_B Z_B$ plane co-incides with the aircraft plane of symmetry. The positive X_B axis points forward. The positive Y_B axis points to the right. The positive Z_B axis points downward.

Earth-Fixed Reference
Frame F_E

The earth-fixed reference frame F_E is a right-handed system $O_E X_E Y_E Z_E$ of orthogonal axes. The positive Z_E axis points downward. The positive Y_E axis points to the east.

Vehicle-Carried Vertical
Reference Frame F_T

The vehicle-carried vertical reference frame F_T is a right-handed system $O_T X_T Y_T Z_T$ of orthogonal axes. The origin O_T of F_T lies in the aircraft center of gravity. The positive Z_T axis points downward. The positive X_T axis is in the aircraft plane of symmetry and points forward.

I. INTRODUCTION

Since the early 1960s, the Department of Aerospace Engineering of Delft University of Technology has been engaged in the development of methods to derive aircraft performance as well as stability and control characteristics from dynamic flight-test data applying high-accuracy instrumentation techniques [1]. Traditional methods of performance testing are based on measurements in steady-straight flight conditions in which the aircraft experiences neither translational nor angular accelerations.

A limited number of efforts to derive aircraft performance from measurements in nonsteady flight have been reported in the literature [2,3,4], but these have probably suffered from inadequate instrumentation. Stability and control characteristics have been derived from separate dynamic response measurements. An important reduction in flight time may be achieved when all characteristics of interest are derived from measurements in dynamic flight. The corresponding flight-test

technique developed at Delft University of Technology has been
described in [5-14].

A. *FLIGHT-TEST TECHNIQUE*

In general, successful application of dynamic flight-test
techniques depends on a thoughtful combination of the aircraft
to be tested, the flight-test instrumentation system, the signals applied to excite the aircraft, the models selected for
identification, and the procedures devised to analyze test
data.

The flight-test technique developed at Delft University of
Technology hinges on accurate measurement of several inertial
variables, e.g., specific aerodynamic forces and angular rates
and such barometric variables, as total and static air pressure.
This flight-test method includes the following:

1. Utilization of a high-accuracy flight-test instrumentation system, comprising high-quality inertial and barometric
sensors [15,16]

2. Careful calibration of all transducers to be used in
the flight-test instrumentation system [17]

3. Analytic or computer-aided development of optimal maneuver shapes, i.e., optimal time histories for the control
surface deflections required to excite the aircraft so as to
maximize the amount of information in the measurements [6,18,
19]

4. Excitation of the aircraft manually or under servo control (according to the optimal test signals developed) during
test flights flown in fine weather

5. Off-line analysis of the measurements recorded in flight, using advanced state and parameter estimation techniques

B. *FLIGHT-TEST DATA ANALYSIS*

To provide an outline of the procedure used for the analysis of dynamic flight-test data, the following procedure steps are distinguished:

1. Includes transformation of transducer output voltages accurately measured, periodically sampled, digitized, and recorded in flight into physical magnitudes, using the results of laboratory calibrations of the flight-test instrumentation system and applying various corrections

2. Results in accurate reconstruction of the motions of the aircraft in symmetric or asymmetric dynamic flight from the flight-test measurements recorded

3. Directed toward the identification of the aerodynamic model of the aircraft, using the flight-path reconstruction results. This includes specification of the relations between the aircraft's state and control variables and the resulting aerodynamic forces and moments, as well as estimation of the aerodynamic derivatives governing these relations [13,14]

4. Comprises the derivation of aircraft performance as well as stability and control characteristics, either from the aerodynamic model obtained or from the reconstructed state variable time histories [8,9]

In the present work attention is focused on the step 2 of the dynamic flight-test data-analysis procedure. Reconstruction of the motions of the aircraft corresponds to the estimation of the state vector trajectory and is indicated as "flight-path reconstruction" and "compatibility check" in the literature (e.g., [8,20]).

Measurement of aircraft performance characteristics such as polar drag curves or steady-state climb performance, dictates a very accurate flight-path reconstruction. This explains the importance of flight-path reconstruction in dynamic flight-test data analysis [7,21].

C. *ORGANIZATION OF THIS ARTICLE*

Section II is devoted to the description of the dynamic system models used for flight-path reconstruction. Basic principles and concepts underlying the data-reduction procedures are introduced and discussed in Section III. In addition, a survey of the procedures applicable to flight-path reconstruction is given in Section III. Results obtained from processing actual flight-test measurements are presented in Section IV.

II. DEVELOPMENT OF THE SYSTEM MODEL

Aircraft state estimation or flight-path reconstruction can be formulated as being the problem of generating knowledge about the motion of the aircraft from onboard measurements. The solution of this problem starts with the development of a mathematical model representing the kinematics of the motion of the body-fixed reference frame F_B with respect to earth. Because the mathematical model can be interpreted as representing a dynamical system, the model in the sequence is indicated as the system model.

Section II.A starts with the development of a simplified model of aircraft motion with respect to earth. In Section II.B this model is interpreted as representing a dynamical system, and state, input, and observation vectors are defined. Finally, input and observation measurement error models are postulated in Section II.C.

A. EQUATIONS OF MOTION

The equations governing the aircraft's dynamics are first discussed and rearranged, taking account of the measurement techniques applied in dynamic flight testing. The translational dynamics of the aircraft are described by the following equations relating all relevant forces such as the aerodynamic forces X_B, Y_B, and Z_B and the centrifugal and gravitational forces:

$$X_B = m(\dot{V}_{x_B} + q_B V_{z_B} - r_B V_{y_B}) + W \sin \theta, \tag{1}$$

$$Y_B = m(\dot{V}_{y_B} + r_B V_{x_B} - p_B V_{z_B}) - W \cos \theta \sin \varphi, \tag{2}$$

$$Z_B = m(\dot{V}_{z_B} + p_B V_{y_B} - q_B V_{x_B}) - W \cos \theta \cos \varphi. \tag{3}$$

The rotational dynamics of the aircraft are represented by the following equations:

$$L_B = I_x \dot{p}_B + (I_z - I_y) q_B r_B - I_{xz} (\dot{r}_B + p_B q_B), \tag{4}$$

$$M_B = I_y \dot{q}_B + (I_x - I_z) r_B q_B + I_{xz} (p_B^2 - r_B^2), \tag{5}$$

$$N_B = I_z \dot{r}_B + (I_y - I_x) p_B q_B - I_{xz} (\dot{p}_B - r_B q_B). \tag{6}$$

In these relations the effects of rotating propellers or turbines have been neglected. Furthermore, I_{xy} and I_{yz} have been taken equal to zero because the aircraft is assumed to be symmetric.

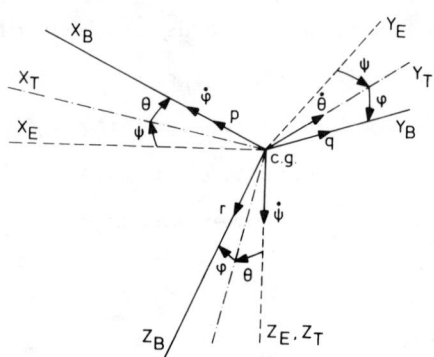

Fig. 1. The body frame of the aircraft of reference F_B relative to the vehicle-carried vertical frame F_T and earth-fixed reference frame F_E.

The orientation of F_B with respect to the earth-fixed reference frame F_E can be expressed in terms of the Euler angles ψ, θ, and φ (Fig. 1). These attitude angles are directly related to the rotation rates p_B, q_B, and r_B according to the following set of kinematic relations:

$$\dot{\psi} = q_B(\sin \varphi/\cos \theta) + r_B(\cos \varphi/\cos \theta), \qquad (7)$$

$$\dot{\theta} = q_B \cos \varphi - r_B \sin \varphi, \qquad (8)$$

$$\dot{\varphi} = p_B + q_B \sin \varphi \tan \theta + r_B \cos \varphi \tan \theta. \qquad (9)$$

It is interesting to note that Eqs. (1)-(9) can be solved by means of numerical integration if (1) the aerodynamic forces X_B, Y_B, and Z_B; (2) the aerodynamic moments L_B, M_B, and N_B; (3) the aircraft weight; (4) the moments and products of inertia; and (5) the initial condition $V_{x_B}(0)$, $V_{y_B}(0)$, $V_{z_B}(0)$, $p_B(0)$, $q_B(0)$, $r_B(0)$, (0), $\varphi(0)$, and $\theta(0)$ are known. Solution of these equations yields time histories of the variables $V_{x_B}(t)$, $V_{y_B}(t)$, $V_{z_B}(t)$, $p_B(t)$, $q_B(t)$, $r_B(t)$, $\psi(t)$, $\varphi(t)$, and $\theta(t)$.

It is even more important to note that the dynamic equations (1)-(3) and the kinematic equations (7)-(9) can be numerically solved independent of Eqs. (4)-(6), which represent the rotational dynamics, if the rotation rates p_B, q_B, and r_B are directly measured. In that case knowledge of the aerodynamic moments and integration of the corresponding equations is no longer necessary.

The aerodynamic forces X_B, Y_B, and Z_B can be written in terms of specific aerodynamic forces A_{x_B}, A_{y_B}, and A_{z_B}:

$$A_{x_B} \triangleq X_B/m, \tag{10}$$

$$A_{y_B} \triangleq Y_B/m, \tag{11}$$

$$A_{z_B} \triangleq Z_B/m. \tag{12}$$

In the hypothetical case of a rigid aircraft, the specific aerodynamic forces are the quantities sensed by accelerometers in the center of gravity of the aircraft. Substitution of Eqs. (10)-(12) into Eqs. (1)-(3) results in the following set of kinematic equations:

$$\dot{V}_{x_B} = A_{x_B} - g_h \sin\theta - q_B V_{z_B} + r_B V_{y_B}, \tag{13}$$

$$\dot{V}_{y_B} = A_{y_B} + g_h \cos\theta \sin\varphi - r_B V_{x_B} + p_B V_{z_B}, \tag{14}$$

$$\dot{V}_{z_B} = A_{z_B} + g_h \cos\theta \cos\varphi - p_B V_{y_B} + q_B V_{x_B}. \tag{15}$$

The position of the aircraft's center of gravity relative to F_E is governed by the following set of equations:

$$\dot{x}_E = V_{x_B} \cos\theta \cos\psi + V_{y_B} (\sin\varphi \sin\theta \cos\psi - \cos\varphi \sin\psi)$$

$$+ V_{z_B} (\cos\varphi \sin\theta \cos\psi + \sin\varphi \sin\psi), \tag{16}$$

$$\dot{y}_E = V_{x_B} \cos\theta \sin\psi + V_{y_B}(\sin\varphi \sin\theta \sin\psi + \cos\varphi \cos\psi)$$

$$+ V_{z_B}(\cos\varphi \sin\theta \sin\psi - \sin\varphi \cos\psi), \qquad (17)$$

$$\dot{z}_E = -V_{x_B} \sin\theta + V_{y_B} \sin\varphi \cos\theta + V_{z_B} \cos\varphi \cos\theta. \qquad (18)$$

Now if the inertial quantities A_{x_B}, A_{y_B}, A_{z_B}, p_B, q_B, and r_B could be measured with absolute accuracy as a function of time, and if the relevant initial conditions were exactly specified, the flight-path reconstruction problem would reduce to integration of the first-order nonlinear differential equations (7)-(9) and (13)-(18). Evidently, the measurements recorded in flight are never completely free of errors. Moreover, in actual flight an initial condition can never be exactly determined. Application of estimation procedures is therefore required to attenuate the effects of these errors.

B. SYSTEM STATE AND OBSERVATION EQUATIONS

The complexity of the multivariable system, represented by Eqs. (7) - (9) and (13) - (18), necessitates its description in terms of vectors and matrices. Consequently, relevant estimation procedures are also described by use of the language and rules of vector-matrix algebra.

Representation of the dynamics of the aircraft in terms of a state vector equation is established to set the stage for the design of the estimation procedure applied for flight-path reconstruction from actual on-board measurements. We introduce the vector-valued quantities

$$\underline{X} \triangleq \text{col}[V_{x_B}, V_{y_B}, V_{z_B}, \psi, \theta, \varphi, x_E, y_E, z_E], \qquad (19)$$

$$\underline{U} \triangleq \text{col}[A_{x_B}, A_{y_B}, A_{z_B}, p_B, q_B, r_B]. \qquad (20)$$

The kinematic relations (7)-(9) and (13)-(18) may be represented together by the following vector differential equation:

$$\dot{\underline{X}}(t) = \underline{f}(\underline{X}(t), \underline{U}(t)), \tag{21}$$

$$\underline{X}(t_0) = \underline{X}(0). \tag{22}$$

The quantity $\underline{X}(t)$ is referred to as the state vector of the dynamic system under consideration. The vector-valued quantity $\underline{U}(t)$ is called the system input signal. The vector function $\underline{f}$ is of the same dimension as $\underline{X}(t)$.

If the state vector $\underline{X}$ and input vector $\underline{U}$ are known, several related quantities such as airspeed V, altitude variation Δh, and sideslip vane angle β_v can be computed according to

$$V = \left(V_{x_B}^2 + V_{y_B}^2 + V_{z_B}^2 \right)^{1/2}, \tag{23}$$

$$\Delta h = -z_E, \tag{24}$$

$$\beta_v = \arctan\left(\frac{V_{y_B} + x_{\beta_B} r_B - z_{\beta_B} p_B}{V_{x_B}} \right) + C_{si} \arctan\left(\frac{V_{y_B}}{V_{x_B}} \right) - C_\beta. \tag{25}$$

Here x_{β_B} and z_{β_B} are the coordinates of the wind vane relative to F_B; C_{si} is the sidewash coefficient; and C_β is the zero shift of the wind vane. The parameters C_β and C_{si} account for the effect of aircraft-induced air-velocity components and should either be given or be estimated from flight-test data (see Section IV). The quantities V, Δh, and β_v are components of a vector-valued system output signal $\underline{M}$:

$$\underline{M} \triangleq \text{col}[V, \Delta h, \beta_v]. \tag{26}$$

Application of high-accuracy instrumentation techniques in dynamic flight testing result in very accurate measurements of the components of the system input vector $\underline{U}$. The components of $\underline{M}$ may also be measured in dynamic flight, although with inherently less accurate barometric and vane-angle transducers.

It is emphasized here that the composition of $\underline{M}$ given in Eq. (26) should be considered merely as a possible example of the output-signal observation configuration. Other variables, such as the angle of yaw ψ, the angle of roll φ, the wind-vane angle of attack α_v, DME-based aircraft-position fixes, or inertial platform measured velocity components may also be considered as system output signals.

Equations (23) - (25) may be written as

$$\underline{M}(t) = \underline{h}(\underline{X}(t), \underline{U}(t)). \tag{27}$$

When Eq. (21) was integrated directly as suggested in Section II.A, calculated output signals could be compared to the measured output signals.

From the deviations between the calculated and the measured outputs, "information" can be extracted concerning the deviations of the calculated or estimated state $\underline{X}(t)$ from the actual state $\underline{X}(t)$. The estimation procedures to be discussed center around optimal use of this information. Equations (21) and (27) are called, respectively, the system state equation and the system observation equation.

C. ERROR MODELS

The output of the sensors used for measurement of the system input- and output-signal components are assumed to be corrupted with time-dependent errors. More precisely the accelerometer and rate gyroscope measurements are assumed to be contaminated with constant bias errors as well as with random measurement errors. The barometric measurements are assumed to be corrupted only with random measurement errors, since

short-circuiting of the pneumatic sensor systems, before and
after each test-flight maneuver, allows for postflight compen-
sation of possible bias errors. This leads to the following
measurement-error models:

$$\underline{U}_m(t) \triangleq \underline{U}(t) - \underline{\lambda} - \underline{w}(t), \tag{28}$$

$$\underline{M}_m(t) \triangleq \underline{M}(t) - \underline{q}(t). \tag{29}$$

The vector-valued quantities $\underline{\lambda}$, $\underline{w}(t)$, and $\underline{q}(t)$ are defined,
respectively, as

$$\underline{\lambda} \triangleq col[\lambda_x, \lambda_y, \lambda_z, \lambda_p, \lambda_q, \lambda_r], \tag{30}$$

$$\underline{w} \triangleq col[w_x, w_y, w_z, w_p, w_q, w_r], \tag{31}$$

$$\underline{q} \triangleq col[q_V, q_{\Delta h}, q_{\beta_V}]. \tag{32}$$

These bias errors are assumed constant, hence

$$\dot{\underline{\lambda}} = 0. \tag{33}$$

The time-dependent random errors $\underline{w}(t)$ and $\underline{q}(t)$ are considered
to be Gaussian white-noise processes having the following sta-
tistical properties.

(1) The time averages are assumed to be equal to zero:

$$E[\underline{w}(t)] = 0, \tag{34}$$

$$E[\underline{q}(t)] = 0, \tag{35}$$

for all t.

(2) The random error variances satisfy

$$E[\underline{w}(t)\underline{w}^T(\tau)] = W(t)\delta_d(t - \tau), \tag{36}$$

$$E[\underline{q}(t)\underline{q}^T(\tau)] = Q(t)\delta_d(t - \tau). \tag{37}$$

Here $\delta_d(t - \tau)$ denotes the Dirac delta function with the pro-
perties $\delta_d(t - \tau) = \infty$ for $t = \tau$; $\delta_d(t - \tau) = 0$ for $t \neq \tau$; and
$\int_{-\infty}^{\infty} \delta_d(t - \tau)d\tau = 1$.

As stated in Section II.A, any estimate of the initial state $\underline{X}(0)$ will deviate from the actual initial state $\underline{X}(0)$. This results in an initial state estimation error $\underline{x}(0)$ defined as

$$\underline{x}(0) = \underline{X}(0) - \hat{\underline{X}}(0). \tag{38}$$

Accurate reconstruction of the motions of the aircraft in flight from the recorded time histories of measurements of the input and observation vectors centers around attenuation of the effects of these measurement errors. In addition, the vector of bias error corrections $\underline{\lambda}$ as well as the parameters C_β and C_{si} will have to be estimated simultaneously when no a priori values are known.

III. FLIGHT-PATH RECONSTRUCTION:
      PRINCIPLES AND METHODS

The algorithms applied to flight-path reconstruction from on-board measurements are discussed in this section. Extremely simple example problems are analyzed to illustrate the principles underlying these algorithms and to introduce fundamental concepts such as nominal trajectory, batch and recursive estimation of parameters or system states, the augmented state vector, and effects of system input and output noise. In these example problems the actual aircraft flight-path reconstruction problem is drastically simplified to exhibit what is measured, what can be directly computed from these measurements, and what is to be estimated by application of statistical principles and related algorithms. The reader who is familiar with basic concepts of estimation theory may want to skip Section III.A.

A. *BASIC SYSTEM DEFINITION*

In this section a description is given of a simple scalar linear dynamic system. This system description is presented to set the stage for the problems to be discussed in the following sections.

Let the scalar variable $X(t)$ denote the state of a one-dimensional linear system with input $U(t)$ and output $M(t)$. The evolution of the linear system state $X(t)$, as a result of the system input $U(t)$ and the corresponding output $M(t)$, are described by the following equations:

$$\dot{X}(t) = AX(t) + BU(t), \tag{39}$$

$$M(t) = CX(t), \tag{40}$$

where A, B, and C are scalar system parameters.

Solution of Eq. (39) obviously requires specification of the initial state:

$$X(t_0) \triangleq X(0). \tag{41}$$

In addition, the input $U(t)$ should be given for $t \geq 0$ (see Fig. 2).

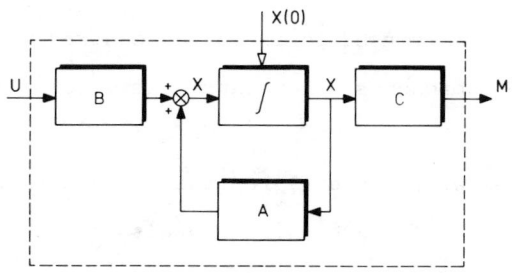

Fig. 2. Block diagram of the scalar linear system $\dot{X} = AX + BU$, $M = CX$.

According to linear system theory, the solution of Eq. (39), i.e., the state of the system as a function of time, may then be computed as follows:

$$X(t) = \exp[A(t - t_0)]X(0) + \int_0^t \exp[A(t - \tau)]BU(\tau)d\tau$$

$$= \Phi(t, t_0)X(0) + \int_0^t \Gamma(t, \tau)U(\tau)d\tau, \qquad (43)$$

where

$$\Gamma(t, \tau) = \exp[A(t - \tau)], \qquad B = \Phi(t, \tau)B. \qquad (44)$$

The output signal is given by Eq. (40). Definitions for the state transition parameter $\Phi(t, t_0)$ and the impulse response parameter $\Gamma(t, \tau)$ occurring in Eq. (43) are apparent from that equation.

Now suppose that the system input U and the system output M are measured and that these measurements are corrupted with errors as follows[1]:

$$U_m(t) = U(t) - \lambda - w(t), \qquad (45)$$

$$M_m(t) = M(t) - q(t), \qquad (46)$$

The subscript m is used to indicate measured magnitudes. The parameter λ represents a constant bias error correction, hence,

$$\dot{\lambda} = 0. \qquad (47)$$

The variables $w(t)$ and $q(t)$ represent measurement noise corrections. Both of these noise signals are assumed to Gaussian, white, and zero mean, hence,

$$E[w(t)] = E[q(t)] = 0 \qquad \text{for all} \quad t. \qquad (48)$$

---

[1]In Eqs. (45), (46), and subsequent expressions the minus sign indicates that λ, $w(t)$ and $q(t)$ are interpreted as corrections to be added to the measurements $U_m(t)$ and $M_m(t)$, respectively [see also Eqs. (28) and (29)].

The second-order moments of the noise statistics are specified as follows:

$$E[w(t)w(\tau)] = W\delta_d(t - \tau), \tag{49}$$

$$E[w(t)q(\tau)] = Q\delta_d(t - \tau). \tag{50}$$

In accordance with present-day flight-test-instrumentation technology, the system input and output signals are assumed to be sampled periodically in time. Consequently, Eqs. (45) and (46) are written as

$$U_m(t_k) = U(t_k) - \lambda - w(t_k), \tag{51}$$

$$M_m(t_k) = M(t_k) - q(t_k), \tag{52}$$

in which t_k denotes a discrete time instant. To simplify the notation only the sequence number k is used instead of the notation t_k to indicate the sample of interest. Discretization implies that the Gaussian noise processes $w(t)$ and $q(t)$ are approximated by the Gaussian random sequences $w(k)$ and $q(k)$, with statistical first- and second-order moments:

$$E[w(k)] = E[q(k)] = 0 \quad \text{for all } k, \tag{53}$$

$$E[w(k)w(l)] = W\delta_{k,l}, \tag{54}$$

$$E[q(k)q(l)] = Q\delta_{k,l}. \tag{55}$$

The Dirac delta function has been replaced by the Kronecker delta function in these expressions, where $\delta_{k,l} = 1$ for $k = l$ and $\delta_{k,l} = 0$ otherwise. If discretization of the input-signal time history $U(\cdot)$ according to

$$U(\tau) = U(k - 1) \quad \text{for } \tau \in [t_{k-1}, t_k)$$

is acceptable, Eqs. (43) and (40) may be replaced by

$$X(k) = \Phi(k, k - 1)X(k - 1) + \Gamma(k, k - 1)U(k - 1), \tag{56}$$

$$M(k) = CX(k). \tag{57}$$

Now the task is set to reconstruct the system state $X(k)$ from the measurements $U_m(i)$ and $M_m(i)$ for $i < k$, $i = k$ or $i > k$. In the following sections further simplifications of the system just delineated are introduced to facilitate the explanation of elementary problems and solution principles relevant to flight-path reconstruction.

*Example 1. Estimation
            of the Initial State*

In this example flight-path reconstruction, or more precisely, estimation of the trajectory of the system state is reduced to calculation of the initial condition $X(0)$ followed by direct integration of the system state equation [see Eqs. (43) and (56)]. For clarity, the linear system model defined earlier is simplified as much as possible. The system parameter A is assumed equal to 0, whereas the parameters B and C are supposed to be equal to 1 (see Fig. 3). Hence,

$$A \triangleq 0, \tag{58}$$

$$B = C \triangleq 1. \tag{59}$$

It follows that the system parameters Φ and Γ satisfy

$$\Phi(k, k - 1) = 1, \tag{60}$$

$$\Gamma(k, k - 1) = t_k - t_{k-1} = T. \tag{61}$$

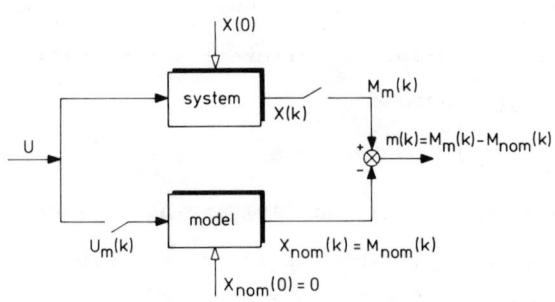

Fig. 3. The simplified system of Example 1.

To simplify even further the estimation problem, the system input U is assumed constant in time. The measurements $U_m(k)$ and $M_m(k)$ are supposed to be free of any errors:

$$U_m(k) = U, \tag{62}$$

$$M_m(k) = M(k). \tag{63}$$

[Compare with Eqs. (45) and (46).]

To solve this almost trivial problem, the concept of nominal quantities is introduced. Nominal quantities are those computed directly from the input-signal measurements recorded. First of all, the nominal system state $X_{nom}(k)$ is computed. In principle, $X_{nom}(0)$ may be given any value. When, for example,

$$X_{nom}(0) = 0, \tag{64}$$

the nominal system state is computed as

$$X_{nom}(k) = X_{nom}(0) + \sum_{j=0}^{k-1} U_m(j)T = \sum_{j=0}^{k-1} U_m(j)T. \tag{65}$$

Since $U_m(k) = U$ for all k, it follows that

$$X_{nom}(k) = UkT. \tag{66}$$

The nominal output-signal observations are

$$M_{nom}(k) = X_{nom}(k) = UkT. \tag{67}$$

The measured output-signal observations are

$$M_m(k) = X(k) = X(0) + UkT. \tag{68}$$

Comparison of $M_m(k)$ with $M_{nom}(k)$ reveals the required information concerning the initial condition $X(0)$ to be (see Fig. 4)

$$m(k) \triangleq M_m(k) - M_{nom}(k) = X(t) - X_{nom}(t)$$

$$= X(0) + UkT - UkT = X(0). \tag{69}$$

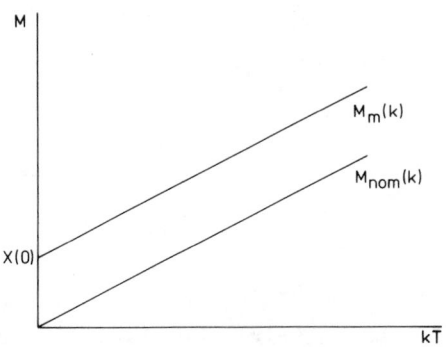

Fig. 4.   Nominal and measured output signal of Example 1.

Consequently, in this very simple deterministic example, the
initial condition X(0) can be determined from a single measurement of the system input U and the system output M(k) for any
value of k.   Evidently, the system state X(k) can now be computed for all k, according to

$$X(k) = X(0) + UkT,$$ (70)

or, in more general terms, by integration of the system input
signal.

To summarize, in this example the flight-path reconstruction problem has been reduced to a straightforward calculation
of the initial state, followed by integration of the system
input signal.   Essential to flight-path reconstruction from
on-board measurements is the notion of nominal quantities and
the fact that comparison of nominal output signals with actual
measurements of the output signal reveals most relevant information concerning the parameters to be estimated.

*Example 2.   Estimation of the Initial
            State from Noise-Corrupted
            Measurements*

The problem discussed previously in Section III.A.1 becomes
slightly more complicated by the assumption that the output-
signal measurements $M_m(k)$ for k = 1, 2, ..., n are corrupted

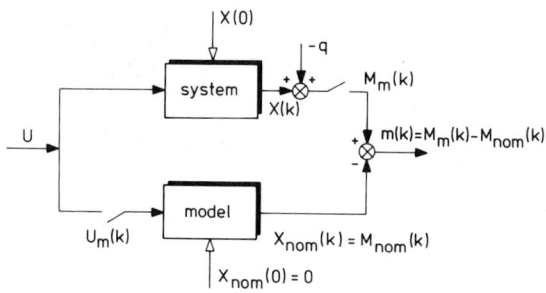

Fig. 5. The system of Example 2, allowing for observation noise.

with random errors $-q(k)$ (see Fig. 5). Now it is no longer desirable to determine $X(0)$ from a single pair of measurements
$U_m(k)$ and $M_m(k)$. To obtain the information required for estimation of $X(0)$, the application of statistical methods is
unavoidable.

When the output-signal measurement errors are additive, Eq.
(63) is replaced by

$$M_m(k) = M(k) - q(k), \tag{71}$$

in which the random errors $-q(k)$ are considered to be an uncorrelated Gaussian sequence [see Eqs. (53) and (55)]. For
state trajectory reconstruction, once again the task is set
first to estimate $X(0)$ from the available measurements $U_m(k)$
and $M_m(k)$ for $k = 0, 1, 2, \ldots, n$. Again, nominal observations
$M_{nom}(k)$ can be computed (see Fig. 6):

$$M_{nom}(k) = X_{nom}(k) = UkT. \tag{67}$$

The output-signal measurements $M_m(k)$ are related to the initial
state $X(0)$ according to

$$M_m(k) = X(k) - q(k) = X(0) + UkT - q(k). \tag{72}$$

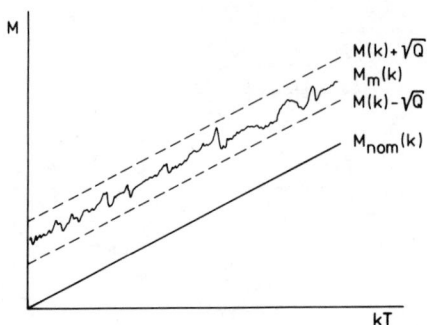

Fig. 6. Nominal and measured output signal of Example 2.

Subtraction of the nominal output-signal observations $M_{nom}(k)$, derived from the input-signal measurements $U_m(k)$, from the actual output-signal measurements $M_m(k)$ yields

$$m(k) = M_m(k) - M_{nom}(k) = X(0) - q(k). \tag{73}$$

In the following, $m(k)$ is referred to as observation residual. Because the random errors $-q(k)$ are supposed to be zero mean [see Eq. (53)], and in order to have a Gaussian, and hence symmetric distribution, it seems reasonable to estimate the initial state $X(0)$ by averaging n residuals $m(k)$. This estimate is written as $\hat{X}(0|n)$ and

$$\hat{X}(0|n) = \frac{1}{n} \sum_{k=1}^{n} m(k). \tag{74}$$

The resulting estimation algorithm is called linear because a linear combination of the available measurements is made. Because $\hat{X}(0|n)$ is derived at once from a batch of n observations $m(k)$, it may also be referred to as a batch algorithm. Alternatively, a recursive algorithm may be derived in which the measurements are processed one by one. From the batch algorithm given previously, it follows that

$$(n + 1)\hat{X}(0|n + 1) - n\hat{X}(0|n + 1) = m(n + 1), \tag{75}$$

hence

$$\hat{X}(0|n + 1) = [n/(n + 1)]\hat{X}(0|n + 1) + (n + 1)^{-1}m(n + 1)$$

$$= [1 - (n + 1)^{-1}]\hat{X}(0|n) + (n + 1)^{-1}m(n + 1).$$

$$(76)$$

Defining the confidence factor;

$$K(n + 1) \triangleq (n + 1)^{-1}, \tag{77}$$

combination of Eqs. (76) and (77) yields

$$\hat{X}(0|n + 1) = (1 - K(n + 1))\hat{X}(0|n)$$

$$+ K(n + 1)m(n + 1), \tag{78}$$

or

$$\hat{X}(0|n + 1) = \hat{X}(0|n) + K(n + 1)$$

$$\times [m(n + 1) - \hat{X}(0|n)]. \tag{79}$$

Because Eqs. (78) and (79) return frequently in the subsequent text, it is highly relevant to discuss these equations in more detail.

The factors $1 - K(n + 1)$ and $K(n + 1)$ that occur in Eq. (78) may be considered as confidence factors or weighting factors. This means that $\hat{X}(0|n + 1)$ is a weighted combination of the preceding estimate $\hat{X}(0|n)$ and the next observation residual $m(n + 1)$. If n is very small, the estimate $\hat{X}(0|n + 1)$ depends on the new measurement as well as on the preceding estimate $\hat{X}(0|n)$. For large n, however, $\hat{X}(0|n + 1)$ is almost entirely determined by $\hat{X}(0|n)$, and the effect of new measurements vanishes as $n \to \infty$. Equation (79) shows that the recursive algorithm may be considered as a kind of prediction-correction scheme. When $m(n + 1)$ is recorded, the best available estimate of $X(0)$ is $\hat{X}(0|n)$. Hence $\hat{X}(0|n)$ can be considered as the best prediction of $m(n + 1)$ obtained by processing all measurements up to and including $m(n)$. The deviation between $m(n + 1)$ and

$\hat{X}(0|n)$ is then used to correct $\hat{X}(0|n)$. Obviously, the relevant information contained in $m(n + 1)$ is corrupted with the measurement error $-q(n + 1)$.

From Eq. (80), recursive expressions may be derived for the propagation in time of the error in the estimate of the initial state. Let

$$e(0|n + 1) \triangleq X(0) - \hat{X}(0 \ n + 1). \tag{80}$$

Equation (80) may then be rewritten as

$$X(0) - \hat{X}(0|n + 1)$$
$$= X(0) - (1 - K(n + 1))\hat{X}(0|n) - K(n + 1)m(n + 1)$$
$$= (1 - K(n + 1))X(0) - (1 - K(n + 1))\hat{X}(0|n)$$
$$+ K(n + 1)q(n + 1). \tag{81}$$

Hence,

$$e(0|n + 1) = (1 - K(n + 1))e(0|n)$$
$$+ K(n + 1)q(n + 1). \tag{82}$$

A measure for the dispersion or scatter in the estimates $\hat{X}(0|n + 1)$ for all n is given by the variance $E[e^2(0|n + 1)]$. Because the estimation error $e(0|n)$ and the observation measurement error $-q(n + 1)$ may be assumed to be uncorrelated, Eq. (83) may be derived from Eq. (82):

$$P(0|n + 1) \triangleq E[e^2(0|n + 1)]$$
$$= (1 - K(n + 1))^2 P(0|n) + K^2(n + 1)Q. \tag{83}$$

Studying the weighted combination of $\hat{X}(0|n)$ and $m(n + 1)$ [see Eq. (78)], computed to obtain the estimate $\hat{X}(0|n + 1)$, the question arises as to what relation should exist between the best or optimal $K(n + 1)$ on the one hand and $P(0|n)$ and Q on the other, to minimize $P(0|n + 1)$. This relation may be derived from the necessary condition

$$\partial P(0|n + 1)/\partial K(n + 1) = -2(1 - K(n + 1))P(0|n)$$
$$+ 2K(n + 1)Q = 0. \tag{84}$$

Because $\partial^2 P(0|n + 1)/\partial K^2(n + 1) \geq 0$, it follows that $P(0|n + 1)$
is minimal if

$$K^0(n + 1) = P(0|n)/[P(0|n) + Q]. \tag{85}$$

Substitution of this expression for the confidence factor
$K(n + 1)$ in the expression for $P(0|n + 1)$ in Eq. (83) yields
the following recursive relation for the variance in the estimation error:

$$P(0|n + 1) = P(0|n)Q/[P(0|n) + Q]$$
$$= (1 + K^0(n + 1))P(0|n). \tag{86}$$

A recursive algorithm for $K^0(n + 1)$, the optimal confidence
factor given $K^0(0)$, can also be derived according to

$$K^0(n + 1) = \frac{P(0|n)}{P(0|n) + Q} = \frac{1}{1 + Q/P(0|n)}$$

$$= \left[1 + \frac{P(0|n - 1) + Q}{P(0|n - 1)}\right]^{-1} = \frac{1}{1 + 1/K^0(n)} \tag{87}$$

$$= \frac{K^0(n)}{K^0(n) + 1}.$$

In contrast to the batch algorithm, application of recursive
algorithms demands for a priori specification of the initial
estimate $\hat{X}(0|0)$ the variance $P(0|0)$ of the errors in an initial
estimate of $X(0)$ as well as the variance Q of the measurement
errors $-q(k)$.

The question may now arise as to which of the two algorithms — the optimal recursive estimation scheme or the batch
algorithm — generates the most accurate estimate of $X(0)$.

After processing $n + 1$ observation residuals $m(k)$, $k = 1$, 2, $\ldots$, $n + 1$, the error in the estimate $\hat{X}(0|n + 1)$ is

$$e(0|n + 1) = X(0) - \hat{X}(0|n + 1)$$

$$= X(0) - \frac{1}{n + 1} \sum_{k=1}^{n+1} m(k) = \frac{1}{n + 1} \sum_{k=1}^{n+1} - q(k),$$

(88)

[see Eq. (74)]. The variance of the estimation error may then be written as

$$P(0|n + 1) = E\left[\left(\frac{1}{n + 1} \sum_{k=1}^{n+1} - q(k)\right)^2\right] = \frac{1}{n + 1} Q.$$ (89)

This leads to $\pm$root mean square estimation error bounds as depicted in Fig. 7.

When applying the batch estimation scheme, no a priori information concerning $X(0)$ is required. In the case of a recursive estimation scheme this corresponds to

$$P(0|0) = \infty.$$ (90)

This leads to the following asymptotic expressions:

$$\lim_{P(0|0)\to\infty} K^0(1) = \lim_{P(0|0)\to\infty} \frac{P(0|0)}{P(0|0) + Q} = 1,$$ (91)

$$\lim_{P(0|0)\to\infty} P(0|1) = \lim_{P(0|0)\to\infty} \frac{P(0|0)Q}{P(0|0) + Q} = Q.$$ (92)

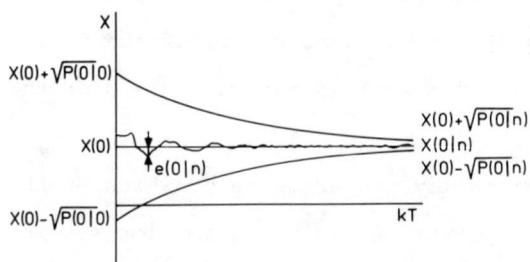

Fig. 7. *Estimate of X(0) and corresponding root mean square error.*

Repeated application of the relation between $P(0|k)$ and
$P(0|k - 1)$ in Eq. (86) for $k = 2, 3, \ldots, n + 1$ results in

$$P(0|n + 1) = (n + 1)^{-1}Q. \tag{93}$$

Hence, when setting $P(0|0) = \infty$, the optimal recursive estima-
tion scheme yields exactly the same estimation accuracy as that
obtained when applying the batch estimation algorithm. A fi-
nite magnitude of $P(0|0)$ affects the error variance $P(0|n)$;
however, as follows from recursive application of Eqs. (86) and
(87) this effect vanishes for $n \to \infty$.

Because the task set was to estimate the system state $X(k)$
after processing a sufficiently large number of measurements
in order to obtain an estimate of the initial state $X(0)$ with
adequate accuracy, the estimate $X(k)$ remains to be computed
according to

$$\hat{X}(k|n + 1) = \hat{X}(0|n + 1) + \sum_{j=0}^{k-1} U_m(j)T$$

$$= \hat{X}(0|n + 1) + UkT. \tag{94}$$

The accuracy of $\hat{X}(k|n + 1)$ may thus be seen to depend in this
particular case only on the accuracy of the initial state esti-
mate $\hat{X}(0|n + 1)$:

$$P(k|n) = E[(X(k) - \hat{X}(k|n))^2] = E[(X(0) - \hat{X}(0|n))^2]$$

$$= P(0|n). \tag{95}$$

To summarize, it should be noted that Example 2 has been
used to introduce statistical estimation concepts that apply
simple averaging of observation residuals $m(k)$ for $k = 1$,
$2, \ldots, n$ between the actual output-signal measurements $M_m(k)$
and the calculated observations $M_{nom}(k)$. In addition, the
concepts of recursive and batch estimation have been exhibited.

*Example 3.   Estimation of Initial
              State and Bias Error*

This example problem differs from Example 2 only in that a
constant bias error is supposed to offset the measurements U_m
of the output signal U in order to facilitate the introduction
of the concept of an augmented vector (see Fig. 8). It should
be noted, however, that here the bias error can be introduced
while conserving model linearity.

The input-signal bias error correction λ was defined in
Eq. (45) as

$$U_m \triangleq U - \lambda. \tag{96}$$

Reconstruction of the time history of the system state X(k)
now requires estimation of the initial state X(0) as well as
estimation of the bias error correction λ. Again, the required information is obtained by comparing nominal observations $M_{nom}(k)$ calculated from the measured input signals $U_m(k)$
with the actual output-signal measurements $M_m(k)$. As in
Examples 1 and 2 (see Fig. 9),

$$M_{nom}(k) = X_{nom}(k) = X_{nom}(0) + U_m kT = (U - \lambda)kT, \tag{97}$$

whereas

$$M_m(k) = X(k) - q(k) = X(0) + UkT - q(k). \tag{98}$$

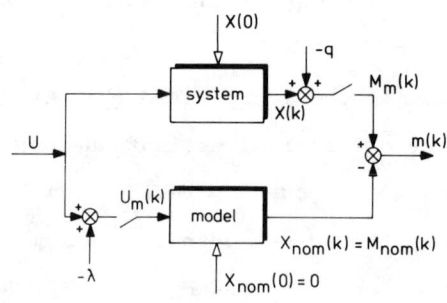

*Fig. 8.   The system of Example 3, taking nto account ob-
servation noise and an input-signal bias error.*

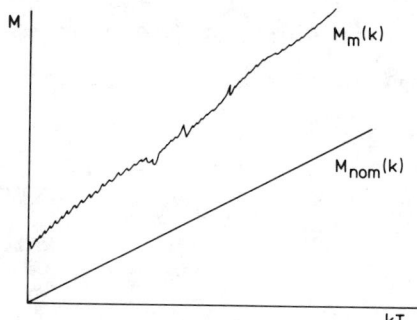

Fig. 9. Nominal and measured output signal of the system of Example 3.

From Eqs. (97) and (98) it follows that

$$m(k) = M_m(k) - M_{nom}(k) = X(0) + \lambda kT - q(k). \tag{99}$$

As components of an augmented parameter vector $\underline{Y}(0)$ the parameters $X(0)$ and λ can now be estimated, processing a batch of, say, n observation residuals $m(k)$, $k = 1, 2, \ldots, n$, with the aid of a statistical estimation algorithm, often referred to as regression analysis. Let

$$\underline{Y}(0) \triangleq \mathrm{col}[X(0), \lambda], \tag{100}$$

$$\underline{r}(k) \triangleq \mathrm{row}[1, kT]. \tag{101}$$

Equation (99) may then be reformulated by writing

$$m(k) = \underline{r}(k)\underline{Y}(0) - q(k). \tag{102}$$

According to regression analysis, the optimal estimate $\hat{\underline{Y}}(0|n)$ minimizing the quadratic cost function

$$J_n = \sum_{k=1}^{n} (m(k) - \underline{r}(k)\hat{\underline{Y}}(0|n))^2 Q^{-1} \tag{103}$$

is then obtained according to (see the Appendix)

$$\hat{\underline{Y}}(0|n) = \left(\sum_{k=1}^{n} \underline{r}^T(k)Q^{-1}\underline{r}(k) \right)^{-1} \sum_{k=1}^{n} \underline{r}^T(k)Q^{-1}m(k). \tag{104}$$

The covariance matrix $P(0|n)$ of the error in the batch estimate $\hat{\underline{y}}(0|n)$ can be derived from Eq. (104), according to

$$\underline{e}(0|n) = \underline{y}(0) - \hat{\underline{y}}(0|n)$$

$$= \underline{y}(0) - \left(\sum_{k=1}^{n} \underline{r}^{T}(k)Q^{-1}\underline{r}(k)\right)^{-1} \sum_{k=1}^{n} \underline{r}^{T}(k)Q^{-1}\underline{m}(k)$$

$$= \underline{y}(0) - \left(\sum_{k=1}^{n} \underline{r}^{T}(k)Q^{-1}\underline{r}(k)\right)^{-1} \sum_{k=1}^{n} \underline{r}^{T}(k)Q^{-1}$$

$$\times (\underline{r}(k)\underline{y}(0) - q(k))$$

$$= \sum_{k=1}^{n} (\underline{r}^{T}(k)Q^{-1}\underline{r}(k))^{-1} \sum_{k=1}^{n} \underline{r}^{T}(k)Q^{-1}q(k). \tag{105}$$

Now,

$$P(0|n) = E[\underline{e}(0|n)\underline{e}^{T}(0|n)]$$

$$= E\left[\left(\sum_{k=1}^{n} \underline{r}^{T}(k)Q^{-1}\underline{r}(k)\right)^{-1} \sum_{k=1}^{n} \underline{r}^{T}(k)Q^{-1}q(k)\right.$$

$$\times \left.\sum_{j=1}^{n} q^{T}(j)Q^{-1}\underline{r}(j)\left(\sum_{k=1}^{n} \underline{r}^{T}(k)Q^{-1}\underline{r}(k)\right)^{-1}\right]. \tag{106}$$

Recalling the fact that $q(k)$ is sequentially uncorrelated, Eq. (55), it follows that

$$P(0|n) = \left(\sum_{k=1}^{n} \underline{r}^{T}(k)Q^{-1}\underline{r}(k)\right)^{-1}. \tag{107}$$

Substitution in Eq. (104) yields

$$\hat{\underline{y}}(0|n) = P(0|n) \sum_{k=1}^{n} \underline{r}^{T}(k)Q^{-1}\underline{m}(k). \tag{108}$$

The recursive regression algorithm, corresponding with the batch algorithm discussed in this example, now follows from

Eq. (104) and comparison of $\hat{Y}(0|n)$ with $\hat{Y}(0|n+1)$ [see also Eq. (75)]:

$$\left(\sum_{k=1}^{n+1} \underline{r}^T(k)Q^{-1}\underline{r}(k)\right)\hat{\underline{Y}}(0|n+1) - \left(\sum_{k=1}^{n} \underline{r}^T(k)Q^{-1}\underline{r}(k)\right)\hat{\underline{Y}}(0|n)$$

$$= \underline{r}^T(n+1)Q^{-1}m(n+1). \tag{109}$$

This expression may be rearranged by using Eq. (107):

$$P(0|n+1)^{-1}\hat{\underline{Y}}(0|n+1)$$

$$- (P(0|n+1)^{-1} - \underline{r}^T(n+1)Q^{-1}\underline{r}(n+1))\hat{\underline{Y}}(0|n)$$

$$= \underline{r}^T(n+1)Q^{-1}m(n+1). \tag{110}$$

Now, rewriting Eq. (110),

$$\hat{\underline{Y}}(0|n+1) = \hat{\underline{Y}}(0|n) + P(0|n+1)\underline{r}^T(n+1)Q^{-1}$$

$$\times [m(n+1) - \underline{r}(n+1)\hat{\underline{Y}}(0|n)] \tag{111}$$

$$\triangleq \hat{\underline{Y}}(0|n) + K(n+1)[m(n+1) - \underline{r}(n+1)\hat{\underline{Y}}(0|n)],$$

or

$$\hat{\underline{Y}}(0|n+1) = (1 - K(n+1)\underline{r}(n+1))\hat{\underline{Y}}(0|n)$$

$$+ K(n+1)m(n+1), \tag{112}$$

where

$$K(n+1) \triangleq P(0|n+1)\underline{r}^T(n+1)Q^{-1}. \tag{113}$$

Using the expressions for $P(0|n)$ and $K(n+1)$, a recursive propagation equation can be derived for the covariance matrix $P(0|n+1)$ of the errors $e(0|n+1)$ in the estimate $\hat{\underline{Y}}(0|n+1)$:

$$P(0|n+1) - P(0|n)$$

$$= -P(0|n+1)\{P(0|n+1)^{-1} - P(0|n)^{-1}\}P(0|n)$$

$$= -P(0|n+1)\left\{\sum_{k=1}^{n+1} \underline{r}^T(k)Q^{-1}\underline{r}(k) - \sum_{k=1}^{n} \underline{r}^T(k)Q^{-1}\underline{r}(k)\right\}P(0|n)$$

(Equation continues)

$$= -P(0|n + 1)\underline{r}^T(n + 1)Q^{-1}\underline{r}(n + 1)P(0|n)$$

$$= -K(n + 1)\underline{r}(n + 1)P(0|n). \tag{114}$$

So,

$$P(0|n + 1) = (1 - K(n + 1)\underline{r}(n + 1))P(0|n). \tag{115}$$

After estimation of $X(0)$ and λ, the time history of the system state may be computed according to

$$\hat{X}(k|n) = \hat{X}(0|n) + (U_m + \hat{\lambda}(0|n))kT. \tag{116}$$

To summarize, Example 3 has been presented to illustrate the introduction of an augmented parameter vector for simultaneous estimation of initial condition and bias error correction. In addition, it has been shown how to correct for a bias error in the measurements. While introducing an augmented parameter vector model linearity was conserved.

Example 4. State Estimation

Example 4 is presented to introduce the notion of recursive estimation of the system state $X(k)$ as a function of time, in contrast to the estimation of only the initial state $X(0)$ (see Refs. [22] and [23]). To set the stage the measurements $U_m(k)$ are assumed to be corrupted with random errors $-w(k)$ (see Fig. 10), whereas now the bias error correction λ is again assumed

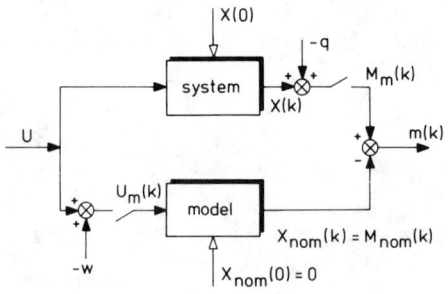

Fig 10. The system of Example 4, allowing for input- and output-signal noise.

to be zero:

$$U_m \triangleq U - w(k). \tag{117}$$

The random errors $-w(k)$ for $k = 0, 1, 2, \ldots$, are considered
to be a zero-mean, sequentially uncorrelated, Gaussian random
sequence [Eqs. (53) and (54)].

Calculation of system state estimates $\hat{X}(k)$ by estimation
of the initial state $X(0)$, followed by integration of the measured input signal $U_m(k)$, does not yield the most accurate results, since integration of $U_m(k)$ obviously implies integration
of the measurement errors $-w(k)$. In fact, a method should now
be devised to attenuate the effects of both the errors $-w(k)$
and $-q(k)$ on the estimates $X(k)$. To this end Kalman and Bucy
developed in the early 1960s recursive algorithms which can be
used to derive an estimate $\hat{X}(k|k)$ from all measurements up to
and including those recorded at time t_k [24]. Application of
these recursive algorithms to the estimation problem in this
example results in attenuation of the errors $-q(k)$ in the
output-signal measurements $M_m(k)$ and in estimation of the effects of the errors $-w(k)$ in the measurements $U_m(k)$ on the
estimates of the system state $X(k)$. Estimation of the latter
effects enables adequate correction. Again, most relevant information is squeezed out of the observation residuals $m(k)$
between the output-signal measurements $M_m(k)$ and the nominal
or computed output signals $M_{nom}(k)$. The nominal output signals
are computed according to

$$M_{nom}(k) = X_{nom}(k) = X_{nom}(k - 1) + U_m(k - 1)T. \tag{118}$$

The actual system state evolves according to

$$X(k) = X(k - 1) + UT. \tag{119}$$

The actual output-signal measurements are again given by

$$M_m(k) = M(k) - q(k) = X(k) - q(k). \tag{120}$$

From Eqs. (118)-(120), it follows for the observation residuals that

$$m(k) = M_m(k) - M_{nom}(k) = X(k) - X_{nom}(k) - q(k)$$

$$= X(k - 1) - X_{nom}(k - 1) + w(k - 1)T - q(k). \tag{121}$$

Recursive substitution of the relations

$$X(i) = X(i - 1) + UT, \tag{122}$$

$$X_{nom}(i) = X_{nom}(i - 1) + U_m(i - 1)T, \tag{123}$$

in the expression derived for m(k) in Eq. (121) yields

$$m(k) = X(0) - X_{nom}(0) + \sum_{i=1}^{k} w(i - 1)T - q(k) \tag{124}$$

$$= X(0) + \sum_{i=1}^{k} w(i - 1)T - q(k), \tag{125}$$

when $X_{nom}(0)$ is set equal to zero. From the assumptions that -w(k) and -q(k) are zero-mean sequentially uncorrelated Gaussian random-error sequences, it follows that a linear combination of these random errors, occurring in Eqs. (124) and (125), may also be considered to be a zero-mean random-error process $\epsilon(k)$, and

$$m(k) = X(0) + \epsilon(k). \tag{126}$$

The variance of $\epsilon(k)$ is not constant but increases with increasing k, since

$$\sigma_\epsilon^2(k) = E[\epsilon^2(k)] = \sum_{i=1}^{k} E[w^2(i - 1)]T^2 + E[q^2(k)]$$

$$= kWT^2 + Q. \tag{127}$$

Application of regression analysis algorithms to the problem of estimating $X(0)$ from a batch of observation residuals $m(1)$, $m(2)$, ..., $m(n)$ under this condition requires introduction of a weighting factor $\sigma_\epsilon^2(k)$ in the cost function of Eq. (103) such that

$$J_n = \sum_{n=1}^{n} \frac{(m(k) - \hat{X}(0|n))^2}{\sigma_\epsilon^2}.$$ (128)

The necessary condition for J_n to be a minimum leads again to an estimator for $X(0)$ as in Example 1. In the present example, this result is of much less value. The reason for this is that the calculation of $\hat{X}(k|n)$ through integration as in Eq. (94) would lead to unacceptable errors due to the measurements errors $-w(i)$ in $U_m(i)$. An optimal estimate of $X(k)$ can be obtained, however, by applying a recursive estimation algorithm, the Kalman filter, designed to take account of both the input-signal and output-signal noise statistics.

To initiate the recursive estimation process it is assumed that an estimate $\hat{X}(k - 1|k - 1)$ of the system state $X(k - 1)$ is derived from measurements up to and including those recorded at t_{k-1}. The task is set to derive the optimal estimate $\hat{X}(k|k)$ from the measurements $U_m(k - 1)$ and $M_m(k)$. First, the most accurate prediction of $X(k)$, i.e., $\hat{X}(k|k - 1)$, is computed according to

$$\hat{X}(k|k - 1) = \hat{X}(k - 1|k - 1) + U_m(k - 1)T.$$ (129)

The error $e(k|k - 1)$ in the prediction $\hat{X}(k|k - 1)$ satisfies

$$\begin{aligned} e(k|k - 1) &\triangleq X(k) - \hat{X}(k|k - 1) \\ &= X(k - 1) + UT - \hat{X}(k - 1|k - 1) \\ &\quad - UT + w(k - 1)T \\ &= e(k - 1|k - 1) + w(k - 1)T. \end{aligned}$$ (130)

The variance of the error in the prediction $\hat{X}(k|k - 1)$ hence equals

$$P(k|k - 1) = E[e^2(k|k - 1)]$$
$$= P(k - 1|k - 1) + E[e(k - 1|k - 1)w(k - 1)T]$$
$$+ T^2W. \tag{131}$$

In Eq. (131), $e(k - 1|k - 1)$ is defined as

$$e(k - 1|k - 1) = X(k - 1) - \hat{X}(k - 1|k - 1). \tag{132}$$

It may be deduced from the above that $\hat{X}(k - 1|k - 1)$ depends on the sequences $q(0)$, $q(1)$, ..., $q(k - 1)$ and $w(0)$, $w(1)$, ..., $w(k - 2)$. Therefore,

$$E[e(k - 1|k - 1)w(k - 1)] = 0, \tag{133}$$

$$P(k|k - 1) = P(k - 1|k - 1) + T^2W. \tag{134}$$

The discrete-time Kalman filter algorithm, which shows great similarity to the recursive algorithm delineated in Example 2, compared to $\hat{X}(k|k - 1)$ generates a more accurate estimate of $X(k)$ according to

$$\hat{X}(k|k) = \hat{X}(k|k - 1) + K(k)[m(k) - \hat{X}(k|k - 1)], \tag{135}$$

or equivalently,

$$\hat{X}(k|k) = (1 - K(k))\hat{X}(k|k - 1) + K(k)m(k). \tag{136}$$

Next, we must derive an expression for the confidence factor $K(k)$, which is often referred to as the filter gain. The optimal filter gain should be derived so as to minimize the variance of the error in the estimate $\hat{X}(k|k)$.

From Eq. (135) we may easily derive the fact that the error in the optimal estimate $\hat{X}(k|k)$ satisfies:

$$e(k|k) \triangleq X(k) - \hat{X}(k|k)$$
$$= (1 - K(k))e(k|k - 1) + K(k)q(k). \tag{137}$$

Hence,

$$P(k|k) = (1 - K(k))^2P(k|k - 1) + K^2(k)Q. \tag{138}$$

If $K^0(k)$ is the optimal gain, then the necessary condition for $P(k|k)$ to be minimized is

$$\partial P(k|k)/\partial K(k) = 0, \tag{139}$$

resulting in

$$K^0(k) = P(k|k - 1)/[P(k|k - 1) + Q]. \tag{140}$$

A remarkable similarity between this expression for $K^0(k)$ and the corresponding expression for the confidence factor in Example 2 should be observed. It should not therefore surprise us that the similarity also holds for the error-propagation equation obtained when substituting the expression for the optimal filter gain $K^0(k)$ in Eq. (138), which yields

$$P^0(k|k) = P(k|k - 1)Q/[P(k|k - 1) + Q]. \tag{141}$$

In summary, Example 4 has been presented to show that flight-path reconstruction from on-board measurements in the presence of input-signal errors $-w(t)$ and random errors in the output-signal observations $-q(t)$ cannot be achieved simply by estimation of the initial state with recursive or batch regression analysis procedures followed by integration of $U_m(t)$. Instead, the Kalman filter algorithm was introduced to obtain and estimate directly $\hat{X}(k|k)$ from all measurements $U_m(i - 1)$ and $M_m(i)$ for $i = 1, 2, \ldots, k$. A disadvantage of the Kalman filter is that only the last estimate $\hat{X}(n|n)$ is based on all available measurements. Smoothing, a recursive algorithm working backward in time, allows calculation of $\hat{X}(k|n)$ starting from $\hat{X}(n|n)$ (see Ref. [24]).

B. *THE AUGMENTED-SYSTEM MODEL*
    OF FLIGHT-PATH RECONSTRUCTION

The augmented-system model representing the motions of the aircraft in nonsteady flight in relation to the measured input signals $\underline{U}_m(t)$ and the measured output signals $\underline{M}_m(t)$ is compiled

in this section, using the results presented in Section II.
The model to be developed is required for flight-path reconstruction from measurements of the system input and output signals recorded in flight.

Applying the definitions of the system state vector $\underline{X}(t)$
and the system input $\underline{U}(t)$ [see Eqs. (19) and (20)], a set of
kinematic relations is represented by the nonlinear state
equation:

$$\underline{\dot{X}}(t) = \underline{f}(\underline{X}(t), \underline{U}(t)), \tag{142}$$

with initial state

$$\underline{X}(t_0) = \underline{X}(0). \tag{143}$$

The relation between the system state $\underline{X}(t)$, the system input
signal $\underline{U}(t)$, and the system output signal $\underline{M}(t)$ is given in
terms of Eq. (27):

$$\underline{M}(t) = \underline{h}(\underline{X}(t), \underline{U}(t)). \tag{144}$$

The models relating the observations $\underline{U}_m(t)$ to $\underline{U}(t)$ and $\underline{M}_m(t)$
to $\underline{M}(t)$ are presented in Eqs. (28) and (29) and as

$$\underline{U}_m(t) = \underline{U}(t) - \lambda - \underline{w}(t), \tag{145}$$

$$\underline{M}_m(t) = \underline{M}(t) - \underline{q}(t). \tag{146}$$

The definitions for the vector-valued error corrections λ,
$\underline{w}(t)$, and $\underline{q}(t)$ are given in Eqs. (30)-(32). As expressed by
Eq. (33),

$$\dot{\lambda} = 0, \tag{147}$$

the bias errors in the input-signal measurements are assumed
constant in time. The errors $-\underline{w}(t)$ and $-\underline{q}(t)$, respectively,
in the measurements $\underline{U}_m(t)$ of the input signals and $\underline{M}_m(t)$ of the
output signals are considered to be Gaussian white-noise pro-
cesses. The first- and second-order moments of these stochastic processes are represented by Eqs. (34)-(37). Combination

of Eq. (21) with Eq. (28) yields

$$\dot{\underline{X}}(t) = \underline{f}(\underline{X}(t), \underline{U}_m(t), \underline{\lambda}, \underline{w}(t)). \tag{148}$$

Combining Eqs. (27)-(29), the following output-signal observation equation is obtained:

$$\underline{M}_m(t) = \underline{h}(\underline{X}(t), \underline{U}_m(t), \underline{\lambda}, \underline{w}(t)) - \underline{q}(t). \tag{149}$$

1. *State and Observation Eqaution*
    for the Augmented System Model

As shown in Section III.C, many methods can be applied to flight-path reconstruction from on-board measurements. Selection of methods, models, and algorithms may depend on many technical and operational aspects:

1.   The quality of the sensors used for measurement of the components of the vector-valued input signal $\underline{U}(t)$, which determines whether the input-signal noise $-\underline{w}(t)$ may be neglected or not

2.   The technical layout and quality of the instrumentation-system output signals, which determines whether or not account should be taken of possible bias errors in these measurements

3.   The availability of knowledge concerning the statistics of the errors in the input-signal measurements $\underline{U}_m(t)$ and the output-signal measurements $\underline{M}_m(t)$

4.   The corrections to be applied to the responses of those sensors used to measure the system output signals; correction of these responses may or may not require taking account of the measured system inputs $\underline{U}_m(t)$ and, consequently, of the errors in these measurements [see, for example, Eq. (25)]

5.   The type of maneuver, i.e., the types of motion (steady or nonsteady, symmetric or asymmetric) to be reconstructed. With respect to the effect of the type of maneuver on the

mechanization of flight-path reconstruction, the following re-
mark is made. Experience has shown that measurements of, for
example, static pressure, recorded during highly nonsteady test-
flight sections, preferably should not be granted to much weight
owing to considerable perturbations resulting from the non-
stationarity of the airflow around the aircraft.

As shown in Example 3, flight-path reconstruction in the
presence of biased measurements $\underline{U}_m(t)$ demands estimation of
corrections for these bias errors. Estimation of these corrections may be performed simultaneously with estimation of
either the initial state $\underline{X}(0)$ or the system state at time t_k,
$\underline{X}(k)$. To facilitate simultaneous estimation of the initial
state $\underline{X}(0)$ or the system state $\underline{X}(k)$ and the corrections λ for
the bias errors in the input-signal measurements, the augmented
system model is introduced with the augmented system state:

$$\underline{Y} \triangleq \text{col}[\underline{X}(t), \lambda]$$

$$= \text{col}[V_{x_B}, V_{y_B}, V_{z_B}, \psi, \theta, \varphi, x_E, y_E, z_E, \lambda_x, \lambda_y, \lambda_z, \lambda_p, \lambda_q, \lambda_r].$$

$$(150)$$

The composition of the augmented system and of the pertaining
augmented system state $\underline{Y}(t)$ given should be considered as merely
a possible configuration. Many other parameters, such as the
sidewash coefficient C_{si}, the upwash coefficient C_{up} of an
angle of attack wind vane, horizontal and vertical constant
wind velocities, wind gradients, and sensor drift-rate coefficients, may be incorporated in $\underline{Y}(t)$. Some of the constant bias
error corrections [see Eq. (150)] and the horizontal position
coordinates x_E and y_E may have to be dropped from $\underline{Y}(t)$, depending on the composition of the output signal $\underline{M}(t)$ and possible
inherent unobservability problems. Bias error corrections may

be dropped altogether from the augmented state vector when highly accurate inertial sensors are used during short maneuvers.

The state equation of the augmented system model is obtained by joining Eqs. (21) and (33) and by using the augmented state vector $\underline{Y}$:

$$\dot{\underline{Y}} = \underline{f}'(\underline{Y}(t), \underline{U}_m(t), \underline{w}(t)) = \underline{f}'(\underline{Y}(t), \underline{U}'(t)), \tag{151}$$

where

$$\underline{U}'(t) \triangleq \underline{U}_m(t) + \underline{w}(t). \tag{152}$$

The vector-valued quantity $\underline{U}'(t)$ has mathematical significance but no particular physical meaning. The observation equation corresponding to the augmented system model now reads

$$\underline{M}_m(t) = \underline{h}'(\underline{Y}(t), \underline{U}'(t)). \tag{153}$$

In Section III.A utilization of nominal observations $\underline{M}_{nom}(t)$, derived from nominal state vector trajectories $\underline{X}_{nom}(t)$, has been exhibited.

Again and again it has been shown that significant information concerning the parameter or state vector quantities to be estimated can be extracted from the deviations between the nominal observations $\underline{M}_{nom}(t)$ and the actual observations $\underline{M}_m(t)$. In addition, we note that many estimation procedures require linearization of the nonlinear state and observation equations presented earlier with respect to a nominal trajectory $\underline{Y}_{nom}(t)$, computed from the recorded input-signal measurements $\underline{U}_m(t)$ by integration of the nominal state equation:

$$\dot{\underline{Y}}_{nom}(t) = \underline{f}'(\underline{Y}_{nom}(t), \underline{U}'_{nom}(t)) = \underline{f}'(\underline{Y}_{nom}(t), \underline{U}_m(t)), \tag{154}$$

for given

$$\underline{Y}_{nom}(0) \triangleq \text{col}[\underline{X}_{nom}(0), \underline{\lambda}_{nom}] = \text{col}[\underline{X}_{nom}(0), 0]. \tag{155}$$

Here,

$$\underline{U}'_{nom}(t) \triangleq \underline{U}_m(t_{k-1}) \quad \text{for} \quad t \in [t_{k-1}, t_k),$$

$$k = 1, 2, \ldots, n. \tag{156}$$

2. *Linearization and Discretization
   of the Model*

Many statistical procedures applied to state trajectory reconstruction from measurements of dynamic system inputs and outputs require linear relations between the system state and the input-signal noise as well as between the output signal observations and the system state. The demand for linearity of these relations stems from the fact that Gaussian stochastic processes remain Gaussian under linear transformations. Under these statistical conditions, application of linear estimation procedures yields optimal results. For clarity it is recalled here that an estimation procedure is called linear if the estimate is computed linearly from the available measurements. Gaussian random variables are so significant to statistical estimation because specification of the first-order (the average) and second-order (the variance) moments implies complete specification of all higher order moments.

Linearization of the augmented-system state equation and the corresponding observation equation [see Eqs. (151) and (153)], derived in this section, is based on the introduction of state perturbation corrections

$$\underline{y}(t) \triangleq \underline{Y}(t) - \underline{Y}_{nom}(t), \tag{157}$$

and input-signal perturbation corrections

$$\underline{u}(t) \triangleq \underline{U}'(t) - \underline{U}'_{nom}(t) = \underline{U}'(t) - \underline{U}_m(t) = \underline{w}(t), \tag{158}$$

followed by differentiation of the augmented state and observation equations [Eqs. (151) and (153)], with respect to the nominal trajectory $\underline{Y}_{nom}(t)$ and the corresponding nominal input signals $\underline{U}'_{nom}(t)$,

$$\dot{\underline{y}}(t) = \underline{f}'(\underline{Y}(t), \underline{U}'(t)) - \underline{f}'(\underline{Y}_{nom}(t), \underline{U}'_{nom}(t))$$

$$\cong (\partial \underline{f}'/\partial \underline{Y})_{\underline{Y}_{nom}, \underline{U}'_{nom}} \underline{y}(t) + (\partial \underline{f}'/\partial \underline{U}')_{\underline{Y}_{nom}, \underline{U}'_{nom}} \underline{w}(t),$$

(159)

and subsequently of the observation equation,

$$\underline{m}(t) \triangleq \underline{M}_m(t) - \underline{M}_{nom}(t) = \underline{M}(t) - \underline{q}(t) - \underline{M}_{nom}(t)$$

$$= \underline{h}'(\underline{Y}(t), \underline{U}'(t)) - \underline{h}'(\underline{Y}_{nom}(t), \underline{U}'_{nom}(t)) - \underline{q}(t)$$

$$\cong (\partial \underline{h}'/\partial \underline{Y})_{\underline{Y}_{nom}, \underline{U}'_{nom}} \underline{y}(t)$$

$$+ (\partial \underline{h}'/\partial \underline{U}')_{\underline{Y}_{nom}, \underline{U}'_{nom}} \underline{w}(t) - \underline{q}(t).$$
(160)

The initial condition of the linearized state equation in many cases is not known and is therefore set equal to zero:

$$\underline{y}(0) \triangleq 0.$$
(161)

It is numerically interesting to note that the rows of the matrix $\partial \underline{f}'/\partial \underline{Y}$ corresponding to $\partial \underline{\lambda}/\partial \underline{Y}$ are equal to zero.

The matrix $\partial \underline{f}'/\partial \underline{Y}$ has the following structure:

$$\frac{\partial \underline{f}'}{\partial \underline{Y}} = \begin{bmatrix} \dfrac{\partial \dot{v}_{x_B}}{\partial v_{x_B}} & \dfrac{\partial \dot{v}_{x_B}}{\partial v_{y_B}} & \cdots & \dfrac{\partial \dot{v}_{x_B}}{\partial \lambda_r} \\[2ex] \dfrac{\partial \dot{z}_E}{\partial v_{x_B}} & \dfrac{\partial \dot{z}_E}{\partial v_{y_B}} & \cdots & \dfrac{\partial \dot{z}_E}{\partial \lambda_r} \\[2ex] & \vdots & & \\[1ex] \dfrac{\partial \dot{\lambda}_r}{\partial v_{x_B}} & \dfrac{\partial \dot{\lambda}_r}{\partial v_{y_B}} & \cdots & \dfrac{\partial \dot{\lambda}_r}{\partial \lambda_r} \end{bmatrix} = \begin{bmatrix} \dfrac{\partial \dot{v}_{x_B}}{\partial v_{x_B}} & \dfrac{\partial \dot{v}_{x_B}}{\partial v_{y_B}} & \cdots & \dfrac{\partial \dot{v}_{x_B}}{\partial \lambda_r} \\[2ex] \dfrac{\partial \dot{z}_E}{\partial v_{x_B}} & \dfrac{\partial \dot{z}_E}{\partial v_{y_B}} & \cdots & \dfrac{\partial \dot{z}_E}{\partial \lambda_r} \\[2ex] 0 & 0 & & 0 \\[1ex] 0 & 0 & \cdots & 0 \end{bmatrix}.$$
(162)

The linearized state and observation equations previously presented may be symbolically represented by

$$\dot{\underline{y}}(t) = A(t)\underline{y}(t) + B(t)\underline{w}(t),$$
(163)

$$\underline{m}(t) = C(t)\underline{y}(t) + D(t)\underline{w}(t) - \underline{q}(t). \tag{164}$$

Under the conditions characterizing the problem of flight-path reconstruction from flight-test data, it can be shown that the contributions of the term $D(t)\underline{w}(t)$ in $\underline{m}(t)$, resulting from the correction of the sideslip vane measurements β_v for the effects of the angular velocities, are negligibly small if the sensors used for measurement of the components p_B, q_B, and r_B of $\underline{U}(t)$ have high accuracy [see Eqs. (20) and (25)]. The linearized observation equation therefore reduces to

$$\underline{m}(t) = C(t)\underline{y}(t) - \underline{q}(t). \tag{165}$$

In high-accuracy instrumentation and data-logging systems, the input signals $\underline{U}(t)$ and the output signals $\underline{M}(t)$ are periodically sampled in time. Moreover, digital computers are used to perform all computations required for flight-path reconstruction and further reduction of flight-test data. Therefore the discret e-time versions of the linearized state and observations are required. These discrete-time versions relate the state perturbation correction $\underline{y}(k)$ and the observation perturbation correction $\underline{m}(k)$ to the quantities $\underline{y}(k - 1)$ and $\underline{w}(k - 1)$. The discrete time versions of Eqs. (163) and (164) are

$$\underline{y}(k) = \Phi(k, k - 1)\underline{y}(k - 1) + \Gamma(k, k - 1)\underline{w}(k - 1), \tag{166}$$

$$\underline{m}(k) = C(k)\underline{y}(k) - \underline{q}(k). \tag{167}$$

The transition matrix $\Phi(k, k - 1)$ is derived from the linear system matrix $A(t)$, approximating $A(t)$ for t in $[t_{k-1}, t_k)$ by $A(t_{k-1})$, according to

$$\Phi(k, k - 1) = \exp[A(t_{k-1})(t_k - t_{k-1})]$$

$$= I + A(t_{k-1})(t_k - t_{k-1})$$

$$+ \left[A^2(t_{k-1})(t_k - t_{k-1})^2/2!\right] + \cdots . \tag{168}$$

There are other methods available for computation of the transition matrix such as numerical integration of the so-called fundamental equation (see Ref. [25]). Truncation of the Taylor series expansion, used for computation of the transition matrix of Eq. (168), may in practice be governed by arguments of computation efficiency or by limitation of the truncation error, based on any possible matrix norm.

The control distribution matrix $\Gamma(k, k - 1)$ is derived from $B(t_{k-1})$ according to

$$
\begin{aligned}
\Gamma(k, k - 1) &= \int_{t_{k-1}}^{t_k} \Phi(t_k, \tau) B(\tau) d\tau \cong \int_{t_{k-1}}^{t_k} \Phi(t_k, \tau) B(t_{k-1}) d\tau \\
&= B(t_{k-1})(t_k - t_{k-1}) \\
&\quad + \left[A(t_{k-1}) B(t_{k-1})(t_k - t_{k-1})^2/2! \right] \\
&\quad + \left[A^2(t_{k-1}) B(t_{k-1})(t_k - t_{k-1})^3/3! \right] + \cdots .
\end{aligned}
\tag{169}
$$

Combining the discrete-time linear state equation with the linear observation equation, the following relations may be derived for the linearized observations $\underline{m}(k)$ [Eq. (167)]:

$$
\begin{aligned}
\underline{m}(k) &= C(k)\Phi(k, k - 1)\underline{y}(k - 1) \\
&\quad + C(k)\Gamma(k, k - 1)\underline{w}(k - 1) - \underline{q}(k),
\end{aligned}
\tag{170}
$$

or

$$
\begin{aligned}
\underline{m}(k) &= C(k)\Phi(k, 0)\underline{y}(0) \\
&\quad + C(k) \sum_{i=1}^{k} \Phi(k, i)\Gamma(i, i - 1)\underline{w}(i - 1) - \underline{q}(k).
\end{aligned}
\tag{171}
$$

It should be noted that Eq. (171) drastically simplifies if the contributions to $\underline{m}(k)$ governed by the perturbations $\underline{w}(i - 1)$, for $i = 1, 2, \ldots, k$, may be neglected for all k. This implies that the stochastic linear system, represented by Eqs. (166)

and (167), is approximated by a deterministic system with stochastic output-signal observations. This simplification yields
the following relation:

$$\underline{m}(k) = C(k)\Phi(k,\ 0)\underline{y}(0) - \underline{q}(k).$$ (172)

This simplified relation shows considerable similarity to the
observation equations underlying regression analysis as shown
in Example 3.

C. FLIGHT-PATH RECONSTRUCTION ALGORITHMS

A survey of algorithms used for estimation of flight-path
variables and corrections for instrumental bias errors from
measurements of system input and output signals is presented
here. Depending on the characteristics of the flight-path reconstruction problem facing the flight-test engineer, he may
choose from these algorithms the one that suits him best. The
selection may be affected by the characteristics of the underlying dynamic system model, the flight-test instrumentation
system to be used and the type of maneuvers to be reconstructed.
Moreover, economic factors such as computing time required,
available somputer capacity, available standard software, and
available skill and experience may have to be taken into
account when selecting the algorithm to be applied.

1. Weighted Least Squares Estimation

If the accelerometers used to measure the specific aero-
dynamic forces A_{x_B}, A_{y_B}, and A_{z_B}, and the rate gyroscopes used
to measure the angular rates p_B, q_B, and r_B are of sufficiently
high accuracy, it may be acceptable to neglect the measurement
errors $-\underline{w}(k)$ in the measurements $\underline{U}_m(k)$ [26]. The observations
$\underline{m}(k)$ [see Eq. (160)] are in that case linearly related to the
initial state perturbation corrections $\underline{y}(0)$ of Eq. (172).

Definition of

$$R(k) \triangleq C(k) \Phi(k, 0),$$ (173)

and substitution into Eq. (172) yields

$$\underline{m}(k) = R(k)\underline{y}(0) - \underline{q}(k).$$ (174)

This linear relation closely resembles Eq. (102) in Example 3.

The estimate minimizing the quadratic cost function

$$J_n = \sum_{k=1}^{n} (\underline{m}(k) - R(k)\underline{y}(0))^T Q^{-1} (\underline{m}(k) - R(k)\underline{y}(0))$$ (175)

is the weighted least squares estimate (see the Appendix)

$$\hat{\underline{y}}(0|n) = \left(\sum_{k=1}^{n} R^T(k) Q^{-1} R(k) \right)^{-1} \sum_{k=1}^{n} R^T(k) Q^{-1} \underline{m}(k).$$ (176)

The reconstructed state time histories then follow from

$$\hat{\underline{Y}}(k|n) = \underline{Y}_{nom}(k) + \Phi(k, 0)\hat{\underline{y}}(0|n),$$ (177)

or from recalculation of the nominal state vector trajectory $\underline{Y}_{nom}(k)$ after correction of the nominal initial state according to

$$\hat{\underline{Y}}(0|n) = \underline{Y}_{nom}(0) + \hat{\underline{y}}(0|n).$$ (178)

This process may be repeated until no significant deviations between two consecutive estimates of the initial state are observed. The covariance matrix $P(0|n)$ of the errors in the estimate $\hat{\underline{y}}(0|n)$ and henceforth in the estimate $\hat{\underline{Y}}(0|n)$ may be computed according to

$$P(0|n) = \left(\sum_{k=1}^{n} R^T(k) Q^{-1} R(k) \right)^{-1}.$$ (179)

Alternatively, $\hat{\underline{y}}(0|k)$ may also be obtained by applying a recursive estimation scheme, as in Example 3:

$$\hat{\underline{y}}(0|k) = \hat{\underline{y}}(0|k-1) + K(k)(\underline{m}(k) - R(k)\hat{\underline{y}}(0|k-1)),$$ (180)

where the confidence or gain matrix is given by

$$K(k) = P(0|k) R^T(k) Q^{-1}. \tag{181}$$

The estimation error covariance matrix follows from

$$P(0|k) = (I - K(k) R(k)) P(0|k - 1). \tag{182}$$

2. *The Kalman Filter Algorithm*

The presence of nonneglectable input-signal-measurement errors calls for the application of a Kalman filter, as in Example 4. Let the linearized system dynamics related to the propagation of the corrections for the state perturbations be given by the following equations:

$$\underline{y}(k) = \Phi(k, k - 1) \underline{y}(k - 1) + \Gamma(k, k - 1) \underline{w}(k - 1), \tag{183}$$

$$\underline{m}(k) = C(k) \underline{y}(k) - \underline{q}(k). \tag{184}$$

The Kalman filter algorithm is then represented by the following set of recursive equations:

$$\hat{\underline{y}}(k|k - 1) = \Phi(k, k - 1) \hat{\underline{y}}(k - 1|k - 1), \tag{185}$$

$$P(k|k - 1) = \Phi(k, k - 1) P(k - 1|k - 1) \Phi^T(k, k - 1)$$
$$+ \Gamma(k, k - 1) W \Gamma^T(k, k - 1). \tag{186}$$

The gain matrix is given by

$$K(k) = P(k|k - 1) C^T(k) (C(k) P(k|k - 1) C^T(k) + Q)^{-1}. \tag{187}$$

It should be remarked here that the covariance matrices of the random errors $-\underline{w}(k - 1)$ and $-\underline{q}(k)$ may be time dependent. In that case W and Q in Eqs. (186) and (187) are replaced, respectively, by the time-dependent covariance matrices $W(k - 1)$ and $Q(k)$. The optimal estimate of the correction $\underline{y}(k)$ for the state perturbation is then obtained from:

$$\hat{\underline{y}}(k|k) = \hat{\underline{y}}(k|k - 1) + K(k) (\underline{m}(k) - C(k) \hat{\underline{y}}(k|k - 1)). \tag{188}$$

The corresponding covariance matrix of the errors in $\hat{\underline{y}}(k|k)$ is computed according to

$$P(k|k) = (I - K(k) C(k)) P(k|k - 1). \tag{189}$$

Finally, actual estimates of the system state $\underline{y}(k)$ are obtained
from

$$\hat{\underline{Y}}(k|k) = \underline{Y}_{nom}(k) + \hat{\underline{y}}(k|k). \tag{190}$$

For actual application of the Kalman filter to flight-path reconstruction, a so-called extended form of the Kalman filter
is applied. The extended Kalman filter differs from the original filter in that the nominal state vector trajectory
$\underline{Y}_{nom}(k)$ is not computed entirely before mechanizing the Kalman
filter algorithm. Instead, for $k = 1, 2, \ldots,$ the nominal state
vector quantity $\underline{Y}_{nom}(k)$ is computed by integration of the nonlinear state equation [Eq. (151)] by using the previous optimal
estimate $\hat{\underline{Y}}(k - 1|k - 1)$ as an initial condition. Hence,

$$\underline{Y}_{nom}(k) = \hat{\underline{Y}}(k|k - 1) = \hat{\underline{Y}}(k - 1|k - 1)$$

$$+ \int_{t_{k-1}}^{t_k} \underline{f}(\underline{Y}(\tau), \underline{U}_m(k - 1)) d\tau. \tag{191}$$

The observation perturbation correction is in this case computed according to

$$\underline{m}(k) = \underline{M}_m(k) - \underline{h}(\hat{\underline{Y}}(k|k - 1)). \tag{192}$$

The optimal estimate now follows from

$$\hat{\underline{Y}}(k|k) = \hat{\underline{Y}}(k|k - 1) + K(k)\underline{m}(k), \tag{193}$$

since $\hat{\underline{y}}(k|k - 1) = 0$.

The result of the extended Kalman filter as just described
is a sequence of state vector estimates $\hat{\underline{Y}}(k|k)$. Clearly, a
sequence of estimates $\hat{\underline{Y}}(k|n)$ can be more accurate because each
estimate depends on all available measurements. The sequential
generation, backward in time, of the latter sequence of estimates is called smoothing. Although in this article attention
is devoted only to "fixed-interval smoothing," it should be
remarked here that there exist three types of smoothing

algorithms: (1) "fixed-point smoothing," which yields an esti-
mate $\hat{\underline{Y}}(k|j)$, for k = constant and $j \geq k$; (2) "fixed-lag
smoothing," which yields an estimate $\hat{\underline{Y}}(k|k + j)$ for fixed j;
and (3) "fixed-interval smoothing," which yields $\hat{\underline{Y}}(k|n)$, where
n is fixed.

The initial condition of the extended fixed-interval-
smoothing algorithm is $\hat{\underline{Y}}(n|n)$, as generated by the extended
Kalman filter. Working backward in time, this procedure cor-
rects the estimate $\hat{\underline{Y}}(k|k)$ obtained from the Kalman filter al-
gorithm according to

$$\hat{\underline{Y}}(k|n) = \hat{\underline{Y}}(k|k) + K_s(k) [\hat{\underline{Y}}(k + 1|n) - \hat{\underline{Y}}(k + 1|k)], \qquad (194)$$

where k = n - 1, n - 2, ..., 0. The gain matrix $K_s(k)$ of the
smoothing algorithm is given by

$$K_s(k) = P(k|k) \Phi^T(k + 1, k) P^{-1}(k + 1|k), \qquad (195)$$

and the covariance matrix of the errors in $\hat{Y}(k|n)$ becomes

$$P(k|n) = P(k|k) + K_s(k) [P(k + 1|n)$$
$$- P(k + 1|k)] K_s^T(k). \qquad (196)$$

Applications of the Kalman filter to the flight-path reconstruction problem are reported in Refs. [8,11,26]. Instead of
the conventional Kalman approach to discrete filtering and
smoothing, the so-called square-root information filter-smoother
algorithms may be applied. Although algebraically equivalent
to the Kalman algorithms, the square-root filters may achieve
improved numerical accuracy [27,28].

3. *Maximum Likelihood Estimation*

Another method applicable to the flight-path reconstruction
problem is the so-called maximum likelihood algorithm. If the
system and observation models are linear, the maximum

likelihood estimation theory results in an algorithm identical to the weighted least-square algorithm discussed in Section III.C.1. The maximum likelihood method, however, differs from the estimation procedures discussed in that the models of the dynamic system and the output-signal observations may be non-linear. Furthermore, the maximum likelihood algorithm when applied to a nonlinear estimation problem is to be implemented as an iterative batch-processing algorithm. This implies that an estimate of the parameters of interest, obtained after processing a certain batch of measurements, is improved by repeated application of the same algorithm to the same batch of measurements, using the previous estimate as a reference condition.

The following assumptions underly the application of the maximum likelihood theory:

1. The system and observation models are assumed highly accurate

2. The input signal noise, i.e., $-\underline{w}(k)$ in the measurements $\underline{U}_m$ is assumed negligibly small

3. The observation noise $-\underline{q}(k)$ is assumed to be zero mean and Gaussian; however, the variance $Q(k)$ is assumed unknown, although usually assumed to be constant, thus $Q(k) = Q$.

Application of the maximum likelihood algorithm to flight-path reconstruction is directed again toward extraction of an estimate $\hat{\underline{Y}}(0|n)$ from the observation residuals:

$$\underline{m}(k) = \underline{M}_m(k) - \underline{M}_{nom}(k), \qquad k = 1, 2, \ldots, n. \qquad (197)$$

In addition to estimating the initial state, the maximum likelihood algorithm can be used to estimate the variance Q of the observation noise $-\underline{q}(k)$.

Each iteration cycle of the maximum likelihood procedure comprises the following steps:

1. An estimate of the initial state $\underline{Y}(0)$ is extracted manually or by any other means from the flight-test data (see Section IV) or is obtained from the previous iteration.

2. An estimate of the covariance matrix Q is extracted either from instrument-calibration data or obtained from the previous iteration. In practice, the first estimate of Q is set equal to the identity matrix I.

3. Using the available estimate of the initial state $\underline{Y}(0)$ as an initial condition, the dynamic system state equations

$$\dot{\underline{Y}}_{nom} = \underline{f}(\underline{Y}_{nom}(t), \underline{U}_m(t)) \tag{198}$$

are numerically integrated to obtain a nominal state vector trajectory $\underline{Y}_{nom}(k)$, $k = 1, 2, \ldots, n$. Nominal observations $\underline{M}_{nom}(k)$ are then derived from $\underline{Y}_{nom}(k)$:

$$\underline{M}_{nom}(k) = \underline{h}(\underline{Y}_{nom}(k), \underline{U}_m(k)). \tag{199}$$

4. The nominal output signals $\underline{M}_{nom}(k)$ are then compared to the measured output signals $\underline{M}_m(k)$ to obtain the observation residuals $\underline{m}(k)$ [see Eq. (195)].

5. The maximum likelihood estimates of $\underline{Y}(0)$ and the covariance matrix Q are those values of $\hat{\underline{Y}}(0|n)$ and $\hat{Q}$ which maximize the conditional probability density function

$$p(\underline{m}(1), \underline{m}(2), \ldots, \underline{m}(n)|\underline{Y}(0), Q). \tag{200}$$

To obtain a maximum of the conditional probability density function, one is required to formulate the sensitivity of $\underline{m}(k)$, $k = 1, 2, \ldots, n$, to variations in $\hat{\underline{Y}}(0|n)$. In Eq. (197) only $\underline{M}_{nom}(k)$ depends on $\underline{Y}_{nom}(0)$. We have to formulate therefore the sensitivity of $\underline{M}_{nom}(k)$ for variations in $\underline{Y}_{nom}(0)$. With

Eqs. (198) and (199) the following sensitivity relations are found:

$$\underline{M}_{nom}(\underline{Y}_{nom}(0) + \underline{y}(0), \underline{U}_m(t))$$

$$\cong \underline{M}_{nom}(\underline{Y}_{nom}(0), \underline{U}_m(t))$$

$$+ [\partial \underline{M}(\underline{Y}(0), t)/\partial \underline{Y}(0)]_{\underline{Y}_{nom}, \underline{U}'_{nom}} \underline{y}(0), \tag{201}$$

or

$$\underline{M}_{nom}(\underline{Y}_{nom}(0) + \underline{y}(0), \underline{U}_m(t)) - \underline{M}_{nom}(\underline{Y}_{nom}(0), \underline{U}_m(t))$$

$$\cong (\partial \underline{h}/\partial \underline{Y})_{\underline{Y}_{nom}, \underline{U}'_{nom}} [\partial \underline{Y}/\partial \underline{Y}(0)]_{\underline{Y}_{nom}, \underline{U}'_{nom}} \underline{y}(0)$$

$$= C(t) [\partial \underline{Y}/\partial \underline{Y}(0)]_{\underline{Y}_{nom}, \underline{U}'_{nom}} \underline{y}(0)$$

$$\underline{\triangle} C(t) S(t) \underline{y}(0), \tag{202}$$

in which $\underline{U}'_{nom}$ is defined as in Eq. (156). The elements of the matrix product $C(t)S(t)$ are referred to as the sensitivity coefficients. The elements of the matrix $S(t)$ can be found by solving a differential propagation equation for $S(t)$ with respect to time. This differential equation may be derived as follows:

$$\frac{d}{dt}[S(t)] = \frac{d}{dt}\left(\frac{\partial \underline{Y}}{\underline{Y}(0)}\right)_{\underline{Y}_{nom}, \underline{U}'_{nom}}$$

$$= \left(\frac{\partial}{\partial \underline{Y}(0)} \frac{d\underline{Y}}{dt}\right)_{\underline{Y}_{nom}, \underline{U}'_{nom}} = \left(\frac{\partial}{\partial \underline{Y}(0)} \underline{f}(\underline{Y}, \underline{U})\right)_{\underline{Y}_{nom}, \underline{U}'_{nom}}$$

$$= \left(\frac{\partial \underline{f}(\underline{Y}, \underline{U})}{\partial \underline{Y}} \frac{\partial \underline{Y}}{\partial \underline{Y}(0)}\right)_{\underline{Y}_{nom}, \underline{U}'_{nom}} = \left(\frac{\partial \underline{f}(\underline{Y}, \underline{U})}{\partial \underline{Y}}\right)_{\underline{Y}_{nom}, \underline{U}'_{nom}} S(t).$$

$$\tag{203}$$

Hence,

$$\dot{S}(t) = A(t) S(t), \tag{204}$$

in which $A(t)$ is identical to the matrix of derivatives in

Eq. (163). It is clear that the initial condition is

$$S(0) = [\partial \underline{Y}/\partial \underline{Y}(0)]_{\underline{Y}_{nom}(0), \underline{U}'_{nom}(0)} = I. \tag{205}$$

The estimation procedure is now initiated by specifying an initial estimate $\underline{Y}_{nom}(0)$ and Q.

The likelihood function of the observation residuals $\underline{m}(1)$, $\underline{m}(2), \ldots, \underline{m}(n)$, is defined as

$$L(\underline{Y}_{nom}(0), Q) \triangleq p(\underline{m}(1), \underline{m}(2), \ldots, \underline{m}(n) | \underline{Y}_{nom}(0), Q). \tag{206}$$

From Eqs. (197) and (198) it follows that $\underline{m}(k)$ depends only on $\underline{Y}_{nom}(0)$ and $\underline{q}(k)$. Because $\underline{q}(k)$ is a sequentially uncorrelated random sequence, Eq. (202) may be written as a product of conditional densities according to

$$L(\underline{Y}_{nom}(0), Q) = \prod_{k=1}^{n} p(\underline{m}(k)\underline{Y}_{nom}(0), Q)$$

$$= \frac{1}{(2\pi|Q|)^{n/2}} \exp\left\{-\frac{1}{2} \sum_{k=1}^{n} [\underline{m}^T(k)Q^{-1}\underline{m}(k)]\right\}$$

$$\tag{207}$$

To obtain the required maximum likelihood estimates of $\underline{Y}(0)$ and Q, $L(\underline{Y}_{nom}(0), Q)$ should be maximized. Maximizing $L(\underline{Y}_{nom}(0), Q)$ is equivalent to maximizing $\ln L(\underline{Y}_{nom}(0), Q)$:

$$\ln L(\underline{Y}_{nom}(0), Q) = -\frac{1}{2} \sum_{k=1}^{n} \underline{m}^T(k)Q^{-1}\underline{m}(k) - \frac{n}{2} \ln |Q| + C, \tag{208}$$

in which C is a normalizing constant. According to Ref. [29], maximization of $\ln L(\underline{Y}_{nom}(0), Q)$ with respect to $\underline{Y}_{nom}(0)$ and Q is replaced by maximization of $\ln L(\underline{Y}_{nom}(0) + \underline{y}(0), Q)$ with respect to $\underline{y}(0)$ and Q.

With $\underline{y}(0)$, an improved estimate of $\underline{m}(k)$, denoted by $\underline{m}_1(k)$ follows directly from Eqs. (197) and (198):

$$\underline{m}_1(k) \triangleq \underline{M}_m(k) - \underline{M}_{nom}(\underline{Y}_{nom}(0) + \underline{y}(0), k)$$

$$\cong \underline{m}(k) - C(k)S(k)\underline{y}(0). \tag{209}$$

Substitution of $\underline{m}_1(k)$ in Eq. (208) yields

$$\ell n \ L(\underline{Y}_{nom}(0) + \underline{y}(0), Q)$$

$$= -\frac{1}{2} \sum_{k=1}^{n} [\underline{m}(k) - C(k)S(k)\underline{y}(0)]^T Q^{-1}$$

$$\times [\underline{m}(k) - C(k)S(k)\underline{y}(0)] - \frac{1}{2} n \ \ell n \ |Q| + C. \tag{210}$$

The estimates $\hat{\underline{y}}(0|n)$ and $\hat{Q}$ follow from the necessary conditions

$$\left(\frac{\partial \ \ell n \ L[\underline{Y}_{nom}(0) + \underline{y}(0), Q]}{\partial \underline{y}(0)}\right)_{\hat{\underline{y}}(0|n), \hat{Q}} = [0, 0, \ldots, 0],$$
$$\tag{211}$$

$$\left(\frac{\partial \ \ell n \ L[\underline{Y}_{nom}(0) + \underline{y}(0), Q]}{\partial Q}\right)_{\hat{\underline{y}}(0|n), \hat{Q}} = 0 \cdot I. \tag{212}$$

With Eq. (210) the following results are obtained:

$$\hat{\underline{y}}(0|n) = \left\{\sum_{k=1}^{n} [S^T(k)C^T(k)\hat{Q}^{-1}C(k)S(k)]\right\}^{-1}$$

$$\times \left\{\sum_{k=1}^{n} [C(k)S(k)\hat{Q}^{-1}\underline{m}(k)]\right\}, \tag{213}$$

$$\hat{Q} = \frac{1}{n} \sum_{k=1}^{n} [\underline{m}(k) - C(k)S(k)\hat{\underline{y}}(0|n)]$$

$$\times [\underline{m}(k) - C(k)S(k)\hat{\underline{y}}(0|n)]^T. \tag{214}$$

The estimated correction $\hat{\underline{y}}(0|n)$ is used to obtain the maximum likelihood estimate

$$\hat{\underline{Y}}(0|n) = \underline{Y}_{nom}(0) + \hat{\underline{y}}(0|n). \tag{215}$$

Flight-path reconstruction is then achieved by integrating the nonlinear augmented state equation [Eq. (154)], using $\hat{\underline{Y}}(0|n)$ as an initial condition. A performance index J_n is computed as

$$J_n = |Q|. \tag{216}$$

If the difference between two consecutive magnitudes of J_n is greater than a prespecified magnitude, the estimate $\hat{\underline{Y}}(0|n)$ is used to initiate the next iteration cycle.

The covariance matrix $P(0|n)$ of the errors in the estimate $\hat{\underline{Y}}(0|n)$ is defined as

$$P(0|n) \triangleq E\{[\hat{\underline{Y}}(0|N) - E(\hat{\underline{Y}}(0|N))][\hat{\underline{Y}}(0|N) - E(\hat{\underline{Y}}(0|N))]^T\}. \tag{217}$$

The following remarks can be made concerning various properties of the maximum likelihood estimation results:

1. Under certain general conditions, maximum likelihood estimates are consistent and asymptotically efficient. This implies that the accuracy of $\hat{\underline{Y}}(0|n)$ will approach the maximum accuracy achievable if the number of measurements increases:

$$P(0|n) = F^{-1}. \tag{218}$$
$$\scriptstyle n\to\infty$$

The matrix F^{-1} is a lower bound of the covariance matrix $P(0|n)$ and is called the Cramèr-Rao lower bound. The matrix F denotes the so-called Fisher information matrix defined as

$$F \triangleq E\left\{\left(\frac{\partial \ln L(\underline{Y}_{nom}(0) + \underline{y}(0), Q)}{\partial \underline{y}(0)}\right)^T_{\underline{Y}(0),Q} \right.$$
$$\left. \times \left(\frac{\partial \ln L(\underline{Y}_{nom}(0) + \underline{y}(0), Q)}{\partial \underline{y}(0)}\right)_{\underline{Y}(0),Q}\right\}. \tag{219}$$

An approximation to this information matrix can be calculated according to

$$F = \sum_{k=1}^{n} S^T(k) C^T(k) \hat{Q}^{-1} C(k) S(k). \tag{220}$$

This matrix also appears in Eq. (213). The information matrix
is usually calculated by substituting the best estimates of Q
and $S(k)$ after the iteration scheme for $\hat{\underline{Y}}(0|n)$ has converged.

2. In practice, $P(0|n)$ is usually approximated by setting

$$P(0|n) = F^{-1}, \tag{221}$$

even if $n \ll \infty$. With $P(0|n)$ it is simple to approximate the
covariance matrix of $\hat{\underline{Y}}(k|n)$, indicated by $P(k|n)$ according to

$$P(k|n) = S(k) P(0|n) S^T(k). \tag{222}$$

3. In cases where the information matrix [as defined in
Eq. (220)] is ill-conditioned, numerical difficulties may arise
in the matrix inversion as required in Eq. (213). If the information matrix F is singular, one or more components of $\underline{Y}(0)$
cannot be estimated from the available measurements and subsequent elimination is then demanded. If the matrix F is ill-
conditioned but still invertable, then the iterative estimation
scheme may diverge. This difficulty may be circumvented by
resorting to alternative optimization algorithms or by modifying
the information matrix, eliminating the smallest eigenvalues
and corresponding eigenvectors. Numerical details of maximum
likelihood estimates are discussed in Ref. [30].

4. The iteration scheme presented earlier is also referred
to as the quasi-linearization method or modified Newton-Raphson
in the literature. In aircraft dynamic-response analysis, the
algorithm is frequently applied to the problem of estimating
the stability derivatives and initial conditions of the linearized equations of motion. From the previous discussion it

follows that the algorithm may equally well be applied to the
flight-path reconstruction problem ([7,11]).

5. In cases where the process noise can not be neglected
it is still possible to derive an expression for the likelihood function and the corresponding maximum likelihood estimates. However, the resulting optimization problem becomes
more difficult to solve. Analytical and numerical details are
discussed in Ref. [18].

IV. EXPERIMENTAL RESULTS

Several flight-test programs have been carried through for
validation of the dynamic flight-test technique mentioned in
Section I. The experimental aircraft used in these flight-test
programs included a De Havilland DHC-2 low-speed piston-engined
aircraft of Delft University of Technology [5,6,21,31], a high-
subsonic Hawker Hunter mk VII of the National Aerospace Laboratory (NLR) [11,12], and a Fokker F-28 twin-engined transport
aircraft [28]. All flight-test programs had in common that
very high accuracy instrumentation techniques were employed,
and utmost care was devoted to instrumentation calibrations
[15-17,32].

In Section IV some experimental results are presented of
flight-path reconstruction of dynamic symmetrical-asymmetrical
flight-test maneuvers with the DHC-2 "Beaver" experimental aircraft, resulting from application of the extended Kalman filter
and smoothing algorithms discussed earlier. These flight-test
maneuvers were executed via a three-axis electrohydraulic servo
system for the exact implementation of a priori calculated elevator, rudder, and aileron input signals. In the course of
the flight-test program, which was carried through in

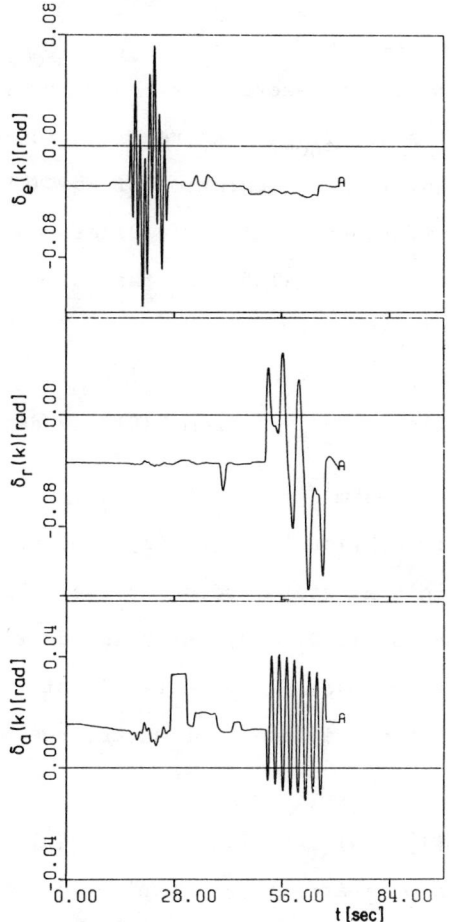

Fig. 11. Elevator, rudder, and aileron input signals of dynamic flight-test maneuver.

cooperation with the Deutsche Forschungs- und Versuchsanstalt fur Luft- und Raumfahrt (DFVLR) and NLR, five different types of test signals were evaluated [31].

Each flight-test maneuver consisted of a longitudinal and a lateral-directional part. After a short period of steady horizontal flight, a 10-sec elevator test signal was implemented, followed after a short delay by a 16-sec aileron-rudder signal. Figure 11 shows one of the test signals of the flight-test program.

A. INITIAL CONDITION COMPUTATION

The subject of this section is the calculation of (1) the nominal initial state $\underline{Y}_{nom}(0)$ as required for nominal flight-path computation, i.e., integration of Eq. (154), and (2) the initial error covariance matrix $P(0|0)$ as required for the solution of Eqs. (186) and (189), of the Kalman filtering and smoothing algorithms.

The initial perturbation correction, i.e., the correction for the deviation of $\underline{Y}_{nom}(0)$ from $\underline{Y}(0)$, is defined as

$$\underline{y}(0) \triangleq \underline{Y}(0) - \underline{Y}_{nom}(0) \tag{223}$$

[see Eq. (157)]. Obviously, $\underline{y}(0)$ is unknown. Consequently, the estimated initial state vector of the Kalman filter [Eq. (185)] is taken as $\hat{\underline{y}}(0|0) = 0$, which is identical to $\hat{\underline{Y}}(0|0) = \underline{Y}_{nom}(0)$. The error made is reflected in statistical terms by the elements of the initial error covariance matrix $P(0|0)$ defined as

$$P(0|0) \triangleq E\{[\hat{\underline{y}}(0|0) - \underline{y}(0)][\hat{\underline{y}}(0|0) - \underline{y}(0)]^T\}. \tag{224}$$

It should be noted that accurate estimation of the nominal initial state is required to minimize linearization errors in the linear system matrices $A(t)$ and $B(t)$ occurring in Eq. (162).

No effort has been made yet to estimate the horizontal position coordinates x_E and y_E. These components are therefore dropped from the augmented system state vector $\underline{Y}$, as defined by Eq. (150). The sidewash coefficient C_{si}, however, is now included because it can not be assumed known when processing actual flight-test measurements. This results in

$$\underline{Y}(t) \triangleq \text{col}[\underline{X}(t), \underline{\lambda}]$$

$$= \text{col}[V_{x_B}, V_{y_B}, V_{z_B}, \psi, \theta, \varphi, z_E, \lambda_x, \lambda_y, \lambda_z, \lambda_p, \lambda_q, \lambda_r, C_{si}].$$

$$\tag{225}$$

In Eq. (225) $\underline{\lambda}$ does not vary with time since the bias errors are assumed to be constant. This also holds true for C_{si}. Since no a priori knowledge is available concerning the components of $\underline{\lambda}_{nom}(0)$, they are all set equal to zero. The remaining elements of $\underline{Y}_{nom}$ are derived from the equations of motion taking account of the fact that the maneuver starts with steady straight flight defined by

$$\dot{V}_{x_B} = \dot{V}_{y_B} = \dot{V}_{z_B} = p_B = q_B = r_B = 0. \tag{226}$$

The pitch angle $\theta_{nom}(0)$ and roll angle $\varphi_{nom}(0)$ may then be expressed as a function of $A_{x_{B_{nom}}}$, $A_{y_{B_{nom}}}$, $A_{z_{B_m}}$:

$$\theta_{nom}(0) = -\arctan A_{x_B}(0)/A_{z_B}(0) \cong -\arctan A_{x_{B_m}}(0)/A_{z_{B_m}}(0), \tag{227}$$

$$\varphi_{nom}(0) = \arctan A_{y_B}(0)/A_{z_B}(0) \cong \arctan A_{y_{B_m}}(0)/A_{z_{B_m}}(0). \tag{228}$$

These expressions follow directly from Eqs. (13)-(15). An estimate of $\psi_{nom}(0)$ may be derived directly from the corresponding measurement

$$\psi_{nom}(0) = \psi_m(0). \tag{229}$$

In steady rectilinear flights with zero angle of roll, A_{y_B} is equal to zero, as follows directly from Eq. (14). For a symmetric aircraft this implies that the velocity vector $\underline{V}$ lies in the plane of symmetry of the aircraft. Consequently, the lateral velocity component $V_{y_B} = 0$. The angle of sideslip β is therefore also equal to zero. If, however, the airflow is not strictly symmetric relative to the plane of symmetry of the aircraft, the condition that $A_{y_B} = 0$ does not necessarily imply that $V_{y_B} = 0$. An ideal sideslip vane would measure β

and implicitly also V_{Y_B} in this flight condition. However, due
to the asymmetry of the airflow, as mentioned earlier, or the
fact that it may not be possible to position the vane in the
plane of symmetry of the aircraft, an, in principle, unknown
sidewash velocity $-C_\beta$ will be induced. Consequently, it is
impossible, in principle, to distinguish the contributions of
$V_{Y_B}(0)$ and $-C_\beta$ in $\beta_V(0)$.

In the present case this ambiguity was resolved by neglecting asymmetric flow effects resulting in

$$V_{Y_{B_{nom}}}(0) = V_{Y_B}(0) \underset{=}{\Delta} 0, \tag{230}$$

and because then $\beta(0) \underset{=}{\Delta} 0$ it, follows from Eq. (25) that

$$C_\beta = -\beta_V(0) \cong -\beta_{V_m}(0). \tag{231}$$

In horizontal flight with zero roll angle φ, it follows that
$\alpha = \theta$ (see Fig. 12). Consequently,

$$V_{X_{B_{nom}}}(0) \cong V_m(0) \cos \theta_{nom}(0), \tag{232}$$

$$V_{z_{B_{nom}}}(0) \cong V_m(0) \sin \theta_{nom}(0). \tag{233}$$

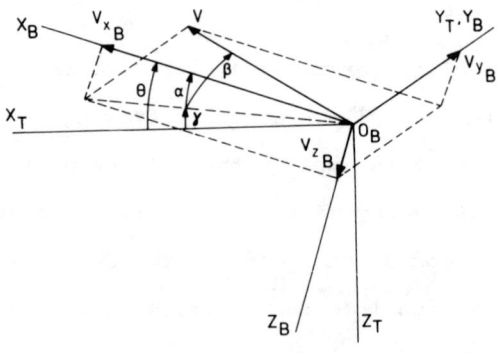

Fig. 12. Relation between pitch angle θ and angle of attack α when roll angle $= 0$.

Finally, by definition

$$z_{E_{nom}}(0) \triangleq 0. \tag{234}$$

Errors in $\hat{\underline{y}}(0|0)$ and therefore also in $\hat{\underline{y}}(0|0)$ result from (1) random measurement errors, (2) constant bias errors in the accelerometer and rate gyroscope measurements, and (3) deviations of the actual flight condition from a horizontal steady rectilinear flight condition. From a quantification of these errors, a reasonable value can be calculated for the a priori variance matrix $P(0|0)$ [13,14].

B. FLIGHT-TEST RESULTS

Numerical values of the elements of the diagonal matrices W and Q are shown in Table I. These values are based on instrumentation calibrations. The flight-path reconstruction results presented here are related to the observation configuration described in Section II.B with system output-signal

Table I. Square Roots of Diagonal Elements of W and Q Matrices

Input signal	$[W_{i,i}]^{1/2}$	Unit	Output signal	$[Q_{i,i}]^{1/2}$	Unit
A_{x_B}	0.32×10^{-2}	m/sec^2	V	0.30	m/sec
A_{y_B}	0.14×10^{-2}	m/sec^2	Δh	0.40	m
A_{z_B}	0.56×10^{-2}	m/sec^2	β	0.15×10^{-1}	rad
p_B	0.56×10^{-4}	rad/sec			
q_B	0.56×10^{-4}	rad/sec			
r_B	0.56×10^{-4}	rad/sec			

components V, Δh, and β_v. Yaw angle ψ and two DME signals
which were also measured in the instrumentation system have not
yet been included in the observation model.

A first result of the application of the extended Kalman
filter and smoothing algorithms, confirmed later by simulation
studies, showed that this observation model resulted in several
"nearly" or even fully unobservable components of the augmented
state vector as defined in Eq. (225). For these augmented
state vector components, the a priori accuracy as expressed by
the relevant diagonal elements of the initial variance matrix
$P(0|0)$ will hardly or not at all improve while processing the
available measurements. This implies that these components may
as well be set equal to zero and, consequently, be eliminated
in the augmented state vector. In the present case the unob-
servable augmented state vector components were ψ, λ_x, λ_y, and
λ_r. The remaining components define an augmented state vector
of lower dimension:

$$\underline{Y}(t) = \text{col}[V_{x_B}, V_{y_B}, V_{z_B}, \theta, \varphi, z_E, \lambda_z, \lambda_p, \lambda_q, C_{si}]. \qquad (235)$$

The numerical values of the corresponding diagonal elements of
the initial covariance matrix are listed in Table II. The
values pertaining to the bias error corrections are based again
on the results of laboratory calibrations.

Figures 13 and 14 show the extended Kalman filter estimates
of the bias error corrections λ_z, λ_q, and λ_p and sidewash coefficient C_{si}. In Figs. 15-17, the extended Kalman smoother
estimates of the output-signal components as well as the corresponding observation residuals are presented.

Table II. Square Roots of Diagnonal Elements of the P(0|0)
Matrix

Quantity	$[P_{ii}(0\|0)]^{1/2}$	Unit	Quantity	$[P_{ii}(0\|0)]^{1/2}$	Unit
V_{x_B}	0.209	m/sec	λ_z	0.04	m/sec^2
V_{y_B}	0.500	m/sec	λ_p	0.001	rad/sec
V_{z_B}	0.030	m/sec	λ_q	0.001	rad/sec
φ	0.031	rad	λ_r	0.001	rad/sec
θ	0.0003	rad	c_{si}	0.1	—
z_E	0.42	m			

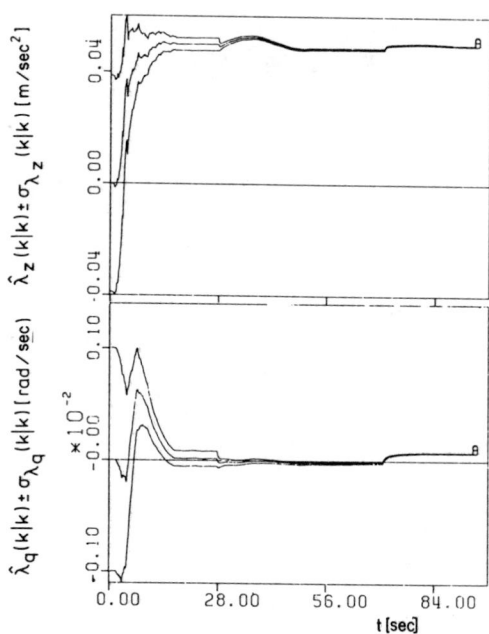

Fig. 13. Extended Kalman filter estimates of bias error
corrections λ_z and λ_q.

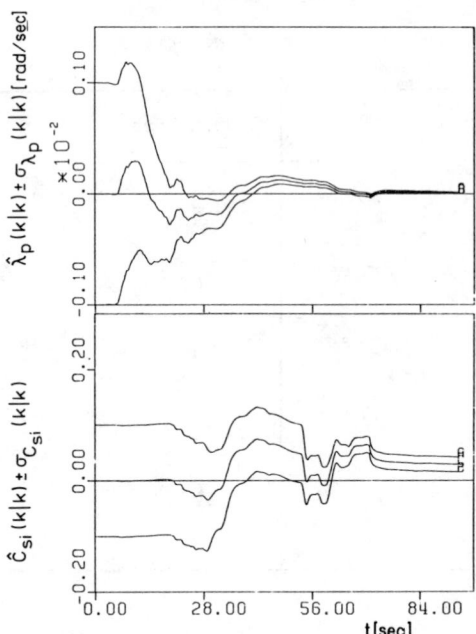

Fig. 14.   Extended Kalman filter estimates of bias error correction λ_p and sidewash C_{si}.

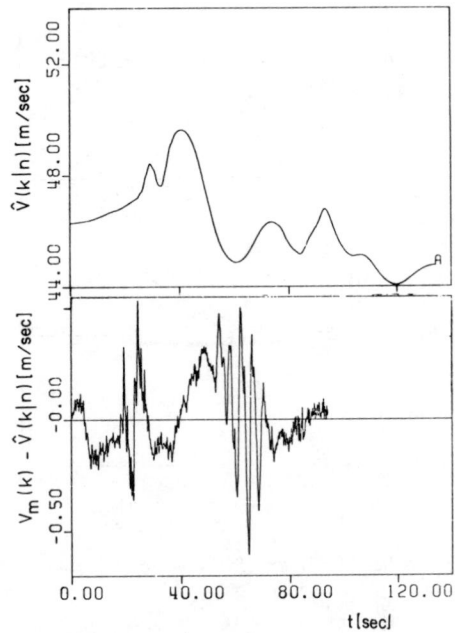

Fig. 15.   Extended Kalman smoother estimate of airspeed V with observation measurement residual.

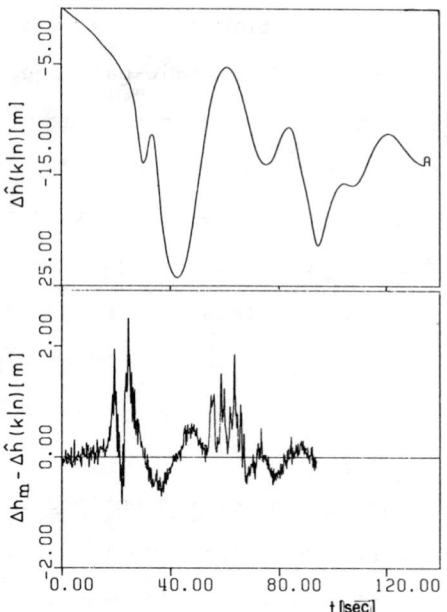

*Fig. 16. Extended Kalman smoother estimate of altitude
variation Δh with observation measurement residual.*

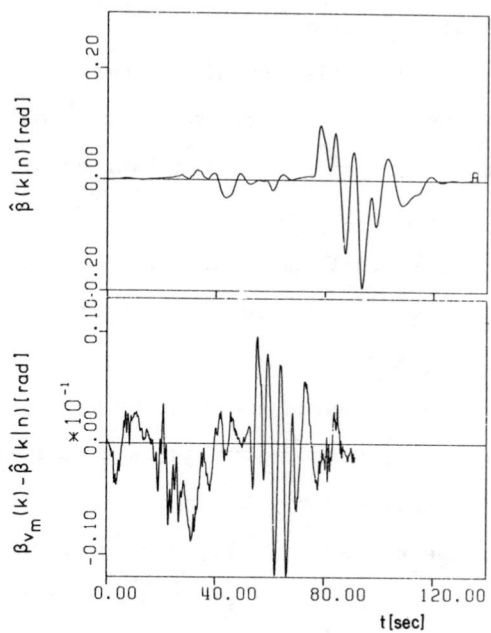

*Fig. 17. Extended Kalman smoother estimate of sideslip
angle β with observation measurement residual.*

With regard to application of the extended Kalman filter and smoothing algorithms for flight-path reconstruction of dynamic maneuvers of the flight-test program mentioned here, more detailed results can be found in Ref. [33]. In Ref. [11], the performance of the maximum likelihood estimation and the extended Kalman filter and smoother algorithms have been compared. In Ref. [26], the weighted least squares solution is approximated by solving a series of regression problems.

V. CONCLUDING REMARKS

Accurate reconstruction of the motions of the aircraft relative to the surrounding air mass is of paramount importance for the deduction of aircraft performance, stability, and control characteristics from measurements in dynamic flight-test maneuvers. This article in particular addresses the flight-path reconstruction problem.

The mathematical models and algorithms used for estimation of the time histories of aircraft state variables from measurements, corrupted with random and constant errors, are discussed in considerable detail in Sections II and III. Some experimental results obtained from actual flight-test measurements with the extended Kalman estimation algorithms, are presented in Section IV.

Perhaps the most important difference in the algorithms for flight-path reconstruction discussed is that system process noise is or is not explicitly taken into account. It has been shown that, in the context of flight-path reconstruction, process noise is directly related to random inertial measurement errors. Consequently, when high-accuracy instrumentation techniques are applied, all algorithms can be expected to yield

closely similar results. Several aspects of flight-path reconstruction fell beyond the scope of the present work, such as
the effect of nonuniform motion of the atmosphere with respect
to earth, the potential benefit of including very accurate geo-
graphic position measurements in the observation model, and
the relation between observability, output-signal observation
configuration, and the geometry of the flight path.

APPENDIX.   DERIVATION OF LINEAR LEAST
            SQUARES BATCH ALGORITHM

Consider the linear model

$$\underline{m}(k) = R(k)\underline{Y} + \underline{\epsilon}(k),$$ (A.1)

in which $\underline{m}(k)$ denotes an n_m-dimensional vector of observations,
$\underline{Y}$ denotes an n_y-dimensional parameter vector, $\underline{\epsilon}(k)$ denotes an
n_m-dimensional model residual that accounts for either obser-
vation-measurement errors or model errors, and $R(k)$ is an
$n_m \times n_y$ matrix of independent variables. When n vector observations are available, a parameter estimate $\underline{\hat{Y}}$ can be calculated,
thus minimizing a quadratic cost function J_n:

$$J_n = \sum_{k=1}^{n} \underline{\hat{\epsilon}}^T(k)Q^{-1}\underline{\hat{\epsilon}}(k),$$ (A.2)

in which Q denotes an $n_m \times n_m$ symmetrical weighting matrix of
full rank and

$$\underline{\hat{\epsilon}}(k) = \underline{m}(k) - R(k)\underline{\hat{Y}}.$$ (A.3)

The necessary condition for J_n to be minimal is

$$\frac{\partial J_n}{\partial \hat{\underline{Y}}} = \frac{\partial J_n}{\partial \hat{\underline{\epsilon}}(k)} \frac{\partial \hat{\underline{\epsilon}}(k)}{\partial \hat{\underline{Y}}} = \frac{\partial \left(\sum\limits_{k=1}^{n} \hat{\underline{\epsilon}}^T(k) Q^{-1} \hat{\underline{\epsilon}}(k) \right)}{\partial \hat{\underline{\epsilon}}(k)} \Bigg/ \frac{\partial \hat{\underline{\epsilon}}(k)}{\partial \hat{\underline{Y}}}$$

$$= \sum_{k=1}^{n} \left\{ 2\hat{\underline{\epsilon}}^T(k) Q^{-1} \frac{\partial \hat{\underline{\epsilon}}(k)}{\partial \hat{\underline{\epsilon}}(k)} \right\} \frac{\partial \hat{\underline{\epsilon}}(k)}{\partial \hat{\underline{Y}}} = [0, 0, \ldots, 0], \qquad (A.4)$$

since Q^{-1} is symmetric. With Eq. (A.3) and $[\partial R(k)\hat{\underline{Y}}/\partial \hat{\underline{Y}}] = R(k)$, it follows that

$$\partial \hat{\underline{\epsilon}}(k)/\partial \hat{\underline{Y}} = -R(k). \qquad (A.5)$$

Because $\partial \hat{\underline{\epsilon}}(k)/\partial \hat{\underline{\epsilon}}(k) = I$, Eq. (A.4) can be written as

$$\frac{\partial J_n}{\partial \hat{\underline{Y}}} = -2 \sum_{k=1}^{n} \hat{\underline{\epsilon}}^T(k) Q^{-1} R(k) = [0, 0, \ldots, 0]. \qquad (A.6)$$

Transposing both sides results in

$$-2 \sum_{k=1}^{n} R^T(k) Q^{-1} \hat{\underline{\epsilon}}(k) = -2 \sum_{k=1}^{n} R^T(k) Q^{-1} (\underline{m}(k) - R(k)\hat{\underline{Y}})$$

$$= \begin{bmatrix} 0 \\ 0 \\ \vdots \\ 0 \end{bmatrix}, \qquad (A.7)$$

which are called the normal equations. If there exists a single solution, it can be calculated according to

$$\hat{\underline{Y}} = \left(\sum_{k=1}^{n} R^T(k) Q^{-1} R(k) \right)^{-1} \sum_{k=1}^{n} R^T(k) Q^{-1} \underline{m}(k), \qquad (A.8)$$

and is indicated as the least squares estimate of Y. In the case of scalar observations, $n_m = 1$, $R(k)$ becomes a $1 \times n_y$ row vector and Q is a scalar. Then Eq. (A.8) reduces to

$$\hat{\underline{Y}} = \left(\sum_{k=1}^{n} R^T(k) R(k) \right)^{-1} R^T(k) m(k).$$ (A.9)

Introduction of a matrix $R^T = [R^T(1), R^T(2), \ldots, R^T(n)]$ and a vector $\underline{m}^T = [m(1), m(2), \ldots, m(n)]$ allows us to write Eq. (A.9) in the form

$$\hat{\underline{Y}} = (R^T R)^{-1} R^T \underline{m},$$ (A.10)

which is the well-known regression estimate.

REFERENCES

1. O. H. GERLACH, *Delft Univ. Technol. Dept. Aerosp. Eng. Rep.* No. VTH-117 (in Dutch with English summary) (1964).

2. F. HUSSENOT, *Tech. Sci. Aeronaut. 6*, 38-49 (1950).

3. ANONYMOUS, *International Civil Aviation Organization Circ.* No. 16-AN/13 (1951).

4. H. B. KLOPFENSTEIN, *American Institute of Aeronautics and Astronautics Pap.* No. 65-211 (1965).

5. O. H. GERLACH, *Soc. Automotive Eng. Natl. Business Aircraft Met. Paper* No. 700236 (1970).

6. O. H. GERLACH, *Advisory Group of Aerospace Research and Development* CP-85 (VTH-163) (1971).

7. J. A. MULDER, *International Federation of Automatic Control Symp. Identification Syst. Parameter Estimation*, 1131-1145 (1973).

8. H. L. JONKERS, *Delft Univ. Technol. Dept. Aerosp. Eng. Rep.* No. VTH-162 (1976).

9. H. L. JONKERS, and J. A. MULDER, *Cong. Int. Council Aeronaut. Sci. 10th Pap.* No. 76-46 (1976).

10. H. L. JONKERS, and J. A. MULDER, *AIAA Aircraft Syst. Technol. Meet. American Institute of Aeronautics and Astronautics Pap.* No. 76-897 (1976).

11. J. A. MULDER, *Advisory Group of Aerospace Research and Development* CP-172 (1975).

12. J. A. MULDER, *Advisory Group of Aerospace Research and Development - Flight Mechanics Panel Symp. Flight Test Tech.*, 11-1—11-30 (1976).

13. J. A. MULDER, H. L. JONKERS, J. J. HORSTEN, J. H. BREEMAN, and J. L. SIMONS, *Advisory Group of Aerospace Research and Development Lecture Ser.* No. 104, 5-1–5-39 (1979).

14. J. A. MULDER, H. L. JONKERS, J. J. HORSTEN, J. H. BREEMAN, and J. L. SIMONS, *Advisory Group for Aerospace Research and Development Lecture Ser.* No. 104, 5-40–5-87 (1979).

15. K. VAN WOERKOM, *Delft Univ. Technol. Dept. Aerosp. Eng. Rep.* No. LR-308 (1981).

16. K. VAN WOERKOM, *Delft Univ. Technol. Dept. Aerosp. Eng.*, to be published.

17. R. J. A. W. HOSMAN, *Advisory Group for Aerospace Research and Development* CP-172 (1974).

18. D. E. STEPNER, and R. E. MEHRA, NASA R-2200 (1973).

19. J. A. MULDER, "Design and Evaluation of Dynamic Flight Test Maneuvers," Dissertation, Department of Aerospace Engineering, Delft University of Technoogy, The Netherlands, 1984, to be published.

20. V. KLEIN, and J. R. SCHIESS, NASA TN D-8514, Langley Research Center (1977).

21. J. A. MULDER, and J. G. DEN HOLLANDER, *Soc. Automotive Eng. Natl. Business Aircraft Meet. Pap.* No. 810597 (1981).

22. S. L. FAGIN, *IEEE Int. Convention Rec.* Part I, Session 30 216-240 (1964).

23. Y. C. HO, "Introduction to Probability, Random Processes and Estimation Theory with Aerospace Applications," *AIAA Recorded Lect. Ser.*

24. R. E. KALMAN, and R. S. BUCY, *J. Basic Eng. 83*, 95-107 (1961).

25. A. P. SAGE, and J. L. MELSA, "Estimation Theory with Application to Communications and Control," McGraw-Hill, New York, 1971.

26. R. J. A. W. HOSMAN, *Delft Univ. Technol. Dept. Aerosp. Eng. Rep.* No. VTH-156 (1971).

27. J. J. BIERMAN, "Factorization Methods for Discrete Sequential Estimation," Academic Press, New York, 1977.

28. J. H. BREEMAN, and J. L. SIMONS, "Evaluation of a Method to Extract Performance Data from Dynamic Maneuvers for a Jet Transport Aircraft," *Congr. Int. Council Aeronaut. Sci. 11th* (1978).

29. R. D. GROVE, R. L. BOWLES, and S. C. MAYHEW, *NASA* TN D-6735 (1972).

30. M. K. GUPTA, and R. K. MEHRA, *IEEE Trans. Autom. Control* *AO-19*, 774-783 (1974).

31. H. C. GARRETSON III, Deutsche Forschungs- und Versuchsanstalt für Luft- und Raumfahrt Braunschweig (1978).

32. J. H. BREEMAN, K. VAN WOERKOM, H. L. JONKERS, and J. A. MULDER, *AGARD Lecture Ser.* No. 104, 4-1—4-22 (1979).

33. J. J. HORSTEN, H. L. JONKERS, and J. A. MULDER, *Delft Univ. Technol. Dept. Aerosp. Eng. Rep.* No. LR-280 (1979).

Optimal Filtering
and Control Techniques
for Torpedo–Ship Tracking Systems

REINHART LUNDERSTAEDT

RUDOLF KERN

Department of Mechanical Engineering
University of the German Armed Forces
Hamburg, Federal Republic of Germany

I. INTRODUCTION

In addition to aeronautics and astronautics, interception problems often occur in naval engineering with regard to the guidance of naval vehicles. By "interception" one generally

Copyright © 1984 by Academic Press, Inc.
All rights of reproduction in any form reserved.
ISBN 0-12-012721-0

means the "hard" interaction of two moving objects, e.g., a
target and a pursuer. This means mathematically that at a prescribed or algorithm-fixed (guidance law) point in time, the
position coordinates of the two moving vehicles have to be
equal.

An example of well-known interception trajectory is the
dogfight curve, where a pursuer is controlled in such a way
that his forward axis is always directed on the target [1].
This classical trajectory has the advantage of being generally
very accurate but the disadvantage of a long interception time,
especially in such cases where the velocities of the target and
the pursuer are nearly equal [2]. The various so-called inter-
ception-course procedures [3], including that of proportional
navigation [4], are therefore more favorable. These procedures
make use of information about the motion of the target in ad-
vance in order to hit it at a prescribed collision point. The
disadvantage of these procedures (especially at the end of the
motion) is that stability problems occur in the guidance law,
thus causing deterioration in accuracy or complete cutoff of
interception [5]. Modern time-domain procedures of control
theory, such as quadratic optimal control, are now used in only
a few cases for the layout of guidance laws for interception
trajectories [6,7]. This is due to the relatively high mathematical expenditure, which is troublesome particularly in solving problems of real time.

Depending on the particular problem, time- and fuel- or
energy-optimal algorithms offer modern solutions for interception trajectories. By their use, the spectrum of the most important applications for the problem of torpedo-ship interception is covered. Due to the complexity of the equations of

motion, especially those of the pursuer, guidance laws cannot be given analytically in this case. If one wants to exclude numerical procedures, which is useful for basic declarations and necessary for real-time tasks, then one obviously has to simplify the mathematical model of the pursuer. A reasonable restriction is the consideration of only planar motions, such as they occur for an torpedo-surface ship interception. If one emphasizes, furthermore, energy-optimal points of view, then one obtains linear equations of motion, which are derived from the kinetics of mass points with on the average only small variations of the velocity of the pursuer. Based on this model, it is easy to apply optimal control theory.

In order to use optimal control laws, it is necessary to measure the position and velocity coordinates of the target (ship) and of the pursuer (torpedo). In both cases stochastic disturbances occur. Because of the linear mathematical model of the pursuer, the disturbances that occur to it can be handled in a simple way by using a Kalman filter. For the target, however, the task is not so easy because the target trajectory is generally not known and has to be generated from the measurement data. Because the measurement equation is nonlinear and a linearization about a nominal trajectory is not possible, one has at first a nonlinear filter problem [8] for the target trajectory. However, this problem can be reduced to an extended Kalman filter so that this measurement task can also be handled by linear theory.

II. MATHEMATICAL MODEL

Starting from a planar motion for the target and the pursuer and assuming the validity of the kinetics of mass points for the pursuer, the normalized equations of motion for the torpedo can be described in accordance with Fig. 1 in the following form:

$$\dot{x}_1 = x_3 + v_1(x_1, x_2, t), \qquad\qquad x_1(t_0) = x_{10},$$

$$\dot{x}_2 = x_4 + v_2(x_1, x_2, t), \qquad\qquad x_2(t_0) = x_{20},$$

$$\dot{x}_3 = -\epsilon x_3\left(x_3^2 + x_4^2\right)^{1/2} - x_5, \qquad x_3(t_0) = x_{30},$$

$$\dot{x}_4 = -\epsilon x_4\left(x_3^2 + x_4^2\right)^{1/2} - x_6, \qquad x_4(t_0) = x_{40}, \qquad (1)$$

$$\dot{x}_5 = -(1/\beta)x_5 + (k/\beta)u_1 \cos(x_7), \qquad x_5(t_0) = x_{50},$$

$$\dot{x}_6 = -(1/\beta)x_6 + (k/\beta)u_1 \sin(x_7), \qquad x_6(t_0) = x_{60},$$

$$\dot{x}_7 = x_8, \qquad\qquad x_7(t_0) = x_{70},$$

$$\dot{x}_8 = \gamma^{-1}u_2, \qquad\qquad x_8(t_0) = x_{80}.$$

where x_1 and x_2 are the position coordinates of the torpedo; x_3 and x_4 its relative velocities in relation to its flow field; x_5 and x_6 the components T_{x1} and T_{x2} of the thrust T; x_7 the

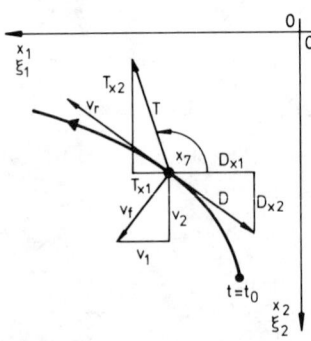

Fig. 1. Kinetics of torpedo.

direction of the thrust in the coordinate systems (x_1, x_2); and x_8 the angular velocity of the torpedo. The components of the flow field are v_1 and v_2; the coefficient in the quadratic drag law D is denoted by ϵ; the moment of inertia of the torpedo about its yaw axis is denoted by γ; the time constant of the engine is β; the amplification of the engine is k. Control variables are introduced: u_1, a measure for the amount of thrust T and u_2, the rudder angle. Time is denoted by t such that $d/dt = (\cdot)$.

The system of Eq. (1) is nonlinear and, in addition, time varying insofar as unstationary flow fields are permitted. It is not amenable to analytic treatment within the scope of an optimization problem. Therefore, the following simplifications are made: (1) $\gamma \to 0$, (2) $\beta \to 0$, (3) $v_f = $ const, and (4) $x_3^2 + x_4^2 \approx 1$; i.e., the dynamics of the angular motion of the torpedo and that of the engine are neglected, and a constant flow field is assumed. Furthermore, the system of Eq. (1) is considered to be normalized in such a way that in addition to v_f the relative velocity v_r of the torpedo can also be regarded approximately as constant in its absolute value. Under these assumptions the system of Eq. (1) becomes

$$
\begin{aligned}
\dot{x}_1 &= x_3 + v_1, & x_1(t_0) &= x_{10}, \\
\dot{x}_2 &= x_4 + v_2, & x_2(t_0) &= x_{20}, \\
\dot{x}_3 &= -\epsilon x_3 - u_1 \cos(u_2), & x_3(t_0) &= x_{30}, \\
\dot{x}_4 &= -\epsilon x_4 - u_1 \sin(u_2), & x_4(t_0) &= x_{40}.
\end{aligned}
\tag{2}
$$

New control variables now are introduced: u_1 as thrust and u_2 as its direction in the coordinate system (x_1, x_2). Finally, taking

$$
\tilde{u}_1 = u_1 \cos(u_2), \qquad \tilde{u}_2 = u_1 \sin(u_2),
\tag{3}
$$

the linear pursuer dynamics follow:

$$
\begin{bmatrix} \dot{x}_1 \\ \dot{x}_2 \\ \dot{x}_3 \\ \dot{x}_4 \end{bmatrix} = \begin{bmatrix} 0 & 0 & 1 & 0 \\ 0 & 0 & 0 & 1 \\ 0 & 0 & -\epsilon & 0 \\ 0 & 0 & 0 & -\epsilon \end{bmatrix} \begin{bmatrix} x_1 \\ x_2 \\ x_3 \\ x_4 \end{bmatrix} + \begin{bmatrix} 0 & 0 \\ 0 & 0 \\ -1 & 0 \\ 0 & -1 \end{bmatrix} \begin{bmatrix} \tilde{u}_1 \\ \tilde{u}_2 \end{bmatrix} + \begin{bmatrix} v_1 \\ v_2 \\ 0 \\ 0 \end{bmatrix} , \qquad (4)
$$

which means that the motion of the torpedo is described by a
linear system having the form

$$
\underline{\dot{x}} = \underline{A} \cdot \underline{x} + \underline{B} \cdot \underline{\tilde{u}} + \underline{v}, \qquad \underline{x}(t_0) = \underline{x}_0 . \qquad (5)
$$

At first it is assumed for the target that its dynamics are
completely known for $t \geq t_0$ and are described by the position
coordinates

$$
\xi_1 = \xi_1(t), \qquad \xi_2 = \xi_2(t), \qquad (6)
$$

and the corresponding velocities

$$
\xi_3(t) = \dot{\xi}_1, \qquad \xi_4(t) = \dot{\xi}_2, \qquad (7)
$$

so that the vector of the relative motion of target (ship) and
pursuer (torpedo) is given by

$$
\underline{e}(t) = \underline{\xi}(t) - \underline{x}(t) . \qquad (8)
$$

III. OPTIMIZATION PROBLEM

In the following, the optimization problem is first formulated. It is then solved and discussed in detail.

A. *FORMULATION*

As mentioned earlier, for a torpedo-ship interception,
time- and energy-optimal solutions are of interest. Therefore
the optimization problem—this means the destination of the control variables $\tilde{u}_1$ and $\tilde{u}_2$ in Eq. (4) and the determination of
the interception time — is formulated in such a way that the

quadratic performance index

$$J(\underline{u}, T) = \frac{1}{2} \underline{e}^T(T) \cdot \underline{P} \cdot \underline{e}(T)$$

$$+ \frac{1}{2} \int_{t_0=0}^{T} \left\{ \delta + (1 - \delta)\left[\underline{e}^T(t) \cdot \underline{Q} \cdot \underline{e}(t)\right.\right.$$

$$\left.\left. + u_1^2(t)\right]\right\} dt \qquad (9)$$

is minimized. In Eq. (9) the weighting factor δ takes into account the interception time T. It is defined for

$$0 \leq \delta < 1. \qquad (10)$$

Consequently, one obtains for $\delta = 0$ pure energy-optimal and for $\delta \to 1$ pure time-optimal solutions. Decidedly, the latter case should be excluded because of the linearity of the optimization problem. The weighting matrices $\underline{P}$ and $\underline{Q}$ in Eq. (9) are given by

$$\underline{P} = \begin{bmatrix} p_{11} & 0 & 0 & 0 \\ 0 & p_{22} & 0 & 0 \\ 0 & 0 & 0 & 0 \\ 0 & 0 & 0 & 0 \end{bmatrix}, \quad \underline{Q} = \begin{bmatrix} q_{11} & 0 & 0 & 0 \\ 0 & q_{22} & 0 & 0 \\ 0 & 0 & 0 & 0 \\ 0 & 0 & 0 & 0 \end{bmatrix}, \qquad (11)$$

where for $\underline{P}$, the interception condition $x_1(T) = \xi_1(T)$, $x_2(T) = \xi_2(T)$ occurs, and for $\underline{Q}$ a weighting of the control deviation $\underline{e}(t)$ is realized. Consequently, it is $p_{11} > 0$, $p_{22} > 0$ and $q_{11} \geq 0$, $q_{22} \geq 0$.

B. SOLUTION

For the solution of the optimization problem, in what follows the absolute value of the thrust $u_1(t)$ is considered to be unconstrained. This means that the control variables $\tilde{u}_1(t)$ and $\tilde{u}_2(t)$ are unconstrained also. One can therefore use the methods of the calculus of variations, since application of

the maximum principle is superfluous. In this way, the linearity of the optimization problem is guaranteed.

From the calculus of variations [9], one obtains for the
optimal control variables

$$\tilde{u}_1^* = -(1 - \delta)^{-1}\psi_3(t, T), \qquad \tilde{u}_2^* = -(1 - \delta)^{-1}\psi_4(t, T), \quad (12)$$

where ψ_3 and ψ_4 are components of the adjoint vector $\underline{\psi}$ belong-
ing to $\underline{x}$. Using Eq. (12) it follows from Eq. (3):

$$u_1^* = -(1 - \delta)^{-1}\left[\psi_3^2(t, T) + \psi_4^2(t, T)\right]^{1/2},$$

$$\tan\left(u_2^*\right) = \psi_4(t, T)/\psi_3(t, T). \tag{13}$$

The optimal interception time T^* can be computed from

$$H(T) - \frac{\partial}{\partial T}\left[\frac{1}{2}\,\underline{e}^T(T) \cdot \underline{P} \cdot \underline{e}(T)\right] = 0, \tag{14}$$

where H is the Hamiltonian. The condition (14) is equivalent
to the demand

$$\partial J(T)/\partial T = 0. \tag{15}$$

The explicit solution of the optimization problem now requires
the determination of the adjoint vector $\underline{\psi}(t)$. If this is done,
then the optimal control variables are explicitly known from
Eq. (12), the optimal interception time T^* can be computed from
Eq. (14), and the open-loop control system [Eqs. (4) and (12)]
can be fed back. First, we present a special case.

1. *Special Case* $\underline{Q} = \underline{0}$

In the special case $\underline{Q} = \underline{0}$, where no weighting of the control deviation $\underline{e}(t)$ in the performance index [Eq. (9)] happens,
the determination of $\underline{\psi}(t)$, and by implication the solution of
the optimization problem, is very simple. One obtains for the
optimal thrust and for the optimal direction of the thrust in

the coordinate system (x_1, x_2)

$$u_1^*(t, T) = -\frac{K(T)}{\epsilon(1 - \delta)}[1 - e^{-\epsilon(T-t)}],$$

$$\tan\left[u_2^*(T)\right] = \tan[\alpha(T)].$$

(16)

where K and α are constants of integration which follow from the transversality condition

$$\psi_i(T) = p_{ii}[\xi_i(T) - x_i(T)], \qquad i = 1, 2,$$

(17)

and are given by

$$K(T) = -\frac{p}{1 + [p/\epsilon^3(1 - \delta)]k_3(T, T)}\left[k_1^2(T) + k_2^2(T)\right]^{1/2},$$

$$\tan[\alpha(T)] = k_2(T)/k_1(T).$$

(18)

We set $p_{11} = p_{22} = p$ as an interception point equal to both position coordinates. The auxiliary functions $k_1(T), \ldots,$ $k_3(t, T)$ in Eq. (18), introduced as abbreviations, are defined by

$$k_1(T) = x_{10} + v_1T + (x_{30}/\epsilon)(1 - e^{-\epsilon T}) - \xi_1(T),$$

$$k_2(T) = x_{20} + v_2T + (x_{40}/\epsilon)(1 - e^{-\epsilon T}) - \xi_2(T),$$

(19)

$$k_3(t, T) = -1 + \epsilon t + e^{-\epsilon t} + e^{-\epsilon T}[1 - ch(\epsilon t)].$$

As it can be seen from Eq. (16), the optimal thrust is an exponential function with $u_1^*(T, T) = 0$ and its direction is constant, which means that the control procedure works with an optimal deflection angle because the target coordinates $\xi_1(T)$ and $\xi_2(T)$ are considered completely known beforehand. Using Eqs. (16) and (18) for the trajectory of the pursuer, it follows that

$$x_1(t, T) = x_{10} + v_1 t + (x_{30}/\epsilon)(1 - e^{-\epsilon t})$$

$$+ \frac{K \cos(\alpha)}{\epsilon^3(1 - \delta)} k_3(t, T),$$

(20)

$$x_2(t, T) = x_{20} + v_2 t + (x_{40}/\epsilon)(1 - e^{-\epsilon t})$$

$$+ \frac{K \sin(\alpha)}{\epsilon^3(1 - \delta)} k_3(t, T),$$

which results in the special case of vanishing initial velocities $x_{30} = x_{40} = 0$ and the vanishing of the flow field
$v_1 = v_2 = 0$ in

$$[x_2(t, T) - x_{20}]/[x_1(t, T) - x_{10}] = \tan[\alpha(T)].$$ (21)

The optimal trajectory is therefore a straight line. Finally,
the optimal interception time T^* is computed from

$$\frac{1}{2} \delta + \frac{p}{2} \frac{\partial}{\partial T} \left\{ \frac{k_1^2(T) + k_2^2(T)}{1 + [p/\epsilon^3(1 - \delta)]k_3(T, T)} \right\} = 0,$$ (22)

which has to be done numerically.

An example for the solution derived is shown in Fig. 2.
In this example, the target trajectory $\xi_1(t) = (1 + v_1)t$,
$\xi_2(t) = 0$ for $t \leq 0$, and $\xi_1(t) = \sin(t) + v_1 t$, $\xi_2(t) = \cos(t) -$
$1 + v_2 t$ for $t > 0$ is assumed. Thus the trajectory is a circle
which is displaced by the velocities v_1 and v_2 of the flow
field. For the pursuer the initial conditions $x_{10} = 0$,
$x_{20} = 1$, and $x_{30} = x_{40} = 0$ are valid. The drag coefficient is
$\epsilon = 1$. Furthermore, it is $v_1 = 0, 1$; $v_2 = 0$, and $p = 10^3$. The
aim of the investigation is to look for the influence of the
weighting factor δ which is responsible for the optimal interception time T^*. In Fig. 2 the target trajectory is pointed
out, and for different values of δ the interception trajectories

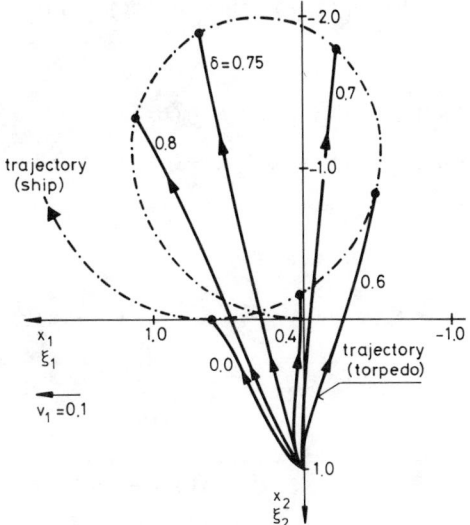

Fig. 2. Optimal interception for torpedo-ship (open loop).

are given. The configuration states that by an appropriate
choice of δ, any combination of time- and energy-optimal pursuer trajectories can be realized.

The solution so far includes only the open-loop control
problem. If a closed-loop control is needed, then the theory
given in [9] yields, for $p_{11} = p_{22} = p$,

$$\begin{bmatrix} \tilde{u}_1^*(t, T) \\ \tilde{u}_2^*(t, T) \end{bmatrix} = \frac{1}{1 - \delta}$$

$$\times \begin{bmatrix} r_{1/2}(T - t) & 0 & r_{3/4}(T - t) & 0 \\ 0 & r_{1/2}(T - t) & 0 & r_{3/4}(T - t) \end{bmatrix}$$

$$\times \begin{bmatrix} x_1(t) \\ x_2(t) \\ x_3(t) \\ x_4(t) \end{bmatrix} - \begin{bmatrix} g_1(t, T) \\ g_2(t, T) \end{bmatrix}, \tag{23}$$

which means that the optimal closed-loop control law has the
form

$$\tilde{u}^*(t, T) = \underline{R}(T - t) \cdot \underline{x}(t) - \underline{g}(t, T). \tag{24}$$

It therefore consists of a linear feedback and additionally of
a feed-forward control. The time-varying coefficients of the
control law in Eq. (23) are given by

$$r_{1/2}(T - t) = (1/\epsilon)[1 - e^{-\epsilon(T-t)}]r(T - t),$$

$$r_{3/4}(T - t) = (1/\epsilon^2)[1 - e^{-\epsilon(T-t)}]^2 r(T - t),$$

$$r(T - t) = \frac{p}{1 + [p/\epsilon^3(1 - \delta)]k_3(T - t)}, \tag{25}$$

$$k_3(T - t) = -1 + \epsilon(T - t) + e^{-\epsilon(T-t)}[2 - ch(\epsilon(T - t))],$$

and for the feed-forward control, one obtains

$$g_1(t, T) = r_{1/2}(T - t)\xi_1(T) - v_1 r_{3/4}(T - t),$$

$$g_2(t, T) = r_{1/2}(T - t)\xi_2(T) - v_2 r_{3/4}(T - t), \tag{26}$$

which follows from the solution of a Riccati matrix differential
equation [9]. The control law depends on the final coordinates
$\xi_1(T)$ and $\xi_2(T)$ of the target because of the feed-forward control $\underline{g}(t, T)$. These therefore have to be known. The coefficients of the controller are pure time functions, which can be
calculated off-line. For $p \gg 1$, they are nearly independent
of δ, which means that the weighting of the interception time
T is only needed for calculating itself from Eq. (9).

2. *General Case* $\underline{Q} \neq \underline{0}$

In the general case $\underline{Q} \neq \underline{0}$, the conditions [Eq. (12)] are
the same as before. Of course for the determination of the
adjoint functions $\psi_3(t)$ and $\psi_4(t)$ distinctions are now necessary because the optimization problem contains for the

eigenvalues λ_i of the system the characteristic equations

$$\lambda_i^4 - \epsilon^2\lambda_i^2 + q_{ii} = 0, \qquad i = 1, 2. \tag{27}$$

Therefore the three cases

$$(1) \quad q_{ii} < \epsilon^4/4; \qquad (2) \quad q_{ii} = \epsilon^4/4; \qquad (3) \quad q_{ii} > \epsilon^4/4, \tag{28}$$

have to be treated separately. Because the procedure is identical in all three cases, we shall select Case 3, which is of
special interest in practice. All further considerations are
referred to this case. Using the abbreviations

$$\gamma_i = \left\{\frac{1}{2}\left[(q_{ii})^{1/2} + \frac{1}{2}\epsilon^2\right]\right\}^{1/2},$$

$$\omega_i = \left\{\frac{1}{2}\left[(q_{ii})^{1/2} - \frac{1}{2}\epsilon^2\right]\right\}^{1/2}, \tag{29}$$

which are the absolute values of the real and imaginary parts
of the eigenvalues λ_i, $i = 1, 2$, and the auxiliary functions
given in Appendix A, one obtains

$$\psi_{i+2}(t, T) = -\left[m_{1i}(t) - \frac{m_{1i}(T)}{m_{0i}(T)} m_{0i}(t)\right]\psi_{i0}$$

$$- (1 - \delta)q_{ii}f_{i+2}(t, T). \tag{30}$$

The constants of integration ψ_{i0}, $i = 1, 2$ are related to

$$\psi_{i0} = -\frac{pk_{1i}(T) + (1 - \delta)q_{ii}k_{2i}(T)}{1 + [p/(1 - \delta)]\ell_{1i}(T) + q_{ii}\ell_{2i}(T)}, \tag{31}$$

with the connection

$$\psi_{10} = K \cos \alpha, \qquad \psi_{20} = K \sin \alpha, \tag{32}$$

to the constants K and α used previously. In Eq. (31) it is
assumed again that $p_{11} = p_{22} = p$, where the abbreviations
$k_{1i}(T)$, $k_{2i}(T)$, $\ell_{1i}(T)$, and $\ell_{2i}(T)$ are assigned in Appendix A.
In Eq. (30) it is now important, because of Eqs. (A.4) and
(A.8), that the solution of the optimization problem depends

not only on the final state $[\xi_1(T), \xi_2(T)]$ of the target, but also on the whole target trajectory for $0 \leq t \leq T$.

Using Eqs. (30) and (12) it is now possible to integrate the equations of motion [Eq. (4)]. By this the trajectory of the torpedo is known. The analytic expressions for $x_1(t)$ and $x_2(t)$ are not given here: in this connection the reader is referred to [10]. In order to show the influence of the weighting matrix $\underline{Q}$, we refer to the example treated previously. It is shown in Fig. 3 with a parameter variation of $q_{11} = q_{22} = q$, the time of interception T is thereby fixed. It is evident from the graphs that by using the parameter q completely different interception trajectories can be realized. This is important in order to make an attack on the ship with the torpedo, for instance, from behind.

The discussion until now is valid only for the open-loop control problem. If one wants to use a closed-loop control, which is generally necessary, one is led again to Eq. (23).

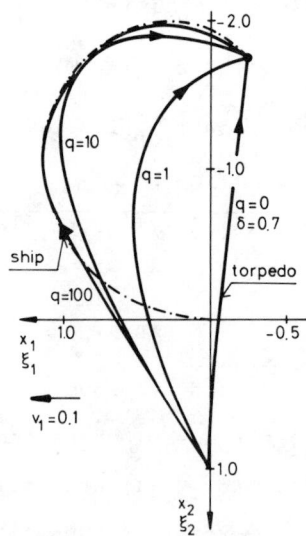

Fig. 3.   Optimal interception for $\underline{Q} \neq \underline{0}$.

The coefficients of the controller are now given by

$$r_i(T - t) = -\frac{d_{1i}(T - t)}{d_{0i}(T - t)} r_{0i}(T - t)$$

$$- (1 - \delta)q_{ii} \frac{d_{2i}(T - t)}{d_{0i}(T - t)},$$

(33)

$$r_{i+2}(T - t) = \left[\frac{d_{1i}(T - t)}{d_{0i}(T - t)}\right]^2 r_{0i}(T - t) + (1 - \delta)q_{ii}$$

$$\times \left[\frac{c_{3i}(T - t)d_{0i}(T - t) + d_{1i}(T - t)d_{2i}(T - t)}{d_{0i}^2(T - t)}\right],$$

$$i = 1, 2.$$

The abbreviation $r_{0i}(T - t)$ is assigned in Appendix B. The
functions $c_{3i}, \ldots, d_{2i}$ in Eq. (33) follow from Appendix A, if
one there replaces the argument t by t - T. In contrast to
Eq. (25) the coefficients of the controller now have stationary
values not equal to zero, which can be seen from

$$\bar{r}_i = \lim_{T \to \infty} r_i(T - t) = (1 - \delta)(q_{ii})^{1/2},$$

$$\bar{r}_{i+2} = \lim_{T \to \infty} r_{i+2}(T - t) = (1 - \delta)\left\{-\epsilon + \left[\epsilon^2 + 2(q_{ii})^{1/2}\right]^{1/2}\right\},$$

$$i = 1, 2.$$

(34)

The stationary value of the coefficient r_i is not influenced at
all by the drag, and the stationary value of the coefficient
r_{i+2} is influenced only slightly, since $0 \leq \epsilon \leq 1$ can be assumed
for physical reasons and, furthermore, Eq. (28) deals with Case
3. Moreover, the slight influence of ϵ on r_i and r_{i+2} is not
only valid for the stationary values but is also generally
valid for $0 \leq t \leq T \leq \infty$, because it can be seen from a numerical exploitation of Eq. (33). For the feed-forward control

$g_i(t, T)$, $i = 1, 2$ in Eq. (23), we now obtain

$$g_i(t, T) = \frac{h_i(T - t)}{n_i(T - t)}\, \xi_i(T) - [r_{i+2}(T - t) - \Omega_i(T - t)]v_i$$
$$- q_{ii}[r_i(T - t)i_{1i}(t, T) + r_{i+2}(T - t)i_{2i}(t, T)$$
$$+ (1 - \delta)i_{4i}(t, T)], \tag{35}$$

with the auxiliary functions $h_i(T - t)$ and $\Omega_i(T - t)$ given in Appendix B. The integrals i_{ji} ($j = 1, 2, 4$; $i = 1, 2$) are as explained in Appendix A, but now with t as the lower and T as the upper integration limit. From Eq. (35) it is evident that for the feed-forward control analogous to the open-loop control, the whole target trajectory has to be known for $0 \le t \le T$. It is not sufficient to know only the final values $[\xi_1(T), \xi_2(T)]$.

The control law [Eq. (23)] with the coefficients of Eq. (33) and the feed-forward control [Eq. (35)] produces good results. It has, however, the disadvantage that it cannot be used for unknown target trajectories. One therefore has to try a suboptimal solution to eliminate this disadvantage. For this purpose one develops the integrals of Eqs. (A.4) and (A.8) given in Appendix A by partial integration up to the accelerations $\ddot{\xi}_i(T)$, $i = 1, 2$. The feed-forward control [Eq. (35)] then goes over in

$$g_i(t, T) = r_i(T - t)\xi_i(t) + r_{i+2}(T - t)$$
$$\times [\xi_{i+2}(t) - v_i] + R_i(t, T), \tag{36}$$

with the residual function

$$R_i(t, T) = (1 - \delta)\epsilon[\xi_{i+2}(t) - \xi_{i+2}(T)]$$
$$- \Omega_i(T - t)[\xi_{i+2}(T) - v_i] + r_i(T - t)I_{1i}(t, T)$$
$$+ r_{i+2}(T - t)I_{2i}(t, T) - (1 - \delta)q_{ii}I_{4i}(t, T). \tag{37}$$

This residual function depends on the integrals given in
Appendix B and on an additional expression resulting from the
hydrodynamic drag. For ships as targets it is certain that
their accelerations are small. Thus far it has been permissible
to neglect these integrals the influence of which for $t \to T$ in
any way vanishes. Furthermore, it can be proven that the additive expressions of higher order that depend on ϵ are also
small. Using this it is possible to put

$$R_i(t, T) = 0, \tag{38}$$

for $0 \leq t \leq T$, and the optimal control law can be replaced by
the suboptimal expression

$$\tilde{u}_i^*(T - t) = -(1 - \delta)^{-1}\{r_i(T - t)[\xi_i(t) - x_i(t)]$$

$$+ r_{i+2}(T - t)[\xi_{i+2}(t) - x_{i+2}(t) - v_i]\},$$

$$i = 1, 2. \tag{39}$$

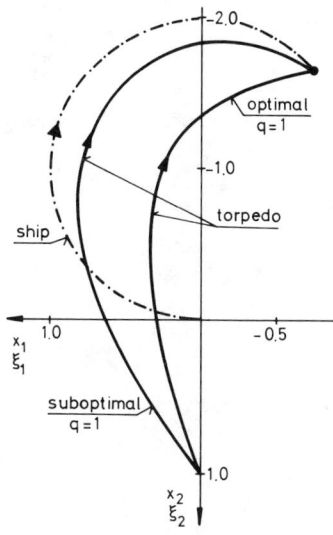

Fig. 4. Optimal and suboptimal interception.

In the suboptimal control law no preliminary knowledge of the
target trajectory is necessary. Only the actual target coordinates for the specific time t are needed.

For the example treated earlier, but now with $v_i = 0$,
$i = 1, 2$, and $\epsilon = 0, 5$, Fig. 4 shows the difference between
the optimal and the suboptimal solutions. One sees that the
optimal solution is comparable within a certain scope to the
proportional navigation, the suboptimal solution to a dog-fight
curve.

An additional example is given in Fig. 5, where the target
with the coordinates $\xi_1(t) = (1 + v_1)t$ and $\xi_2(t) = 0$ for $t \geq 0$
is pursued. Variation parameter is again $q_{11} = q_{22} = q$ and
q_{11}, q_{22}, respectively. Figure 5 points out that in the

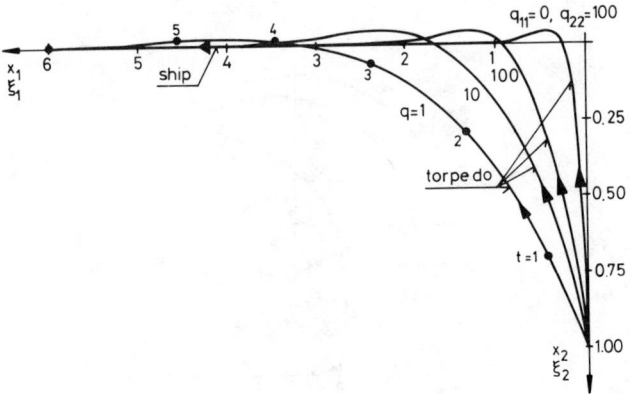

Fig. 5. Suboptimal interception trajectories.

suboptimal case as well as different pursuer trajectories are realizable by an appropriate choice of q_{11} and q_{22}, which is extremely important for tactical considerations.

IV. OPTIMAL FILTERING

For further development, the general suboptimal control law [Eq. (39)] is used. Under this control law, the state variables x_1, ..., x_4 of the pursuer (torpedo) and the state variables ξ_1, ..., ξ_4 of the target (ship) occur. If one uses in accordance with Eq. (8) relative coordinates, then the number of state variables is reduced from eight to four, but in this case no inertial assignments for the coordinates are possible. It does not matter which consideration is used. In any case, the state variables contained in the control law have to be measured. These measurements are subject to stochastic errors so that one has to develop appropriate filter algorithms.

A. *KALMAN FILTER FOR THE TORPEDO*

As shown in Eq. (4) the torpedo can be described by linear state equations. One can assume that in these state equations no stochastic parts occur if, for example, no influence of wave motions of the flow field is considered. With a proportional working measurement device one obtains the measurement equation

$$\begin{bmatrix} y_1 \\ y_2 \end{bmatrix} = \begin{bmatrix} 1 & 0 & 0 & 0 \\ 0 & 1 & 0 & 0 \end{bmatrix} \begin{bmatrix} x_1 \\ x_2 \\ x_3 \\ x_4 \end{bmatrix} + \underline{w}(t), \qquad (40)$$

if only the position coordinates are measured, and

$$
\begin{bmatrix} y_1 \\ y_2 \\ y_3 \\ y_4 \end{bmatrix} = \begin{bmatrix} 1 & 0 & 0 & 0 \\ 0 & 1 & 0 & 0 \\ 0 & 0 & 1 & 0 \\ 0 & 0 & 0 & 1 \end{bmatrix} \begin{bmatrix} x_1 \\ x_2 \\ x_3 \\ x_4 \end{bmatrix} + \underline{w}(t), \tag{41}
$$

if the velocity coordinates are measured as well; i.e., in general,

$$
\underline{y} = \underline{C} \cdot \underline{x} + \underline{w}(t). \tag{42}
$$

Assuming that $\underline{w}$ in Eq. (42) is a normal distributed white-noise process with $E[\underline{w}(t)] = \underline{0}$, then for Eqs. (4) and (40) a linear Kalman filter can be outlined. In accordance with the theoretical background given in [11], for this case

$$
\dot{\hat{\underline{x}}} = \underline{A} \cdot \hat{\underline{x}} + \underline{K}(t) \cdot [\underline{y} - \underline{C} \cdot \hat{\underline{x}}] + \underline{d},
$$
$$
\hat{\underline{x}}(0) = \hat{\underline{x}}_0 = E[\underline{x}(0)] \tag{43}
$$

is valid. Here is for reduction $\underline{d} = \underline{B} \cdot \tilde{\underline{u}} + \underline{v}$ the deterministic part in Eq. (5). The Kalman matrix $\underline{K}(t)$ now follows from

$$
\underline{K}(t) = \underline{P}(t) \cdot \underline{C}^T \cdot \underline{R}^{-1}, \tag{44}
$$

where the covariance matrix $\underline{P}(t)$ obeys the Riccati matrix differential equation

$$
\dot{\underline{P}} = \underline{A} \cdot \underline{P} + \underline{P} \cdot \underline{A}^T - \underline{P} \cdot \underline{C}^T \cdot \underline{R}^{-1} \cdot \underline{C} \cdot \underline{P},
$$
$$
\underline{P}(0) = E\left[(\hat{\underline{x}}_0 - \underline{x}_0) \cdot (\hat{\underline{x}}_0 - \underline{x}_0)^T\right]. \tag{45}
$$

Furthermore, it is

$$
E[\underline{w}(t) \cdot \underline{w}(t + \tau)] = \underline{R} \cdot \delta(\tau) \tag{46}
$$

for the noise process in Eq. (42). If one assumes for $\underline{P}(0)$ and $\underline{R}$ diagonal matrices of the form

$$
\underline{P}(0) = \text{diag}\left(\sigma_{ii}^2\right), \qquad \underline{R} = \text{diag}\left(r_{ii}^2\right), \qquad i = 1, \ldots, 4, \tag{47}
$$

and postulates additionally that there are equal stochastic
qualities for the position coordinates on the one hand and
equal stochastic qualities for the velocity coordinates on the
other, which means that

$$\sigma_{11} = \sigma_{22}, \qquad \sigma_{33} = \sigma_{44}, \tag{48}$$

$$r_{11} = r_{22}, \qquad r_{33} = r_{44}, \tag{49}$$

then the filter can be given analytically in a simple way. One
obtains

$$
\begin{bmatrix} \dot{\hat{x}}_1 \\ \dot{\hat{x}}_2 \\ \dot{\hat{x}}_3 \\ \dot{\hat{x}}_4 \end{bmatrix} =
\begin{bmatrix}
-k_{11} & 0 & 1 - k_{13} & 0 \\
0 & -k_{22} & 0 & 1 - k_{24} \\
-k_{31} & 0 & -(\epsilon + k_{33}) & 0 \\
0 & -k_{42} & 0 & -(\epsilon + k_{44})
\end{bmatrix}
\begin{bmatrix} \hat{x}_1 \\ \hat{x}_2 \\ \hat{x}_3 \\ \hat{x}_4 \end{bmatrix}
$$

$$
+ \begin{bmatrix}
k_{11}y_1 + k_{13}y_3 \\
k_{22}y_2 + k_{24}y_4 \\
k_{31}y_1 + k_{33}y_3 \\
k_{42}y_2 + k_{44}y_4
\end{bmatrix} + \underline{d}, \tag{50}
$$

and because of Eqs. (48) and (49),

$$k_{11} = k_{22}, \qquad k_{13} = k_{24}, \qquad k_{33} = k_{44}, \qquad k_{31} = k_{42}. \tag{51}$$

The explicit expressions of the elements k_{ij} ($i = 1, \ldots, 4$;
$j = 1, \ldots, 4$) are assigned in Appendix C. If the velocities
x_3 and x_4 are not measured, then $y_3 = y_4 \equiv 0$ is valid. From
this follows $k_{13} = k_{33} \equiv 0$. In order to consider this case,
the auxiliary variable Δ is introduced in the elements k_{ij} in
Appendix C. If the velocities x_3 and x_4 are measured then it
is set $\Delta = 1$; otherwise it is $\Delta = 0$. By this the stated expressions are valid for both cases. In the elements of the
Kalman matrix it is evident that these are not explicitly

dependent on the relations of Eq. (47) but depend only on the
ratios

$$\rho^2 = (r_{11}/\sigma_{11})^2, \qquad \lambda^2 = (r_{11}/\sigma_{33})^2, \qquad \gamma^2 = (r_{11}/r_{33})^2.$$

$$(52)$$

This simplifies considerably the choice of appropriate values,
especially for $\underline{P}(0)$.

B. PASSIVE TRACKING

For the designation of the target data needed in the control law, i.e., the state variables $\xi_1(t), \ldots, \xi_4(t)$, one can
go two different ways. On the one hand, the variables
$\xi_1(t), \ldots, \xi_4(t)$ can be determinated directly by active bearing measurements (active tracking). On the other hand, it is
also possible to use only passive bearing measurements (passive
tracking). From this one obtains the relative coordinates
$e_1(t), \ldots, e_4(t)$. Because the state vector $\underline{x}$ is known from
Section IV.A, in this case the inertial coordinates $\xi_1(t), \ldots,$
$\xi_4(t)$ can also be stated by Eq. (8). In further developments
this method is used.

For passive target tracking it is useful to change from a
continuous consideration to a discrete one. This has advantages
for the algorithms to be developed. For this it is necessary
to write the equations of motion [Eq. (4)] of the torpedo in a

discrete form. Using the sampling time T_0, that means

$$
\begin{bmatrix} x_1[(k+1)T_0] \\ x_2[(k+1)T_0] \\ x_3[(k+1)T_0] \\ x_4[(k+1)T_0] \end{bmatrix} =
\begin{bmatrix}
1 & 0 & \epsilon^{-1}[1-\exp(-\epsilon T_0)] & 0 \\
0 & 1 & 0 & \epsilon^{-1}[1-\exp(-\epsilon T_0)] \\
0 & 0 & \exp(-\epsilon T_0) & 0 \\
0 & 0 & 0 & \exp(-\epsilon T_0)
\end{bmatrix}
$$

$$
\times \begin{bmatrix} x_1(kT_0) \\ x_2(kT_0) \\ x_3(kT_0) \\ x_4(kT_0) \end{bmatrix} + \underline{\tilde{c}}(kT_0), \tag{53a}
$$

$$
\underline{\tilde{c}}(kT_0) =
\begin{bmatrix}
\epsilon^{-2}[1-\exp(-\epsilon T_0) - \epsilon T_0] & 0 \\
0 & \epsilon^{-2}[1-\exp(-\epsilon T_0) - \epsilon T_0] \\
\epsilon^{-1}[\exp(-\epsilon T_0) - 1] & 0 \\
0 & \epsilon^{-1}[\exp(-\epsilon T_0) - 1]
\end{bmatrix}
$$

$$
\times \begin{bmatrix} \tilde{u}_1(kT_0) \\ \tilde{u}_2(kT_0) \end{bmatrix} + \begin{bmatrix} v_1 T_0 \\ v_2 T_0 \\ 0 \\ 0 \end{bmatrix}, \qquad k = 0, 1, 2, \ldots .
$$

$$\tag{53b}$$

The dynamics of the torpedo are consequently described by the general linear difference equation

$$
\underline{x}[(k+1)T_0] = \underline{\phi}(t_0) \cdot \underline{x}(kT_0)
$$

$$
+ \underline{\tilde{B}}(T_0) \cdot \underline{\tilde{u}}(kT_0) + \underline{v}(T_0). \tag{54}
$$

For the target (ship) the considerations have to be improved in some details.

Case 1. Assuming that the ship is moving with constant velocity, the discrete equations of motion for the target are given by

$$
\begin{bmatrix}
\xi_1[(k+1)T_0] \\
\xi_2[(k+1)T_0] \\
\xi_3[(k+1)T_0] \\
\xi_4[(k+1)T_0]
\end{bmatrix}
=
\begin{bmatrix}
1 & 0 & T_0 & 0 \\
0 & 1 & 0 & T_0 \\
0 & 0 & 1 & 0 \\
0 & 0 & 0 & 1
\end{bmatrix}
\begin{bmatrix}
\xi_1(kT_0) \\
\xi_2(kT_0) \\
\xi_3(kT_0) \\
\xi_4(kT_0)
\end{bmatrix}
$$

$$
+
\begin{bmatrix}
v_1 T_0 \\
v_2 T_0 \\
0 \\
0
\end{bmatrix}, \qquad k = 0, 1, 2, \ldots, \tag{55}
$$

where now $\xi_3(t)$ and $\xi_4(t)$ are relative velocities in accordance with $x_3(t)$ and $x_4(t)$. Inserting Eqs. (53) and (55) into Eq. (8), one obtains as relative motion between the torpedo and the ship:

$$
\begin{bmatrix}
e_1[(k+1)T_0] \\
e_2[(k+1)T_0] \\
e_3[(k+1)T_0] \\
e_4[(k+1)T_0]
\end{bmatrix}
=
\begin{bmatrix}
1 & 0 & T_0 & 0 \\
0 & 1 & 0 & T_0 \\
0 & 0 & 1 & 0 \\
0 & 0 & 0 & 1
\end{bmatrix}
\begin{bmatrix}
e_1(kT_0) \\
e_2(kT_0) \\
e_3(kT_0) \\
e_4(kT_0)
\end{bmatrix}
$$

$$
+ \underline{z}(kT_0), \qquad k = 0, 1, 2, \ldots, \tag{56}
$$

which is identical with the control deviation $\underline{e}(t)$ in Eq. (8). In Eq. (56), $\underline{z}(kT_0)$ is a known control vector given explicitly by

$$
\underline{z}(kT_0) =
\begin{bmatrix}
\epsilon^{-1}[-1 + \exp(-\epsilon T_0) + \epsilon T_0]\left[x_3(kT_0) + \epsilon^{-1}\tilde{u}_1(kT_0)\right] \\
\epsilon^{-1}[-1 + \exp(-\epsilon T_0) + \epsilon T_0]\left[x_4(kT_0) + \epsilon^{-1}\tilde{u}_2(kT_0)\right] \\
[1 - \exp(-\epsilon T_0)]\left[x_3(kT_0) + \epsilon^{-1}\tilde{u}_1(kT_0)\right] \\
[1 - \exp(-\epsilon T_0)]\left[x_4(kT_0) + \epsilon^{-1}\tilde{u}_2(kT_0)\right]
\end{bmatrix}.
$$

$$
\tag{57}
$$

Consequently, the relative motion is described by the general
linear difference equation

$$\underline{e}[(k + 1)T_0] = \underline{F}(T_0) \cdot \underline{e}(kT_0) + \underline{z}(kT_0). \tag{58}$$

Case 2. The assumption of constant ship velocity is in
some cases not fulfilled. Indeed, it is sufficient to restrict
for variable velocity on linear changes. Using this assumption, Eq. (55) becomes

$$
\begin{bmatrix}
\xi_1[(k + 1)T_0] \\
\xi_2[(k + 1)T_0] \\
\xi_3[(k + 1)T_0] \\
\xi_4[(k + 1)T_0] \\
\xi_5[(k + 1)T_0] \\
\xi_6[(k + 1)T_0]
\end{bmatrix}
=
\begin{bmatrix}
1 & 0 & T_0 & 0 & \frac{1}{2}T_0^2 & 0 \\
0 & 1 & 0 & T_0 & 0 & \frac{1}{2}T_0^2 \\
0 & 0 & 1 & 0 & T_0 & 0 \\
0 & 0 & 0 & 1 & 0 & T_0 \\
0 & 0 & 0 & 0 & 1 & 0 \\
0 & 0 & 0 & 0 & 0 & 1
\end{bmatrix}
$$

$$
\times
\begin{bmatrix}
\xi_1(kT_0) \\
\xi_2(kT_0) \\
\xi_3(kT_0) \\
\xi_4(kT_0) \\
\xi_5(kT_0) \\
\xi_6(kT_0)
\end{bmatrix}
+
\begin{bmatrix}
v_1T_0 \\
v_2T_0 \\
0 \\
0 \\
0 \\
0
\end{bmatrix}. \tag{59}
$$

In Eq. (59) the additional state variables ξ_5 and ξ_6 are constant parameters to be identified. The difference equation for
the relative vector $\underline{e}(t)$, which now has the dimensions 6 × 1
follows from Eq. (59) in the same way as in the previous case;
$\underline{F}(T_0)$ is the system matrix from Eq. (59); and the control vector $\underline{z}(kT_0)$ is derived from Eq. (57).

As a measurement equation for the generation of the target
trajectory, which means the destination of the vector $\underline{e}(kT_0)$,
only the bearing angle

$$\beta(kT_0) = \tan^{-1}[e_1(kT_0)/e_2(kT_0)] + \eta(kT_0) \tag{60}$$

is available [8], in which $\eta(kT_0)$ is a white-noise process with normal distribution

$$E[\eta(kT_0)] = 0,$$

$$E[\eta(jT_0)\eta(kT_0)] = \begin{cases} \sigma^2(kT_0); & j = k \\ 0 & ; \quad j \neq k \end{cases}, \quad k = 0, 1, 2, \ldots .$$

(61)

The measurement equation (60) is nonlinear. From this equation in connection with Eq. (58) the vector $\underline{e}(kT_0)$ has to be determined. If one excludes pathological cases [8], this can be done in a deterministic way: i.e., $\eta(kT_0) \equiv 0$ for Eq. (56) with four measurements, and for Eq. (52) with six measurements. A necessary condition for the observability is therefore a sufficient course change of the pursuer in order to have enough information about the target to solve Eq. (58). In the stochastic case as well at least four measurements are needed to solve Eq. (56) and six to solve Eq. (59), respectively, but in general it is better to have more in order to carry out a data preprocessing. The stochastic measurements of Eq. (60) have to be prepared by a dynamic filter. Following the considerations in [8] and [12], one then obtains an extended Kalman filter belonging to Eq. (58) and given in Appendix D. This filter has been derived for the measurement equation [Eq. (56)]; for Eq. (59) it has to be extended correspondingly. The filter algorithm itself follows from a linearization of the measurement equation about the momentary estimation vector $\underline{\hat{e}}(k + 1|k)$. Now, simulations show that the derived algorithm has an inefficient convergence behavior, especially in regard to the covariance matrix $\underline{P}(k|k)$. This is based on the dependence of the linearized measurement vector $\underline{c}(k)$ upon the estimated state vector $\underline{\hat{e}}(k|k - 1)$, by which a feedback occurs in the calculation

of the amplification vector $\underline{k}(k)$, which represents here the
Kalman matrix. A decoupling of state estimation and covariance
matrix calculations can be reached in a simple way if one sub-
mits the measurement equation to a pseudolinearization [8,12].
Then one obtains the following algorithm with better convergence behavior:

Prediction phase:

$$\hat{\underline{e}}(k + 1|k) = \underline{F}(k + 1, k) \cdot \hat{\underline{e}}(k, k) + \underline{z}(k),$$

$$\hat{\underline{P}}(k + 1|k) = \underline{F}(k + 1, k) \cdot \hat{\underline{P}}(k|k) \cdot \underline{F}^T(k + 1, k),$$

$$k = 0, 1, 2, \ldots . \tag{62}$$

Measurement phase:

$$\hat{\underline{c}}^T(k + 1) = [\cos \beta(k + 1), -\sin \beta(k + 1), 0, 0]. \tag{63}$$

Correction phase:

$$\hat{\underline{k}}(k + 1) = \hat{\underline{P}}(k + 1|k) \cdot \hat{\underline{c}}(k + 1)$$

$$\times \{\hat{\underline{c}}^T(k + 1) \cdot \hat{\underline{P}}(k + 1|k) \cdot \hat{\underline{c}}(k + 1)$$

$$+ \sigma^2(k + 1)\}^{-1},$$

$$\hat{\underline{e}}(k + 1|k + 1) = \hat{\underline{e}}(k + 1|k) - \hat{\underline{k}}(k + 1) \cdot \hat{c}^T(k + 1) \tag{64}$$

$$\cdot \hat{\underline{e}}(k + 1|k),$$

$$\hat{\underline{P}}(k + 1|k + 1) = \hat{\underline{P}}(k + 1|k) - \hat{\underline{k}}(k + 1) \cdot \hat{c}^T(k + 1)$$

$$\cdot \hat{\underline{P}}(k + 1|k).$$

Initial condition:

$$\hat{\underline{e}}(0|0) = \underline{0}, \qquad \hat{\underline{P}}(0|0) = \text{diag}\left(\hat{\sigma}_0^2\right). \tag{65}$$

In this algorithm, as well as in the extended Kalman filter in
Appendix D, the argument T_0 is omitted. The variance $\hat{\sigma}_0^2$ in
Eq. (65) can be set in general equal to one.

The algorithm given by Eqs. (62)-(65) is now used to generate the target trajectory $\xi_1(t) = 1t$, $\xi_2(t) \equiv 0$ for $t \geq 0$. The pursuer (torpedo) thereby derives the following identification trajectory with the initial conditions $x_{10} = 1.12$; $x_{20} = 4.50$:

$$0 \leq t \leq 1: \quad x_1(t) = x_{10} = \text{const},$$
$$x_2(t) = -1t + x_{20}.$$

$$1 \leq t \leq 2: \quad x_1(t) = 1(t - 1) + x_{10},$$
$$x_2(t) = x_2(1) = \text{const}.$$

$$2 \leq t \leq 3: \quad x_1(t) = x_1(2) = \text{const},$$
$$x_2(t) = -1(t - 2) + x_2(1).$$

$$3 \leq t \leq 4: \quad x_1(t) = 1(t - 3) + x_1(2),$$
$$x_2(t) = x_2(3) = \text{const}.$$

$$4 \leq t \leq 5: \quad x_1(t) = x_1(4) = \text{const}.$$
$$x_2(t) = -1(t - 4) + x_3(3).$$

Consequently, the pursuer dynamics is assumed to be an ideal rectangular trajectory leading to certain simplifications in Eq. (57). $T_0 = 1/12$ is chosen as the sampling time; i.e., there are 12 scanning steps within each course correction of

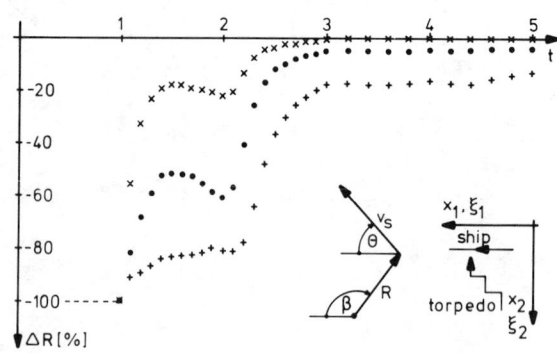

Fig. 6.   Target filter: relative distance error.

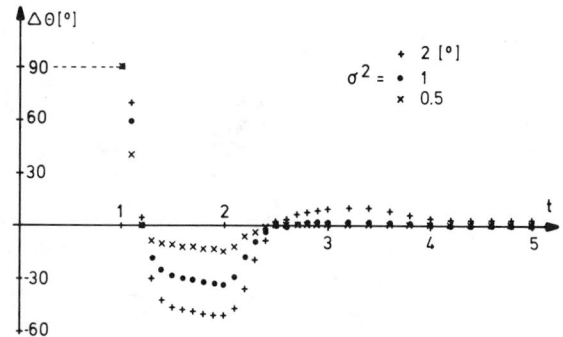

Fig. 7.   Target filter: course-angle error.

the pursuer. By this we get the possibility of preprocessing
in the algorithm [Eqs. (62)-(65)]. The results of the simulation are given in Figs. 6 and 7 for the standard deviations of
the measurement noise of $\sigma = (0.5°; 1°; 2°)$. In Fig. 6 the
relative distance error $\Delta R = (\hat{R} - R)/R$ is drawn in percent with
$R = \left(e_1^2 + e_2^2\right)^{1/2}$. Figure 7 shows the course error $\Delta\Theta = \hat{\Theta} - \Theta$.

From Figs. 6 and 7 it is evident that the filter is not
working until the first course correction of the torpedo. This
is based on the fact that, as mentioned earlier, the problem is
not previously observable [8]. After the first course correction the filter gives usable results, although we cannot over-
look the fact that the filter would diverge without a further
course correction. This is based especially on the parallel
course of torpedo and ship during the second motion interval
$1 \leq t \leq 2$. This divergence can be removed by the second course
correction of the torpedo. After this, stationary accuracy is
nearly reached. The accuracy can be raised in the course by
another course correction.

V. CONTROL AND FILTERING

It is now an obvious step to combine the results of Section III and Section IV in order to carry out a closed-loop control-filter procedure. This is done by a hybrid simulation on the basis of the simulation plan of Fig. 8.

At the top of Fig. 8 the dynamics of the torpedo are given as derived in Section II and inclusive of the Kalman filter of Section IV.A. The target filter is outlined below, and the nonlinear measurement equation for the relative target data and the controllers are given. The two systems — torpedo (pursuer) and ship (target) — are connected by a sample and hold circuit (SH). By this, the continuous system of the torpedo turns mathematically, in a discrete one, as it is needed for the target filter.

For the first phase the target trajectory has to be generated. In this identification phase the switch between the two controllers is turned to the left (1) and the system works in an open loop. If the torpedo has enough information about the

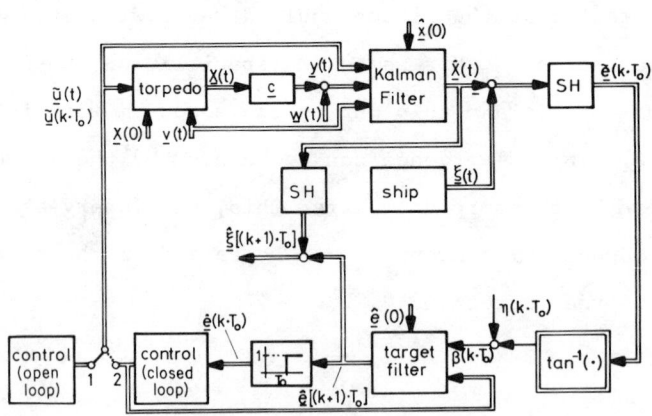

Fig. 8. Simulation plan.

target, the identification phase is finished; i.e., the switch
between the two controllers is turned to the right (2) and the
closed-loop control begins. During this phase the target filter is working in parallel. If an additional time-optimization
is needed, as outlined in Section III, the controller, which
in hardware is a microprocessor or a minicomputer, respectively,
carries out a target trajectory prediction and calculates T^*
from Eq. (14). In this contribution, however, this is not a
central point.

As an example of the simulation procedure the results of
Fig. 9 are given. Referring to the target trajectory of Section III.B.2, where the target is moving on the $x_1 - \xi_1$ axis
with the normalized constant velocity "one" the pursuer is
started with the initial conditions $x_1(0) = 1.12$; $x_2(0) = 4.50$.
The variance of the measurement device is $\sigma^2 = 1°$. In the
identification phase $0 \leq t \leq 3$ the target trajectory is generated. In order to look for geometric influences two different
identification trajectories are chosen. At time $t = 3$ the control phase begins; it is finished at $t = 9$. The controlled
trajectories are comparable to the results in Fig. 5. The

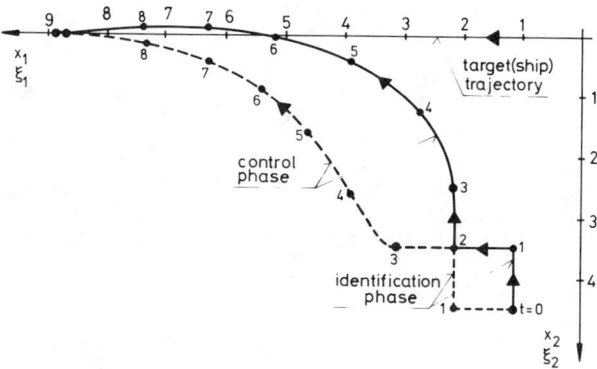

*Fig. 9. Interception trajectories with identification
and control.*

accuracy obtained can be considered satisfactory. In the
$x_1 - \xi_1$ axis, one obtains an absolute error of $\Delta e_1 = -0.15$ for
the upper and $\Delta e_1 = -0.275$ for the lower pursuer trajectory;
in the $x_2 - \xi_2$ axis the corresponding values are $\Delta e_2 = -0.025$
and $\Delta e_2 = 0.03$, respectively.

VI. EXTENSIONS

In Section III a control law was derived which is very
appropriate to any target trajectory. In Sections IV and V,
attention was focused on targets with constant velocities. An

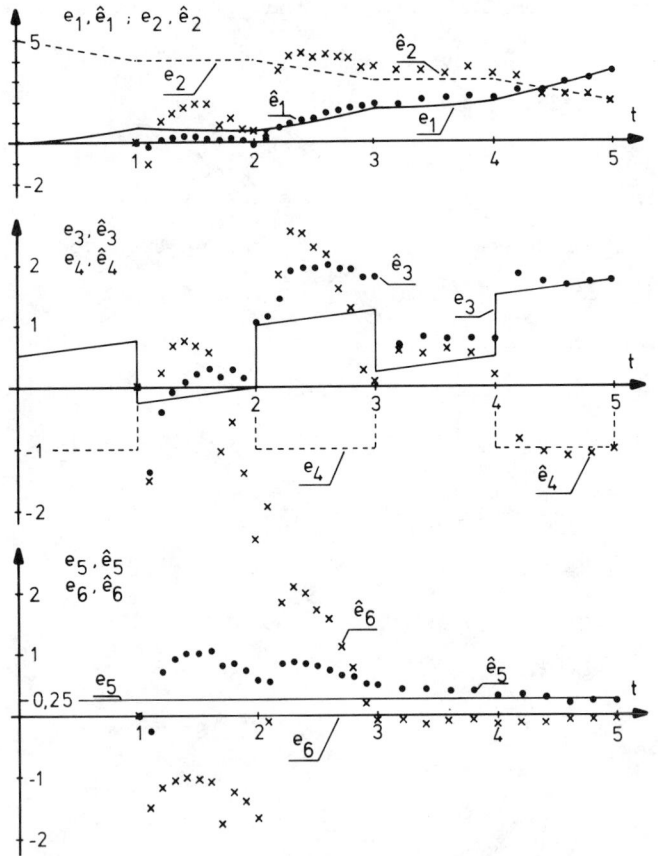

*Fig. 10. Results of target filter for an interception
with variable target velocity.*

extension for linear changes in the target velocity, i.e., constant acceleration, was given by Eq. (59). In order to prove the efficiency of the algorithms derived, including this equation, the example of Section IV.B is taken but now for $\xi_1(t) = 0.5t + 0.125t^2$ and $\xi_2(t) \equiv 0$. Consequently, the ship now has an acceleration of $\ddot{\xi}_1 = 0.25$ in the direction of ξ_1. The identification course of the torpedo is very similar to that given in Section IV.B. The results of this example are shown in Fig. 10. Without any detailed discussion it is obvious that the target filter works efficiently in this case, too.

Future efforts shall be particularly concentrated on targets with any velocity of time. Furthermore, theoretical considerations shall be concentrated in general on the problem of the best identification trajectory. This may be done by an intensive investigation of the observability matrix.

VII. CONCLUSIONS

The contribution presented here deals with control and filter problems for the interception of torpedo-ship situations. The central focus is on engineering applications and not primary theoretical aspects. The restriction to plane motions is not decisive, it is only done to provide formula limitations. All analytic results are prepared such that they can be realized in a simple manner numerically by a microprocessor or a minicomputer, respectively. Some of the calculations can therefore be done off-line, which is appropriate for real-time situations.

APPENDIX A. AUXILIARY FUNCTIONS
             FOR THE OPEN-LOOP CONTROL
             PROBLEM FOR THE CASE $\underline{Q} \neq \underline{0}$

$$c_{0i}(t) = \mathrm{ch}(\gamma_i t)\,\cos(\omega_i t) - \frac{\gamma_i^2 - \omega_i^2}{2\gamma_i \omega_i}\,\mathrm{sh}(\gamma_i t)\,\sin(\omega_i t),$$

$$c_{1i}(t) = \frac{\gamma_i\left(3\omega_i^2 - \gamma_i^2\right)}{2\gamma_i \omega_i\left(\gamma_i^2 + \omega_i^2\right)}\,\mathrm{ch}(\gamma_i t)\,\sin(\omega_i t)$$

$$+ \frac{\omega_i\left(3\gamma_i^2 - \omega_i^2\right)}{2\gamma_i \omega_i\left(\gamma_i^2 + \omega_i^2\right)}\,\mathrm{sh}(\gamma_i t)\,\cos(\omega_i t),$$

$$c_{2i}(t) = (2\gamma_i \omega_i)^{-1}\mathrm{sh}(\gamma_i t)\,\sin(\omega_i t),$$

$$c_{3i}(t) = \frac{\gamma_i}{2\gamma_i \omega_i\left(\gamma_i^2 + \omega_i^2\right)}\,\mathrm{ch}(\gamma_i t)\,\sin(\omega_i t)$$

$$- \frac{\omega_i}{2\gamma_i \omega_i\left(\gamma_i^2 + \omega_i^2\right)}\,\mathrm{sh}(\gamma_i t)\,\cos(\omega_i t).$$

$$\text{(A.1)}$$

$$m_{2i}(t) = c_{2i}(t) + \epsilon c_{3i}(t), \qquad m_{1i}(t) = c_{1i}(t) + \epsilon m_{2i}(t),$$

$$m_{0i}(t) = c_{0i}(t) + \epsilon m_{1i}(t). \tag{A.2}$$

$$f_{i+2}(t, T) = \left[m_{2i}(t) - \frac{m_{2i}(T)}{m_{0i}(T)}\,m_{0i}(t)\right]x_{i0}$$

$$+ \left[c_{3i}(t) - \frac{c_{3i}(T)}{m_{0i}(T)}\,m_{0i}(t)\right](x_{i+2,0} + v_i)$$

$$+ [m_{0i}(t)(c_{0i}(T) - 1) - m_{0i}(T)(c_{0i}(t) - 1)]$$

$$\times \frac{\epsilon}{q_{ii}m_{0i}(T)}\,v_i - \left[i_{4i}(t) - \frac{i_{4i}(T)}{m_{0i}(T)}\,m_{0i}(t)\right].$$

$$\text{(A.3)}$$

$$i_{4i}(t) = \int_0^t m_{1i}(t - \tau)\,\xi_i(\tau)\,d\tau, \qquad i = 1, 2. \tag{A.4}$$

$$d_{2i}(t) = c_{2i}(t) - \epsilon c_{3i}(t), \qquad d_{1i}(t) = c_{1i}(t) - \epsilon d_{2i}(t)$$

$$d_{0i}(t) = c_{0i}(t) - \epsilon d_{1i}(t). \tag{A.5}$$

$$k_{1i}(T) = \left[1 + q_{ii} \frac{c_{2i}(T)m_{2i}(T)}{c_{0i}(T)m_{0i}(T)}\right]x_{i0}$$

$$+ \left[\frac{d_{1i}(T)}{c_{0i}(T)} + q_{ii} \frac{c_{2i}(T)c_{3i}(T)}{c_{0i}(T)m_{0i}(T)}\right](x_{i+2,0} + v_i)$$

$$+ \left[\frac{d_{2i}(T)}{c_{0i}(T)} - (c_{0i}(T) - 1)\frac{c_{2i}(T)}{c_{0i}(T)m_{0i}(T)}\right]\epsilon v_i$$

$$+ \left[\frac{i_{1i}(T)}{c_{0i}(T)} - \frac{c_{2i}(T)i_{4i}(T)}{c_{0i}(T)m_{0i}(T)}\right]q_{ii} - \frac{1}{c_{0i}(T)} \xi_i(T), \tag{A.6}$$

$$k_{2i}(T) = \left[\frac{c_{1i}(T)}{c_{0i}(T)} + q_{ii} \frac{c_{3i}(T)m_{2i}(T)}{c_{0i}(T)m_{0i}(T)}\right]x_{i0}$$

$$+ \left[\frac{d_{2i}(T)}{c_{0i}(T)} + q_{ii} \frac{c_{3i}^2(T)}{c_{0i}(T)m_{0i}(T)}\right](x_{i+2,0} + v_i)$$

$$+ \left[\frac{c_{3i}(T)}{c_{0i}(T)} - (c_{0i}(T) - 1) \frac{c_{3i}(T) - (\epsilon/q_{ii})m_{0i}(T)}{c_{0i}(T)m_{0i}(T)}\right]\epsilon v_i$$

$$- \left[\frac{i_{3i}(T)}{c_{0i}(T)} + q_{ii} \frac{c_{3i}(T)i_{4i}(T)}{c_{0i}(T)m_{0i}(T)}\right].$$

$$\ell_{1i}(T) = \frac{c_{2i}(T)m_{1i}(T)}{c_{0i}(T)m_{0i}(T)} - \frac{c_{3i}(T)}{c_{0i}(T)},$$

$$\tag{A.7}$$

$$\ell_{2i}(T) = \frac{c_{3i}(T)m_{1i}(T)}{c_{0i}(T)m_{0i}(T)}.$$

$$i_{1i}(t) = \int_0^t c_{3i}(t - \tau)\xi_i(\tau)d\tau,$$

$$\tag{A.8}$$

$$i_{3i}(t) = \int_0^t c_{0i}(t - \tau)\xi_i(\tau)d\tau, \qquad i = 1, 2.$$

APPENDIX B. AUXILIARY FUNCTIONS
            FOR THE CLOSED-LOOP CONTROL
            PROBLEM FOR THE CASE $\underline{Q} \neq \underline{0}$

$$r_{0i}(T - t) = [n_i(T - t)]^{-1}$$

$$\times \{p[c_{0i}(T - t)d_{0i}(T - t) + q_{ii}c_{2i}(T - t)d_{2i}(T - t)]$$

$$- (1 - \delta)q_{ii}[c_{1i}(T - t)d_{0i}(T - t)$$

$$+ q_{ii}c_{3i}(T - t)d_{2i}(T - t)]\},$$

$$\hspace{10cm} \text{(B.1)}$$

$$n_i(T - t) = c_{0i}(T - t)d_{0i}(T - t) + [p/(1 - \delta)]$$

$$\times [c_{3i}(T - t)d_{0i}(T - t) - c_{2i}(T - t)d_{1i}(T - t)]$$

$$+ q_{ii}c_{3i}(T - t)d_{1i}(T - t), \quad i = 1, 2.$$

$$h_i(T - t) = -p$$

$$\times \Big\{ c_{0i}(T - t)d_{0i}(T - t)m_{1i}(T - t)$$

$$+ q_{ii}\Big[c_{3i}(T - t)(c_{0i}(T - t)c_{2i}(T - t)$$

$$+ d_{0i}(T - t)m_{2i}(T - t))$$

$$+ c_{3i}(T - t)d_{1i}(T - t)m_{1i}(T - t)$$

$$- c_{2i}(T - t)d_{1i}(T - t)m_{2i}(T - t)$$

$$+ q_{ii}c_{3i}^3(T - t)\Big]\Big\},$$

$$\hspace{10cm} \text{(B.2)}$$

$$\Omega_i(T - t) = \frac{p}{n_i(T - t)}$$

$$\times \Big\{ m_{1i}(T - t)[c_{0i}(T - t)d_{1i}(T - t)$$

$$- c_{1i}(T - t)d_{0i}(T - t)]$$

$$+ q_{ii}\Big[c_{2i}(T - t)c_{3i}(T - t)(d_{1i}(T - t) - c_{1i}(T - t))$$

$$+ c_{3i}^2(T - t)(c_{0i}(T - t)d_{0i}(T - t))\Big]\Big\}$$

[Eq. (B.2) continues]

$$+ \frac{(1 - \delta)q_{ii}}{n_i(T - t)}\Big\{m_{2i}(T - t)[c_{1i}(T - t)d_{0i}(T - t)$$

$$- c_{0i}(T - t)d_{1i}(T - t)]$$

$$+ c_{0i}(T - t)c_{3i}(T - t)$$

$$\times [c_{0i}(T - t) - d_{0i}(T - t)]$$

$$+ q_{ii}c_{3i}^2(T - t)[c_{1i}(T - t) - d_{1i}(T - t)]\Big\}$$

$$+ (1 - \delta)\epsilon[c_{0i}(T - t) - 1], \quad i = 1, 2.$$

$$I_{1i}(t, T) = \int_t^T c_{1i}(t - \tau)\ddot{\xi}_i(\tau)d\tau,$$

$$I_{2i}(t, T) = \int_t^T c_{0i}(t - \tau)\ddot{\xi}_i(\tau)d\tau, \tag{B.3}$$

$$I_{4i}(t, T) = \int_t^T [c_{3i}(t - \tau) - (\epsilon/q_{ii})c_{0i}(t - \tau)]$$

$$\times \ddot{\xi}_i(\tau)d\tau, \quad i = 1, 2.$$

APPENDIX C. ELEMENTS OF THE KALMAN
MATRIX $\underline{K}(t)$ FOR THE TORPEDO
INTERTIAL FILTER

$$N(t) = \epsilon^{-4}[2 - 2ch(\epsilon t) + \epsilon t sh(\epsilon t)]$$

$$+ (\rho^2/\epsilon^3)[2 - ch(\epsilon t) + (\epsilon t - 1)e^{\epsilon t}] \tag{C.1}$$

$$+ \lambda^2 t e^{\epsilon t} + (\rho\lambda)^2 e^{\epsilon t} + \Delta(\gamma^2/\epsilon)(t + \rho^2)sh(\epsilon t).$$

$$k_{11}(t) = [N(t)]^{-1}\{\epsilon^{-3}[-2 + ch(\epsilon t) + (1 + \epsilon t)e^{-\epsilon t}]$$

$$+ 2(\rho/\epsilon)^2[-1 + ch(\epsilon t)]$$

$$+ \lambda^2 e^{\epsilon t} + \Delta(\gamma^2/\epsilon)sh(\epsilon t)\},$$

$$k_{13}(t) = \Delta\gamma^2 k_{31}(t), \tag{C.2}$$

$$k_{31}(t) = [N(t)]^{-1}\{\epsilon^{-2}[1 - (1 + \epsilon t)e^{-\epsilon t}] + (\rho^2/\epsilon)[1 - e^{-\epsilon t}]\},$$

$$k_{33}(t) = \Delta[\gamma^2/N(t)](\rho^2 + t)e^{-\epsilon t}.$$

$$k_{22}(t) = k_{11}(t), \qquad k_{24}(t) = k_{13}(t),$$

$$k_{42}(t) = k_{31}(t), \qquad k_{44}(t) = k_{33}(t).$$

(C.3)

APPENDIX D. EXTENDED KALMAN FILTER FOR THE RELATIVE MOTION OF TORPEDO AND SHIP

Prediction phase:

$$\hat{\underline{e}}(k + 1|k) = \underline{F}(k + 1, k)\hat{\underline{e}}(k|k) + \underline{z}(k),$$

$$\underline{P}(k + 1|k) = \underline{F}(k + 1, k) \cdot \underline{P}(k|k) \cdot \underline{F}^T(k + 1, k),$$

(D.1)

$$k = 0, 1, 2, \ldots .$$

Measurement phase:

$$\underline{c}^T(k + 1) = \frac{\partial h}{\partial \underline{e}}\bigg|_{\underline{e}=\hat{\underline{e}}(k+1|k)}$$

$$= \left[\frac{\cos \beta(k + 1|k)}{\hat{R}(k + 1|k)}, -\frac{\sin \beta(k + 1|k)}{\hat{R}(k + 1|k)}, 0, 0\right],$$

$$h = \tan^{-1}[e_1/e_2], \qquad \hat{R}^2 = \hat{e}_1^2 + \hat{e}_2^2.$$

(D.2)

Correction phase:

$$\underline{k}(k + 1) = \underline{P}(k + 1|k) \cdot \underline{c}(k + 1)$$

$$\times \{\underline{c}^T(k + 1) \cdot \underline{P}(k + 1|k)$$

$$\cdot \underline{c}(k + 1) + \sigma^2(k + 1)\}^{-1},$$

$$\hat{\underline{e}}(k + 1|k + 1) = \hat{\underline{e}}(k + 1|k) + \underline{k}(k + 1)$$

$$\times \{\beta(k + 1) - h[\hat{\underline{e}}(k + 1|k)]\},$$

$$\underline{P}(k + 1|k + 1) = \underline{P}(k + 1|k) - \underline{k}(k + 1)$$

(D.3)

$$\cdot \underline{c}^T(k + 1) \cdot \underline{P}(k + 1|k).$$

Initial condition:

$$\underline{P}(0|0) = \text{diag}\left(\sigma_0^2\right), \quad \sigma_0^2 \gg 1; \qquad \hat{\underline{e}}(0|0) = \underline{0}.$$

(D.4)

REFERENCES

1. OPERATIONAL ANALYSIS STUDY GROUP (ed.), "Naval Operations Analysis." Naval Institute Press, Annapolis, Maryland, 1977.

2. R. LUNDERSTÄDT, "Steuerung und Regelung von Unterwasser-fahrzeugen." Preprints of the Workshop "Regelungssynthese im Zustandsraum" of VDI/VDE-Gesellschaft Mess- und Regelungstechnik, Frankfurt (1977).

3. P. GARNELL and D. J. EAST, "Guided Weapon Control Systems." Pergamon, London, 1977.

4. L. A. STOCKUM, "A Study of Guidance Controllers for Homing Missiles." Ph.D. Thesis, Ohio State Univ. (1974).

5. W. HOFMANN, "Ein vereinheitlichter Zugang zur systemati-schen Auslegung von Flugkörper-Lenksystemen am Beispiel der Zweipunktlenkung." Preprints of 105, Wehrtechnisches Symposium "Regelungstechnik-Automatisierungstechnik," Mannheim (1979).

6. N. W. REES, "An Application of Optimal Control Theory to the Guided Torpedo Problem." *Proc. Joint Automatic Control Conference* (1968).

7. W. HOFMANN, "Ein Ansatz zur Verbesserung des Ablagever-haltens von mit Proportionalnavigation gelenkten Flugkörpern durch Einsatz moderner Regelungsverfahren." Preprints of 105, Wehrtechnisches Symposium "Regelung-stechnik-Automatisierungstechnik," Mannheim (1979).

8. V. J. AIDALA, *IEEE Trans. Aerosp. Electron. Syst. AES-15*, 29-39 (1979).

9. M. ATHANS and P. L. FALB, "Optimal Control." McGraw-Hill, New York, 1966.

10. R. LUNDERSTÄDT, "Zur Optimierung ebener Interzeptions-bewegungen." *Regelungstechnik 29*, 255-262, 298-304 (1981).

11. A. E. BRYSON and Y. C. HO, "Applied Optimal Control." Ginn, Waltham, Massachusetts, 1969.

12. M. WEINGART, "Implementierung und Erprobung eines Kalman-Filters für die Zielbahngenerierung aus Peilmessungen." Diploma Thesis, Univ. German Armed Forces, Hamburg (1981).

State Estimation
of Ballistic Trajectories
with Angle-Only Measurements

MICHAEL R. SALAZAR

Nichols Research Corporation
Functional Analysis Directorate
Huntsville, Alabama

LIST OF SYMBOLS

A	Measurement matrix
Δt	Time between measurements
ΔX	Deviations of state from nominal set
$\widehat{\Delta X}$	Estimate of ΔX
ΔY	Deviations of measurements from nominal or calculated set ($\hat{Y}$)

E, E_T	Total energy
ECI	Earth-centered inertial coordinate system
ϵ	Measurement errors
f, g	The f- and g-series values
f_1, f_2, f_3	State vector velocity components
f_4, f_5, f_6	Time derivative of state vector velocity components
G	Gravitational constant
g_1, g_2	Azimuth, elevation measurements
H, H'	Kalman gain matrix
I	Identity matrix
J	Jacobian matrix
KE	Kinetic energy
λ	Marquardt matrix conditioning factor
M	Mass of earth
μ	$G \times M$
PE	Potential energy
Φ	State transition matrix
R	Range
$\dot{R}$	Range rate
$\underline{R}$	Range vector
R_e	Radius of earth
R_S	Center of earth to sensor radius
R_T	Center of earth to target radius
R_x, R_y, R_z	The x, y, z components of $\underline{R}$
$\underline{S}$	Sensor position vector
S, S_1^{-1}	State covariance matrix
S_x, S_y, S_z	The x, y, z components of $\underline{S}$
SCT	Sensor-centered topographic coordinate system
σ_E	1σ uncertainty in E

$\sigma_\theta,\ \sigma_\phi$	1σ measurement uncertainty in $\theta,\ \phi$
$\underline{T}$	Target-position vector
$\underline{T}_0$	Intitial target-position vector
$T_x,\ T_y,\ T_z$	The x, y, z components of $\underline{T}$
$t_n,\ t_{n-1},\ \ldots,\ t_{n-L}$	Time for each measurement
τ	Time step for f and g series
$\theta,\ \phi$	Azimuth, elevation measurements
$\dot{\theta},\ \dot{\phi}$	First time derivative of $\theta,\ \phi$
$\ddot{\theta},\ \ddot{\phi}$	Second time derivative of $\theta,\ \phi$
$u_0,\ p_0,\ q_0$	The f- and g-series terms
$\underline{V}_0$	Initial target-velocity vector
$\underline{V}_R$	Relative velocity vector (target-sensor)
V_S	Sensor-velocity magnitude
$\underline{V}_S$	Sensor-velocity vector
V_T	Target-velocity magnitude
$\underline{V}_T$	Target-velocity vector
W	Measurement covariance matrix
$\dot{X}$	Time derivative of X
$\hat{X}$	Estimate of X
$\underline{XS}$	Sensor state vector
$X,\ \underline{X}$	State vector
$x,\ y,\ z$	State vector-position components
$\dot{x},\ \dot{y},\ \dot{z}$	State vector-velocity components
$x_1,\ x_2,\ x_3$	State vector-position components
$x_4,\ x_5,\ x_6$	State vector-velocity components
$\dot{x}_1,\ \dot{x}_2,\ \dot{x}_3$	Time derivative of state vector-position components
$\dot{x}_4,\ \dot{x}_5,\ \dot{x}_6$	Time derivative of state vector-velocity components
$\dot{x}_R,\ \dot{y}_R,\ \dot{z}_R$	The x, y, z components of $\underline{V}_R$

$\dot{x}_S, \dot{y}_S, \dot{z}_S$ The x, y, z components of $\underline{V}_S$

χ^2 Chi squared

$\hat{Y}$ Estimate of Y

Y, $\underline{Y}$ Measurement vector

*   Normalized

I. INTRODUCTION

This article deals with the problem of state estimation of ballistic trajectories with angle-only measurements. This type of problem becomes difficult when the observer is free-falling and more difficult if the observer is then located in the plane of the observed trajectory. The methods described in this report are very effective against this most difficult case and are superior to existing angle-only tracking filters in terms of stability and tracking performance.

The first method described herein utilizes the methods of Marquardt matrix conditioning [1] and the explicit Jacobian in determining the weighted least squares state estimate for the nonlinear, time-varying, dynamic system of a ballistic trajectory with nonlinear noisy measurements. In this case both the equations of motion and the angle observations are nonlinear functions of the state. This Marquardt least squares (MLS) technique is a nonrecursive or batch process in that all the observations must be processed each time a state estimate is made. The method of incorporating a priori knowledge of the total energy into the MLS algorithm to assis in poor observability problems is also discussed.

The second method takes the explicit Jacobian technique developed for the MLS algorithm and applies it to the recursive Kalman filter equations. This improved Jacobian-Kalman filter formulation together with the MLS for initialization form the complete angle-only tracking algorithm.

This article is organized as follows. Section II provides the background material and the step-by-step development of the equations for the standard weighted least squares batch filter and the MLS algorithms. A discussion of the energy-constraint concept and the application of the explicit Jacobian method to the Kalman filter is included in this section. Section II.A presents the weighted least squares solution for the general nonlinear system problem. This serves as a common basis for both the standard weighted least squares batch filter and the MLS algorithms, which are developed in Sections II.B and II.C, respectively. No attempt is made to make these derivations mathematically rigorous. The energy constraint concept is discussed in Section II.D. The application of the explicit Jacobian technique to the Kalman filter equations is presented in Section II.E. Section III gives the performance results for several ballistic trajectory problem test cases. Section IV contains the conclusions.

II. ALGORITHM DESCRIPTION

A. WEIGHTED LEAST SQUARES CONCEPT

In general, the fundamental problem of concern can be stated as follows. A nonlinear system can be represented by the linearized model matrix equation

$$\Delta Y = A\Delta X + \epsilon, \tag{1}$$

where ΔY is an $n \times 1$ matrix of the deviations of the observations Y from the nominal or calculated set $\hat{Y}$; ΔX is a $k \times 1$ matrix of deviations of the unknown parameters from a nominal (known) set; A is an $n \times k$ $(n > k)$ known matrix; and ϵ is an n × 1 matrix of observation errors (unknown). The problem is stated as follows: given ΔY and A and the linearized model of Eq. (1), find the "best" estimate of ΔX, called $\widehat{\Delta X}$. Once the "best" estimate $\widehat{\Delta X}$ has been determined, the "best" estimate of the unknown parameters $\hat{X}$ may be determined by

$$\hat{X} = X + \widehat{\Delta X}, \tag{2}$$

where X is the known or nominal set of parameters. In this case the "best" estimate is achieved when the weighted least squares criterion is satisfied; that is, when the sum of the squares of the components of the weighted residual vector is minimized. This quantity can be represented by

$$\Delta Y^T W^{-1} \Delta Y, \tag{3}$$

where W is the known n × n measurement covariance matrix. This weighting matrix accounts for the difference in confidence between various observations and the possible correlation between them.

The well-known weighted least squares solution (formulated by Gauss in 1794[1] to this problem is

$$\widehat{\Delta X} = (A^T W^{-1} A)^{-1} A^T W^{-1} \Delta Y, \tag{4}$$

which combined with Eq. (2) yields the total expression

$$\hat{X} = X + (A^T W^{-1} A)^{-1} A^T W^{-1} \Delta Y. \tag{5}$$

Given some initial guess of X, this expression is normally iterated as follows until the current parameter estimates do not vary appreciably from the previous iteration.

[1]*Formulation included only the diagonal terms of W.*

> Evaluate A with X
>
> Calculate $\hat{Y}$ for X
>
> $\Delta Y = Y - \hat{Y}$
>
> $\hat{X} = X + (A^T W^{-1} A)^{-1} A^T W^{-1} \Delta Y$
>
> $X = \hat{X}$

Equation (5) establishes the basis for both the standard weighted least squares batch filter and the Marquardt least squares algorithms. Sections II.B and II.C show how these two different formulations are derived from this common expression, thus demonstrating both their similarities and differences.

B. *STANDARD WEIGHTED LEAST SQUARES BATCH FILTER*

For a given set of observation times $t = t_n$, t_{n-1}, $\cdots$, t_{n-L}, the linearized model represented by Eq. (1) can be expanded to give

$$
\begin{bmatrix} \Delta Y_n \\ \Delta Y_{n-1} \\ \vdots \\ \Delta Y_{n-L} \end{bmatrix} = \begin{bmatrix} A_n \Delta X_n \\ A_{n-1} \Delta X_{n-1} \\ \vdots \\ A_{n-L} \Delta X_{n-L} \end{bmatrix} + \begin{bmatrix} \epsilon_n \\ \epsilon_{n-1} \\ \vdots \\ \epsilon_{n-L} \end{bmatrix}. \tag{6}
$$

For a time-varying "linear" differential equation model, that is,

$$
\frac{d}{dt} \Delta X(t) = F(X(t)) \Delta X(t) \tag{7}
$$

the "approximate" solution is given by

$$
\Delta X(t + \Delta t) = \Delta X(t) + \Delta t F(X(t)) \Delta X(t). \tag{8}
$$

factoring out $\Delta X(t)$,

$$
\Delta X(t + \Delta t) = \{I + \Delta t F(X(t))\} \Delta X(t), \tag{9}
$$

which yields the following transition relation

$$
\Delta X_{n-1} = \Phi_{n-1,n} \Delta X_n. \tag{10}
$$

Note that two approximations were required to arrive at Eq. (10):
(1) the linear differential equation model and (2) its approximate solution. The significance of this will become apparent in the development of the Marquardt weighted least squares algorithm in Section II.C. Using Eq. (10), Eq. (6) can be written as

$$
\begin{bmatrix} \Delta Y_n \\ \Delta Y_{n-1} \\ \vdots \\ \Delta Y_{n-L} \end{bmatrix} = \begin{bmatrix} A_n \Delta X_n \\ A_{n-1} \Phi_{n-1,n} \Delta X_n \\ \vdots \\ A_{n-L} \Phi_{n-L,n} \Delta X_n \end{bmatrix} + \begin{bmatrix} \epsilon_n \\ \epsilon_{n-1} \\ \vdots \\ \epsilon_{n-L} \end{bmatrix}.
\tag{11}
$$

Now, factor out ΔX_n to yield

$$
\begin{bmatrix} \Delta Y_n \\ \Delta Y_{n-1} \\ \vdots \\ \Delta Y_{n-L} \end{bmatrix} = \begin{bmatrix} A_n \\ A_{n-1} \Phi_{n-1,n} \\ \vdots \\ A_{n-L} \Phi_{n-L,n} \end{bmatrix} \Delta X_n + \begin{bmatrix} \epsilon_n \\ \epsilon_{n-1} \\ \vdots \\ \epsilon_{n-L} \end{bmatrix}.
\tag{12}
$$

From Eq. (12) the following matrix is defined:

$$
J_n = \begin{bmatrix} A_n \\ A_{n-1} \Phi_{n-1,n} \\ \vdots \\ A_{n-L} \Phi_{n-L,n} \end{bmatrix}.
\tag{13}
$$

Equation (11) can now be written in the compact form

$$
\Delta Y_{(n)} = J_n \Delta X_n + \epsilon_{(n)}.
\tag{14}
$$

The weighted least squares solution for this linearized model can be written as an extension of Eq. (4) to give

$$
\widehat{\Delta X_n} = \left(J_n^T W_{(n)}^{-1} J_n \right)^{-1} J_n^T W_{(n)}^{-1} \Delta Y_{(n)}.
\tag{15}
$$

Similarly, the final expression becomes

$$
\hat{X}_n = X_n + \left(J^T W_{(n)}^{-1} J_n \right)^{-1} J_n^T W_{(n)}^{-1} \Delta Y_{(n)},
\tag{16}
$$

which can be iterated by the procedure described in Section II.A. Although the iterative solution for this expression essentially represents the standard weighted least squares batch filter, a few more definitions are required before the final algorithm can be presented.

For the ballistic trajectory problem, the state (parameter) vector may be expressed in terms of an earth-centered inertial Cartesian coordinate system:

$$\underline{X} = \begin{bmatrix} x \\ y \\ z \\ \dot{x} \\ \dot{y} \\ \dot{z} \end{bmatrix} = \begin{bmatrix} x_1 \\ x_2 \\ x_3 \\ x_4 \\ x_5 \\ x_6 \end{bmatrix}. \tag{17}$$

For a spherical earth, the exoatmospheric trajectory equations of motion are

$$\underline{\dot{X}} = \begin{bmatrix} \dot{x}_1 \\ \dot{x}_2 \\ \dot{x}_3 \\ \dot{x}_4 \\ \dot{x}_5 \\ \dot{x}_6 \end{bmatrix} = \begin{bmatrix} f_1 \\ f_2 \\ f_3 \\ f_4 \\ f_5 \\ f_6 \end{bmatrix} = \begin{bmatrix} x_4 \\ x_5 \\ x_6 \\ -GMx_1/R^3 \\ -GMx_2/R^3 \\ -GMx_3/R^3 \end{bmatrix}, \tag{18}$$

where G is the gravitational constant, M is the mass of the earth, and R is the magnitude of the position vector.

The first-order Taylor series approximation for the transition matrix Φ defined in Eqs. (10) is

$$\Phi_{n-1,n} = I + F(X(t_n))(t_{n-1} - t_n),$$

$$\Phi_{n-2,n} = I + F(X(t_n))(t_{n-2} - t_n),$$

$$\vdots$$

$$\Phi_{n-L,n} = I + F(X(t_n))(t_{n-L} - t_n),$$

(19)

where the $F(X(t_n))$ matrix is defined by

$$[F(X(t_n))]_{i,j} = \partial f_i / \partial x_j |_{X=X(t_n)}$$

$$= \begin{bmatrix} \partial f_1/\partial x_1 & \cdots & \partial f_1/\partial x_6 \\ \vdots & & \vdots \\ \partial f_6/\partial x_1 & \cdots & \partial f_6/\partial x_6 \end{bmatrix}_{X=X(t_n)},$$

(20)

where f_i are the derivative functions of the state vector $\dot{\underline{X}} = \underline{f}(X)$ defined in Eq. (18). The partial derivatives for Eq. (20) are given in Appendix A. A more accurate method for determining the transition matrix is to use the transition matrix determined for the previous observation time in determining the transition matrix for the current observation time. This reduces the time interval over which the transition matrix must be valid and allows for evaluation of the F matrix with the updated nominal state. This method produces the following set:

$$\Phi_{n-1,n} = I + F(X(t_n))(t_{n-1} - t_n),$$

$$\Phi_{n-2,n} = \{I + F(X(t_{n-1}))(t_{n-2} - t_{n-1})\}\Phi_{n-1,n},$$

$$\vdots$$

$$\Phi_{n-L,n} = \{I + F(X(t_{n+1-L}))(t_{n-L} - t_{n+1-L})\}\Phi_{n+1-L,n+2-L}.$$

(21)

By this method the magnitude of the time interval is restricted to the time between observations. If this time interval is still too large, this propagation technique can be further

applied by subdividing the time step between observations. For
example, if there are m subdivided time steps of h, then the
first expression in Eq. (21) would be

$$\Phi_{n-1,n} = \{I + hF(X(t_n))\}\{I + hF(X(t_n - h))\}$$

$$\times \cdots \times \{I + hF(X(t_n - (m - 1)h))\}. \tag{22}$$

The angle-only observation set for the ballistic trajectory
problem can be defined as

$$\underline{Y} = \begin{bmatrix} \theta \\ \phi \end{bmatrix} = \begin{bmatrix} g_1 \\ g_2 \end{bmatrix} = \begin{bmatrix} \tan^{-1}(R_x/R_y) \\ \tan^{-1}\left[R_z/\left(R_x^2 + R_y^2\right)\right] \end{bmatrix}, \tag{23}$$

where R_x, R_y, and R_z is the relative position vector (target
sensor) in a topographic coordinate system that is defined in
Appendix B. The first-order Taylor series approximation for
the measurement matrix A_n defined in Eq. (6) is

$$[A_n]_{i,j} = [A(X(t_n))]_{i,j} = \partial g_i/\partial x_j\big|_{X=X(t_n)}$$

$$= \begin{bmatrix} \partial g_1/\partial x_1 & \cdots & \partial g_1/\partial x_6 \\ \partial g_2/\partial x_1 & \cdots & \partial g_2/\partial x_6 \end{bmatrix}_{X=X(t_n)} \tag{24}$$

where the g_i terms are the functions relating observations to
states $\underline{Y} = \underline{g}(X)$ defined in Eq. (23). The remaining set
A_{n-1}, ..., A_{n-L} is obtained in similar fashion using
$X(t_{n-1})$, ..., $X(t_{n-L})$. The partial derivatives in Eq. (24) are
given in Appendix B.

The weighting or measurement covariance matrix W, assumed
in this study, is

$$[W]_{i,j} = \begin{bmatrix} \sigma_\phi^2 & 0 \\ 0 & \sigma_\phi^2 \end{bmatrix}, \tag{25}$$

where σ_θ and σ_ϕ represent the one-sigma uncertainties in the

uncorrelated measurement set (θ, ϕ). One final step is required
before the final algorithm can be presented. Instead of building the entire matrix J_n defined in Eq. (13), Eq. (16) may be
rewritten in the following form:

$$\hat{X}_n = X_n + \left\{ \sum_{i=1}^{L+1} J_n^T(i) W_{n+1-i}^{-1} J_n(i) \right\}^{-1}$$

$$\times \left\{ \sum_{i=1}^{L+1} J_n^T(i) W_{n+1-i}^{-1} \Delta Y_{n+1-i} \right\}, \tag{26}$$

where $J_n(1) = A_n$; $J_n(2) = A_{n-1}\Phi_{n-1,n}$, ..., $J_n(L+1) = A_{n-L}\Phi_{n-L,n}$.

All definitions have now been given for the standard
weighted least squares batch filter algorithm which is presented
in Fig. 1 and uses the notation defined in this section. This
is essentially the algorithm described by Lincoln Laboratory [2]
and is basically the Gauss weighted least squares solution defined for the ballistic trajectory problem. The formulation
presented in Fig. 1 requires at least three pairs of angle observations or three cycles of the measurement loop. For this
formulation the nominal state vector is defined at the final
observation point.

In this algorithm some initial state guess is successively
corrected in an attempt to satisfy the weighted least squares
criterion. For observations at times t_n, t_{n-1}, ..., t_{n-L}, this
quantity is

$$\sum_{i=1}^{L+1} (Y_{n+1-i} - \hat{Y}_{n+1-i})^T W_{n+1-i}^{-1} (Y_{n+1-i} - \hat{Y}_{n+1-i}). \tag{27}$$

If the process is converging, this quantity becomes in-
creasingly smaller at a decreasing rate as it asymptotically
approaches its minimum value. The process could thus be

$\dot{X}$ = f(X) EQUATIONS OF MOTION
Y = g(X) EQUATIONS RELATING OBSERVATIONS TO STATES

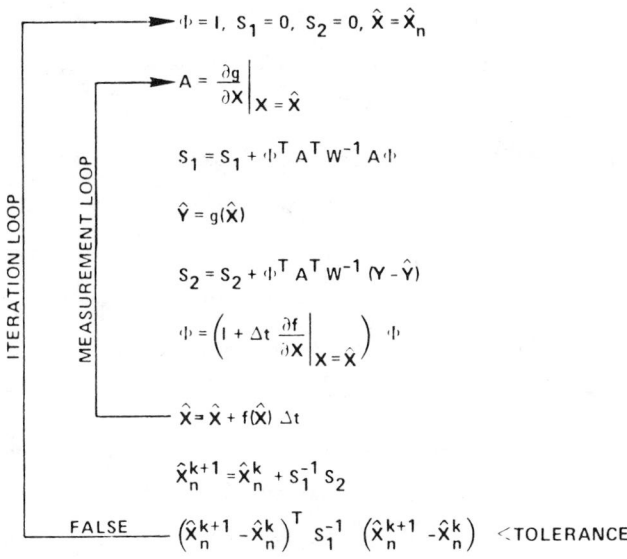

Fig. 1. Standard weighted least squares batch filter algorithm. (For Figs. 1, 3, 5, 6, and 7, see List of Symbols for definitions.)

terminated when this quantity does not decrease significantly from the previous iteration. Another criterion for terminating the iterative process is that shown in Fig. 1:

$$\left(\hat{x}_n^{k+1} - \hat{x}_n^{k}\right)^T s_1^{-1} \left(\hat{x}_n^{k+1} - \hat{x}_n^{k}\right). \tag{28}$$

Note that in this quantity the measurement residual has been replaced by the difference in the nominal state vectors for successive iterations and the measurement covariance has been replaced by the state covariance. If the process is converging, this quantity approaches zero. Therefore, termination of the iterative process would be based on the magnitude of this

quantity rather than on a change in magnitude as in the previous method. In either case the cutoff point is established
at the level at which tracking performance is unaffected.

Because the weighted least squares batch filter is a batch
or nonrecursive process, all of the observations must be processed each time a new measurement pair is added to the observation set. In addition to this requirement, all the measurements of the observation set must be processed for each iteration of the least squares process, as indicated in Fig. 1.
Therefore, termination of the iterative process at the earliest
possible point is essential. The nonrecursive characteristic
of the batch filter and resulting processing requirements tend
to restrict this algorithm to the initialization of track function. For the initialization of track function, the resulting
weighted least squares state estimate and state covariance are
projected forward to the next observation point where they serve
as the initial state and covariance for a Kalman filter algorithm. The weighted least squares algorithm can also be applied
again at this point using these same projected quantities as
initial conditions.

In Section III it is shown that this algorithm does not
always converge. Therefore, a series of tests are made after
each iteration to determine if the process is converging. The
algorithm is determined to be nonconvergent if any of the following tests are true:

1. Too many iterations (15)

2. Position magnitude too large (1.5×10^7 m)

3. Velocity magnitude too large (1.0×10^4 m/sec)

4. Altitude too low (100 m)

The values listed are possible limits for these tests. If the process is nonconvergent, then the pair of angle measurements for the next observation point are added to the observation set, and the weighted least squares algorithm is performed again.

Thus far, the problem of obtaining an initial state guess to begin the iterative least squares process has not been addressed. One possible method for obtaining this initial state guess is that described by Lincoln Laboratory [2] and presented in Fig. 2. For brevity, the derivation of this method is given in Appendix C. In this method the azimuth and elevation are fit with a second-order polynomial for a batch of angle-measurement data (at least three pairs). The azimuth and elevation along with their first and second derivatives are solved for with the polynomial fit at the time of the last observation. These quantities are then used in an iterative set of energy

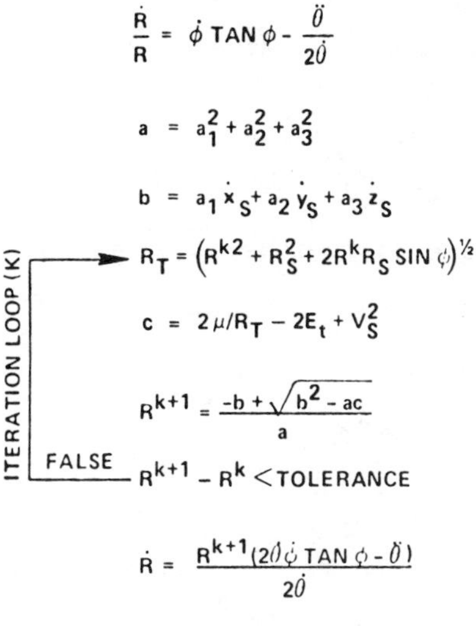

$$\frac{\dot{R}}{R} = \dot{\phi} \, TAN \, \phi - \frac{\ddot{\theta}}{2\dot{\theta}}$$

$$a = a_1^2 + a_2^2 + a_3^2$$

$$b = a_1 \dot{x}_S + a_2 \dot{y}_S + a_3 \dot{z}_S$$

$$R_T = \left(R^{k2} + R_S^2 + 2R^k R_S \, SIN \, \psi\right)^{1/2}$$

$$c = 2\mu/R_T - 2E_t + V_S^2$$

$$R^{k+1} = \frac{-b + \sqrt{b^2 - ac}}{a}$$

$$R^{k+1} - R^k < TOLERANCE$$

ITERATION LOOP (K)

FALSE

$$\dot{R} = \frac{R^{k+1}(2\dot{\theta}\dot{\psi} \, TAN \, \phi - \ddot{\theta})}{2\dot{\theta}}$$

Fig. 2. Initial state guess algorithm.

and geometry equations to solve for the range and range rate, which are in turn used to estimate the initial guess of the state vector for the weighted least squares process.

The iterative energy and geometry equations used to establish the range and range rate require an initial guess of the range and an a priori estimate of the target energy. This set of equations is cycled until the new estimate of the range does not vary from the previous estimate by more than some given tolerance. The selection of energy as a constraint for the initial state guess is made because of its relative constancy over the whole trajectory for a given set of ICBM threats with the same ground range. Whereas the initial state guess from this process is constrained in an energy sense, the final state estimate from the batch filter algorithm (Fig. 1) satisfies the weighted least squares criterion and is therefore independent from the initial state guess and the a priori energy estimate. If this initial state guess algorithm fails to converge, then the pair of angle measurements for the next observation point are added to the observation set and the process is repeated.

C. *MARQUARDT WEIGHTED LEAST SQUARES*

The Marquardt weighted least squares (MLS) technique is derived from the same basic least squares solution as the batch filter algorithm described in Section II.B. For clarity this expression is repeated here

$$\hat{X}_n = X_n + \left(J_n^T W_{(n)}^{-1} J_n \right)^{-1} J_n^T W_{(n)}^{-1} \Delta Y_{(n)}, \tag{29}$$

where

$$
J_n = \begin{bmatrix} A_n \\ A_{n-1}\Phi_{n-1,n} \\ \vdots \\ A_{n-L}\Phi_{n-L,n} \end{bmatrix} . \tag{30}
$$

At this point, one may deviate from the derivation of the previous section by solving for the Jacobian matrix [Eq. (30)]
explicitly, thus eliminating the requirement for a transition
matrix Φ that is based on an approximate solution [Eq. (8)] to
an approximate model [Eq. (7)]. This is accomplished by a different interpretation of the measurement matrix A as defined in
Eq. (24) and Appendix B. Instead of taking the partial derivatives of the functions relating observations to states with
respect to each state for some instant in time

$$
(\partial g(x)/\partial X|_{X=X(t_n),X(t_{n-1}),\ldots,X(t_{n-L})}),
$$

proceed to take the partials of the measurement equations at
times t_n, t_{n-1}, $\ldots$, t_{n-L} with respect to each state at time
t_n:

$$
(\partial g(X)/\partial X_n|_{X=X(t_n),X(t_{n-1}),\ldots,X(t_{n-L})}).
$$

This requires a closed-form solution of the equations of motion,
which is supplied through the f and g series equations for a
free-falling body. Although this approach requires a somewhat
cumbersome development of equations, the final working algorithm shall be shown to be a very precise and rapid method for
calculating the Jacobian matrix explicitly.

For simplicity in this section the observation set is defined somewhat differently from the previous section:

$$\underline{Y} = \begin{bmatrix} \theta \\ \phi \end{bmatrix} = \begin{bmatrix} g_1 \\ g_2 \end{bmatrix} = \begin{bmatrix} \tan^{-1}(R_y/R_x) \\ \sin^{-1}(R_z/|R|) \end{bmatrix}, \tag{31}$$

where

$$\underline{R} = \underline{T} - \underline{S} \quad \text{or} \quad \begin{bmatrix} R_x \\ R_y \\ R_z \end{bmatrix} = \begin{bmatrix} T_x \\ T_y \\ T_z \end{bmatrix} = \begin{bmatrix} S_x \\ S_y \\ S_z \end{bmatrix}, \tag{32}$$

where $\underline{T}$ and $\underline{S}$ are the ECI position vectors of the target and sensor, respectively. Now we proceed to take the partial derivatives of the measurement equations with respect to each state $(x_1, x_2, \ldots, x_6)$ as specified in Eq. (24), keeping in mind the new interpretation of the measurement matrix A:

$$\frac{\partial \theta}{\partial x_i} = \frac{R_x}{R_x^2 + R_y^2} \left(\frac{\partial R_y}{\partial x_i} - \frac{R_y}{R_x} \frac{\partial R_x}{\partial x_i} \right), \tag{33}$$

$$\frac{\partial \phi}{\partial x_i} = \left(|\underline{R}|^2 - R_z^2 \right)^{-1/2} \left(\frac{\partial R_z}{\partial x_i} - \frac{R_z}{|\underline{R}|^2} \right.$$
$$\left. \times \left(R_x \frac{\partial R_x}{\partial x_i} + R_y \frac{\partial R_y}{\partial x_i} + R_z \frac{\partial R_z}{\partial x_i} \right) \right). \tag{34}$$

From Eq. (32), we continue with

$$\partial R_x/\partial x_i = \partial T_x/\partial x_i, \qquad \partial R_y/\partial x_i = \partial T_y/\partial x_i,$$
$$\partial R_z/\partial x_i = \partial T_z/\partial x_i. \tag{35}$$

For some instant in time the dolution at this point would be trivial: $\partial T_x/\partial x_1 = 1$, $\partial T_y/\partial x_2 = 1$, $\partial T_z/\partial x_3 = 1$, and all remaining terms are zero. This would be the desired result for the least squares batch filter described in Section II.B. For the general case of interest, we continue with the f and g series equations for a free-falling body:

$$\underline{T} = f\underline{T}_0 + g\underline{V}_0, \qquad \underline{T}_0 = \begin{bmatrix} x_1 \\ x_2 \\ x_3 \end{bmatrix}_{t=t_n}, \qquad \underline{V}_0 = \begin{bmatrix} x_4 \\ x_5 \\ x_6 \end{bmatrix}_{t=t_n} . \qquad (36)$$

Continuing with Eq. (36),

$$\frac{\partial T_x}{\partial x_i} = x_1 \frac{\partial f}{\partial x_i} + x_4 \frac{\partial g}{\partial x_i}, \qquad i = 2, 3, 5, 6,$$

$$\frac{\partial T_x}{\partial x_1} = x_1 \frac{\partial f}{\partial x_1} + f + x_4 \frac{\partial g}{\partial x_1}, \qquad \frac{\partial T_x}{\partial x_4} = x_1 \frac{\partial f}{\partial x_4} + x_4 \frac{\partial g}{\partial x_4} + g,$$

$$\frac{\partial T_y}{\partial x_i} = x_2 \frac{\partial f}{\partial x_i} + x_5 \frac{\partial g}{\partial x_i}, \qquad i = 1, 3, 4, 6,$$

$$(37)$$

$$\frac{\partial T_y}{\partial x_2} = x_2 \frac{\partial f}{\partial x_2} + f + x_5 \frac{\partial g}{\partial x_2}, \qquad \frac{\partial T_y}{\partial x_5} = x_2 \frac{\partial f}{\partial x_5} + x_5 \frac{\partial g}{\partial x_5} + g,$$

$$\frac{\partial T_z}{\partial x_i} = x_3 \frac{\partial f}{\partial x_i} + x_6 \frac{\partial g}{\partial x_i}, \qquad i = 1, 2, 4, 5,$$

$$\frac{\partial T_z}{\partial x_3} = x_3 \frac{\partial f}{\partial x_3} + f + x_6 \frac{\partial g}{\partial x_3}, \qquad \frac{\partial T_z}{\partial x_6} = x_3 \frac{\partial f}{\partial x_6} + x_6 \frac{\partial g}{\partial x_6} + g.$$

To continue, the eighth-order terms for f and g are introduced. First, however, we define the terms u_0, p_0, and q_0 for simplicity [3].

$$u_0 = \frac{\mu}{|\underline{T}_0|^3} = \frac{\mu}{\left(x_1^2 + x_2^2 + x_3^2\right)^{3/2}},$$

$$p_0 = \frac{\underline{V}_0 \cdot \underline{T}_0}{|\underline{T}_0|^2} = \frac{x_4 x_1 + x_5 x_2 + x_6 x_3}{x_1^2 + x_2^2 + x_3^2}, \qquad (38)$$

$$q_0 = \frac{|\underline{V}_0|^2 - |\underline{T}_0|^2 u_0}{|\underline{T}_0|^2} = \frac{x_4^2 + x_5^2 + x_6^2}{x_1^2 + x_2^2 + x_3^2} - \frac{\mu}{\left(x_1^2 + x_2^2 + x_3^2\right)^{3/2}},$$

where μ is the product of the gravitational constant and mass of the earth. With these terms, f and g can be defined [3] as follows:

$$f = 1 - \frac{1}{2}u_0\tau^2 + \frac{1}{2}u_0p_0\tau^3 + \frac{1}{24}\left(3u_0q_0 - 15u_0p_0^2 + u_0^2\right)\tau^4$$

$$+ \frac{1}{8}\left(7u_0p_0^3 - 3u_0p_0q_0 - u_0^2p_0\right)\tau^5$$

$$+ \frac{1}{720}\left(630u_0p_0^2q_0 - 24u_0^2q_0 - u_0^3 - 45u_0q_0^2 - 945u_0p_0^4\right.$$

$$\left. + 210u_0^2p_0^2\right)\tau^6$$

$$+ \frac{1}{5040}\left(882u_0^2p_0q_0 - 3150u_0^2p_0^3 - 9450u_0p_0^3q_0 + 1575u_0p_0q_0^2\right.$$

$$\left. + 63u_0^3p_0 + 10{,}395u_0p_0^5\right)\tau^7$$

$$+ \frac{1}{40{,}320}\left(1107u_0^2q_0^2 - 24{,}570u_0^2p_0^2q_0 - 2205u_0^3p_0^2\right.$$

$$+ 51{,}975u_0^2p_0^4 - 42{,}525u_0p_0^2q_0^2 + 155{,}925u_0p_0^4q_0$$

$$\left. + 1575u_0q_0^3 + 117u_0^3q_0 - 135{,}135u_0p_0^6 + u_0^4\right)\tau^8,$$

$$\tag{39a}$$

$$g = \tau - \frac{1}{6}u_0\tau^3 + \frac{1}{4}u_0p_0\tau^4 + \frac{1}{120}\left(9u_0q_0 - 45u_0p_0^2 + u_0^2\right)\tau^5$$

$$+ \frac{1}{360}\left(210u_0p_0^3 - 90u_0p_0q_0 - 15u_0^2p_0\right)\tau^6$$

$$+ \frac{1}{5040}\left(3150u_0p_0^2q_0 - 54u_0^2q_0 - 225u_0q_0^2 - 4725u_0p_0^4\right.$$

$$\left. + 630u_0^2p_0^2 - u_0^3\right)\tau^7$$

$$+ \frac{1}{40{,}320}\left(3024u_0^2p_0q_0 - 12{,}600u_0^2p_0^3 - 56{,}700u_0p_0^3q_0\right.$$

$$\left. + 9450u_0p_0q_0^2 + 62{,}370u_0p_0^5 + 126u_0^3p_0\right)\tau^8. \tag{39b}$$

Using Eqs. (38), (39a) and (39b), we continue with

$$\frac{\partial f}{\partial x_i} = \frac{\partial f}{\partial u_0} \frac{\partial u_0}{\partial x_i} + \frac{\partial f}{\partial p_0} \frac{\partial p_0}{\partial x_i} + \frac{\partial f}{\partial q_0} \frac{\partial q_0}{\partial x_i} ,$$

$$\frac{\partial g}{\partial x_i} = \frac{\partial g}{\partial u_0} \frac{\partial u_0}{\partial x_i} + \frac{\partial g}{\partial p_0} \frac{\partial p_0}{\partial x_i} + \frac{\partial g}{\partial q_0} \frac{\partial q_0}{\partial x_i} .$$

(40)

We solve first for the partials of u_0, p_0, and q_0 with respect
to the states

$$\frac{\partial u_0}{\partial x_i} = - \frac{3\mu x_i}{\left(x_1^2 + x_2^2 + x_3^2\right)^{5/2}} , \qquad i = 1, 2, 3,$$

$$\frac{\partial u_0}{\partial x_i} = 0, \qquad i = 4, 5, 6,$$

$$\frac{\partial p_0}{\partial x_i} = \frac{x_{i+3}}{x_1^2 + x_2^2 + x_3^2} - \frac{2x_i (x_4 x_1 + x_5 x_2 + x_2 x_3)}{\left(x_1^2 + x_2^2 + x_3^2\right)^2} , \qquad i = 1, 2, 3,$$

(41)

$$\frac{\partial p_0}{\partial x_i} = \frac{x_{i-3}}{x_1^2 + x_2^2 + x_3^2} , \qquad i = 4, 5, 6,$$

$$\frac{\partial q_0}{\partial x_i} = - \frac{\left(x_4^2 + x_5^2 + x_6^2\right) 2x_i}{x_1^2 + x_2^2 + x_3^2} + \frac{3\mu x_i}{\left(x_1^2 + x_2^2 + x_3^2\right)^{5/2}} , \qquad i = 1, 2, 3,$$

$$\frac{\partial q_0}{\partial x_i} = \frac{2x_i}{x_1^2 + x_2^2 + x_3^2} , \qquad i = 4, 5, 6.$$

And finally solving for the partials of f and g with respect to
u_0, p_0, and q_0:

$$\frac{\partial f}{\partial u_0} = -\frac{1}{2}\tau^2 + \frac{1}{2}p_0\tau^3 + \frac{1}{24}\left(3q_0 - 15p_0^2 + 2u_0\right)\tau^4$$

$$+ \frac{1}{8}\left(7p_0^3 - 3p_0 q_0 - 2u_0 p_0\right)\tau^5$$

$$+ \frac{1}{720}\left(630p_0^2 q_0 - 48u_0 q_0 - 3u_0^2 - 45q_0^2 - 945_0^4 + 420u_0 p_0^2\right)\tau^6$$

(Eq. continues)

$$+ \frac{1}{5040}\Big(1764u_0p_0q_0 - 6300u_0p_0^3 - 9450p_0^3q_0 + 1575p_0q_0^2$$

$$+ 189u_0^2p_0 + 10,395p_0^5\Big)\tau^7$$

$$+ \frac{1}{40,320}\Big(2214u_0q_0^2 - 49,140u_0p_0^2q_0 - 6615u_0^2p_0^2$$

$$+ 103,950u_0p_0^4 - 42,525p_0^2q_0^2 + 155,925p_0^4q_0$$

$$+ 1575q_0^3 + 351u_0^2q_0 + 135,135p_0^6 + 4u_0^3\Big)\tau^8,$$

$$(42a)$$

$$\frac{\partial g}{\partial u_0} = -\frac{1}{6}\tau^3 + \frac{1}{4}p_0\tau^4 + \frac{1}{120}\Big(9q_0 - 45p_0^2 + 2u_0\Big)\tau^5$$

$$+ \frac{1}{360}\Big(210p_0^3 - 90p_0q_0 - 30u_0p_0\Big)\tau^6$$

$$+ \frac{1}{5040}\Big(3150p_0^2q_0 - 108u_0q_0 - 225q_0^2 - 4725p_0^4$$

$$+ 1260u_0p_0^2 - 3u_0^2\Big)\tau^7$$

$$+ \frac{1}{40,320}\Big(6048u_0p_0q_0 - 25,200u_0p_0^3 - 56,700p_0^3q_0$$

$$+ 9450p_0q_0^2 + 62,370p_0^5 + 378u_0^2p_0\Big)\tau^8, \qquad (42b)$$

$$\frac{\partial f}{\partial p_0} = \frac{1}{2}u_0\tau^3 + \frac{1}{24}(-30u_0p_0)\tau^4 + \frac{1}{8}\Big(21u_0p_0^2 - 3u_0q_0 - u_0^2\Big)\tau^5$$

$$+ \frac{1}{720}\Big(1260u_0p_0q_0 - 3780u_0p_0^3 + 420u_0^2p_0\Big)\tau^6$$

$$+ \frac{1}{5040}\Big(882u_0^2q_0 - 9450u_0^2p_0^2 - 28,350u_0p_0^2q_0 + 1575u_0q_0^2$$

$$+ 63u_0^3 + 51,975u_0p_0^4\Big)\tau^7$$

$$+ \frac{1}{40,320}\Big(-49,140u_0^2p_0q_0 - 4410u_0^3p_0 + 207,900u_0^2p_0^3$$

$$- 85,050u_0p_0q_0^2 + 623,700u_0p_0^3q_0$$

$$- 810,810u_0p_0^5\Big)\tau^8, \qquad (42c)$$

$$\frac{\partial g}{\partial p_0} = \frac{1}{4}u_0\tau^4 + \frac{1}{120}(-90u_0p_0)\tau^5 + \frac{1}{360}\left(630u_0p_0^2 - 90u_0q_0 - 15u_0^2\right)\tau^6$$

$$+ \frac{1}{5040}\left(6300u_0p_0q_0 - 18,900u_0p_0^3 + 1260u_0^2p_0\right)\tau^7$$

$$+ \frac{1}{40,320}\left(3024u_0^2q_0 - 37,800u_0^2p_0^2 - 170,100u_0p_0^2q_0\right.$$

$$\left. + 9450u_0q_0^2 + 311,850u_0p_0^4 + 126u_0^3\right)\tau^8, \quad (42d)$$

$$\frac{\partial f}{\partial q_0} = \frac{1}{24}(3u_0)\tau^4 + \frac{1}{8}(-3u_0p_0)\tau^5$$

$$+ \frac{1}{720}\left(630u_0p_0^2 - 24u_0^2 - 90u_0q_0\right)\tau^6$$

$$+ \frac{1}{5040}\left(882u_0^2p_0 - 9450u_0p_0^3 + 3150u_0p_0q_0\right)\tau^7$$

$$+ \frac{1}{40,320}\left(2214u_0^2q_0 - 24,570u_0^2p_0^2 - 85,050u_0p_0^2\right.$$

$$\left. + 155,925u_0p_0^4 + 4725u_0q_0^2 + 117u_0^3\right)\tau^8, \quad (42e)$$

$$\frac{\partial g}{\partial q_0} = \frac{1}{120}(9u_0)\tau^5 + \frac{1}{360}(-90u_0p_0)\tau^6$$

$$+ \frac{1}{5040}\left(3150u_0p_0^2 - 54u_0^2 - 450u_0q_0\right)\tau^7$$

$$+ \frac{1}{40,320}\left(3024u_0^2p_0 - 56,700u_0p_0^3 + 18,900u_0p_0q_0\right)\tau^8.$$

$$(42f)$$

This completes the concept of explicit Jacobian, which is a
fundamentally different approach from the algorithm described
in Section II.B where the transition matrix is required. In
Section II.E the extension of this explicit Jacobian concept to
the Kalman filter formulation is proposed.

The iterative solution method of the weighted least squares
fromulation [Eq. (29)] can be enhanced by the method of Marquardt
matrix conditioning [1]. Basically this method inolves the
addition of a variable factor λ to the diagonal terms of the
state covariance $\left(J_n^T W_{(n)}^{-1} J_n\right)$ to improve the convergence properties of the iterative solution method (Gauss-Newton). This

method results in the following modification of Eq. (29):

$$\hat{X}_n = X_n + \overbrace{\left(J_n^T W_{(n)}^{-1} J_n + \lambda I\right)^{-1}}^{\text{normalized } S_1^*} \overbrace{J_n^T W_{(n)}^{-1} \Delta Y_{(n)}}^{\text{normalized } S_2^*}. \qquad (43)$$
$$\underbrace{\phantom{\left(J_n^T W_{(n)}^{-1} J_n + \lambda I\right)^{-1} J_n^T W_{(n)}^{-1} \Delta Y_{(n)}}}_{\text{denormalized } \Delta X_n}$$

Note that to include the factor λ, a normalizing/denormalizing procedure is required. Using the symbol * to indicate normalized, Eq. (43) can be represented by the following set of equations which define this normalizing procedure:

$$\hat{X}_n = X_n + \Delta X_n, \qquad \Delta X_n^* = \left(S_1^* + \lambda I\right)^{-1} S_2^*,$$

$$\left[S_1^*\right]_{i,j} = S_1(i,j)/[S_1(i,\ i)]^{1/2}[S_1(j,\ j)]^{1/2},$$

$$\left[S_2^*\right]_i = S_2(i)/[S_1(i,\ i)]^{1/2}, \qquad \Delta X_n = \Delta X_n^*/[S_1(i,\ i)]^{1/2}$$

The Marquardt method approaches the Gauss-Newton method (used in the algorithm described in Section II.B) as $\lambda \to 0$ and the method of steepest descent as $\lambda \to \infty$. In addition, the step size increases as $\lambda \to 0$ and decreases as $\lambda \to \infty$. The strategy is to decrease λ if the solution is converging and to increase λ if it is diverging. This method thus has the ability to converge from a distant initial guess and also the ability to converge rapidly once the vicinity of the solution is reached.

As discussed in Section II.B, one can avoid building the entire matrix J_n by formulating Eq. (43) with its algebraic equivalent

$$\hat{X}_n = X_n + \left\{ \left(\sum_{i=1}^{L+1} J_n^T(i) W_{n+1-i}^{-1} J_n(i) \right) + \lambda I \right\}^{-1}$$

$$\times \left\{ \sum_{i=1}^{L+1} J_n^T(i) W_{n+1-i}^{-1} \Delta Y_{n+1-i} \right\}. \qquad (45)$$

where $J_n(1) = A_n, \ J_n(2) = A_{n-1}, \ \ldots, \ J_n(L + 1) = A_{n-L}.$

With the techniques of explicit Jacobian and Marquardt matrix conditioning established, the final working algorithm is now presented in Fig. 3. The algorithm is presented in the same format as the standard weighted least squares batch filter of Section II.B (Fig. 1) so that the two methods can be compared directly. The absence of a transition matrix and the use of Marquardt matrix conditioning are the distinguishing features of this formulation. The use of the closed-form solution for the equations of motion (f and g series) required by the

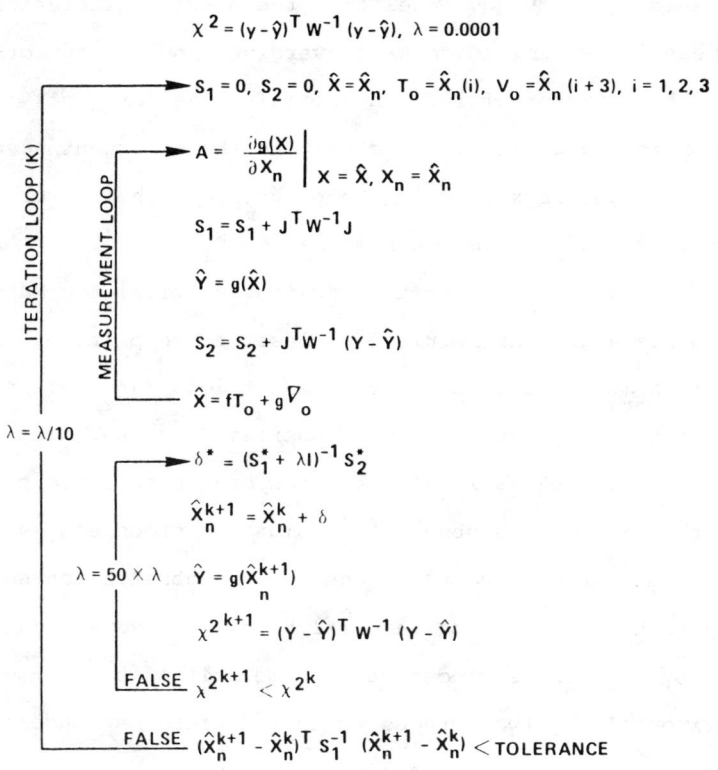

$$X = fT_0 - gV_0 \quad \text{EQUATION OF MOTION}$$
$$Y = g(X) \quad \text{EQUATIONS RELATING OBSERVATIONS TO STATES}$$

$$\chi^2 = (y - \hat{y})^T W^{-1} (y - \hat{y}), \quad \lambda = 0.0001$$

$$S_1 = 0, \; S_2 = 0, \; \hat{X} = \hat{X}_n, \; T_0 = \hat{X}_n(i), \; V_0 = \hat{X}_n(i+3), \; i = 1, 2, 3$$

$$A = \left. \frac{\partial g(X)}{\partial X_n} \right|_{X = \hat{X}, \, X_n = \hat{X}_n}$$

$$S_1 = S_1 + J^T W^{-1} J$$

$$\hat{Y} = g(\hat{X})$$

$$S_2 = S_2 + J^T W^{-1} (Y - \hat{Y})$$

$$\hat{X} = fT_0 + g V_0$$

$\lambda = \lambda/10$

$$\delta^* = (S_1^* + \lambda I)^{-1} S_2^*$$

$$\hat{X}_n^{k+1} = \hat{X}_n^k + \delta$$

$\lambda = 50 \times \lambda$ $\quad \hat{Y} = g(\hat{X}_n^{k+1})$

$$\chi^{2\,k+1} = (Y - \hat{Y})^T W^{-1} (Y - \hat{Y})$$

FALSE $\quad \chi^{2\,k+1} < \chi^{2\,k}$

FALSE $\quad (\hat{X}_n^{k+1} - \hat{X}_n^k)^T S_1^{-1} (\hat{X}_n^{k+1} - \hat{X}_n^k) < \text{TOLERANCE}$

ITERATION LOOP (K)

MEASUREMENT LOOP

Fig. 3. Marquardt weighted least squares with explicit Jacobian.

explicit Jacobian technique is also a key element. Note also
that the partials for the measurement matrix are the partials
of the measurement equations at times t_n, t_{n-1}, ..., t_{n-L} with
respect to each state at time t_n. This means that the R_x, R_y,
R_z terms in Eqs. (33) and (34) are evaluated with the states at
time t_n, t_{n-1}, ..., t_{n-L}, whereas Eqs. (37), (38), and (41) are
evaluated with the state at time $t_n(\hat{X}_n)$.

   The method for varying the Marquardt factor λ is included
in Fig. 2. The chi-squared (χ^2) quantity is the criterion used
to determine if the process is converging. If this quantity
increases from the previous iteration, the process is determined
to be diverging, and the factor λ is increased by a factor of
50. If this quantity decreases from the previous iteration,
the process is determined to be converging, and the factor λ is
decreased by a factor of 10. An initial value for λ of 0.0001
is used in this algorithm. Note that in the divergent case the
process is looped back only to the point where the factor λ is
added, thus using the same quantities for S_1^* and S_2^*.

   Like the batch filter of the previous section, the Marquardt
least squares algorithm requires at least three pairs of angle
observations (three cycles of measurement loop) and defines the
nominal state vector at the final observation point. The
Marquardt least squares is also a batch or nonrecursive process so that all of the observations must be processed each
time a new measurement pair is added to the observation set.
Both algorithms produce the same weighted least squares state
estimate by successively correcting some initial state guess.
A comparison of the two methods would therefore be concerned
with the stability and computation requirements of each.

The Marquardt least squares could use the energy-constrained initial state guess algorithm described in Fig. 2 to supply the initial state guess to begin the iterative process. A less expensive and simplified method can also be used since the Marquardt least squares process is very stable and converges for a very wide range of state estimates. This simplified initial state guess technique is described as follows.

For observation times of t_n, t_{n-1}, ..., t_{n-L}, the following observation set may be defined:

$$\underline{Y}_n = \begin{bmatrix} \theta_n \\ \phi_n \end{bmatrix},$$

$$\underline{Y}_{n-1} = \begin{bmatrix} \theta_{n-1} \\ \phi_{n-1} \end{bmatrix}, \tag{46}$$

$$\underline{Y}_{n-L} = \begin{bmatrix} \theta_{n-L} \\ \phi_{n-L} \end{bmatrix}.$$

Using the first (n - L) and last (n) observation points, the approximate angle rates are

$$\dot{\theta} = (\theta_n - \theta_{n-L})/(t_n - t_{n-L}),$$

$$\dot{\phi} = (\phi_n - \phi_{n-L})/(t_n - t_{n-L}). \tag{47}$$

The velocity magnitude for the target can be approximated by the square root of the sum of its components squared:

$$V = (\dot{R}^2 + (R\dot{\theta})^2 + (R\dot{\phi})^2)^{1/2}. \tag{48}$$

Solving for the range rate yields

$$\dot{R} = V^2 - (R\dot{\theta})^2 - (R\dot{\phi})^2)^{1/2} \tag{49}$$

Given some initial guess for the velocity magnitude V and range (acquisition range R_0), the range rate $\dot{R}$ may be determined by

$$R_n = R_0 + \dot{R}(t_n - t_{n-L}). \tag{50}$$

At this point the working coordinate system of Fig. 4 (X', Y', Z') is introduced so that the state vector in this new system can be computed directly from the quantities R_n, $\dot{R}$, $\dot{\theta}$, and $\dot{\phi}$. The X, Y, Z coordinate system in Fig. 4 is the observer-centered Cartesian coordinate system produced from the difference in the target and sensor [earth-centered inertial coordinate system (ECI)] position vectors ($\underline{X} - \underline{XS}$). The state vector for the target at the final observation point in the working coordinate system can now be computed directly as

$$\underline{X}' = \begin{bmatrix} x_1 \\ x_2 \\ x_3 \\ x_4 \\ x_5 \\ x_6 \end{bmatrix} = \begin{bmatrix} R_n \\ 0 \\ 0 \\ \dot{R} \\ R_n \dot{\theta} \\ R_n \dot{\phi} \end{bmatrix}. \tag{51}$$

The transformation matrix to convert this state vector from the X', Y', Z' system to the X, Y, Z system is

$$C = \begin{bmatrix} \cos\phi\cos\theta & -\sin\theta & -\sin\phi\cos\theta \\ \cos\phi\sin\theta & \cos\theta & -\sin\phi\sin\theta \\ \sin\phi & 0 & \cos\phi \end{bmatrix}_{\substack{\theta=\theta_n \\ \phi=\phi_n}}. \tag{52}$$

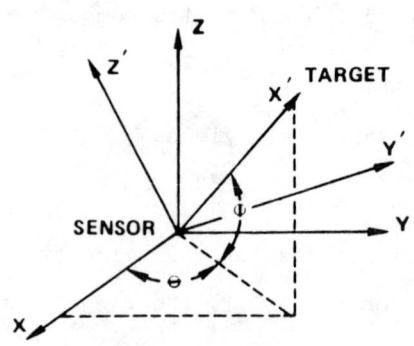

Fig. 4. Working coordinate system (X', Y', Z').

The desired state vector at the final observation point in an ECI coordinate system can now be written

$$\underline{X} = \underline{C}\underline{X}' + \underline{X}\underline{S}. \tag{53}$$

This simplified technique could be used as the primary method for establishing the initial state guess for the Marquardt least squares algorithm, or it could be used as a backup for the energy-constrained iterative algorithm (Fig. 2) when it fails to converge.

D. *ENERGY CONSTRAINT*

As stated earlier, the problem of state estimation of ballistic trajectories with angle-only measurements becomes difficult when the observer is free-falling and more difficult if the observer is then located in the plane of the observed trajectory. Section II.D presents the idea of incorporating an energy constraint into the angle-only tracking algorithms to assist in these poor-observability problems or to enhance the solution of any type of tracking problem in general. Energy is selected as a constraint because of its relative constancy over the whole trajectory for a given set of ICBM threats with the same ground range. The following energy-constraint method can be incorporated into either of the weighted least squares algorithms discussed in Section II.B or Section II.C.

The total energy (per unit mass) of a free-falling body can be expressed as the sum of the kinetic and potential energy

$$E = KE + PE = \frac{1}{2}V^2 - (\mu/R), \tag{54}$$

where V is the velocity magnitude, R is the magnitude of the position vector (earth-centered system), and μ is the product of the gravitational constant and the mass of the earth. The total energy defined in Eq. (54) is incorporated into the

angle-only tracking algorithm by assuming some a priori knowledge of the expected magnitude of this quantity and its associated uncertainty. This expected or mean-energy magnitude can
be considered a pseudomeasurement, thus expanding the measurement set of the Marquardt least squares, for example, to

$$
\underline{Y} = \begin{bmatrix} \theta \\ \phi \\ E \end{bmatrix} = \begin{bmatrix} g_1 \\ g_2 \\ g_3 \end{bmatrix} = \begin{bmatrix} \tan^{-1}(R_y/R_x) \\ \sin^{-1}(R_z/|\underline{R}|) \\ \frac{1}{2}v^2 - (\mu/R) \end{bmatrix}.
\tag{55}
$$

In other words, the a priori mean-energy magnitude serves as
the measured energy, whereas the estimated or calculated energy
is determined from Eq. (54) using the estimated state vector.
Because the total energy is constant over the whole trajectory,
its contribution as a pseudomeasurement is utilized only once
at a single measurement point.

Reformulating the total energy in terms of the target ECI
state vector

$$
E = \frac{x_4^2 + x_5^2 + x_6^2}{2} - \frac{\mu}{\left(x_1^2 + x_2^2 + x_3^2\right)^{1/2}}
\tag{56}
$$

the required partials for the Jacobian matrix (Section II.C) or
the measurement matrix (Section II.B) are obtained:

$$
\partial E/\partial x_i = \mu x_i / \left[\left(x_1^2 + x_2^2 + x_3^2\right)^{1/2}\right], \qquad i = 1, 2, 3,
$$
$$
\partial E/\partial x_i = X_i, \qquad i = 4, 5, 6.
\tag{57}
$$

The measurement covariance matrix is also expanded to accom-
modate the expanded measurement set

$$
[W]_{i,j} = \begin{bmatrix} \sigma_\theta^2 & 0 & 0 \\ 0 & \sigma_\phi^2 & 0 \\ 0 & 0 & \sigma_E^2 \end{bmatrix}.
\tag{58}
$$

The uncertainty in the energy could be input directly or computed from a uniform distribution given some maximum and minimum energy values:

$$\sigma_E^2 = \left[(E_{max} - E_{min})^2 \right]/12. \qquad (59)$$

Because the energy-constraint concept is based on some a priori energy estimate, the resulting least squares state estimate will be biased if this assumed a priori energy is different from the actual energy. Furthermore, the bias will increase as this difference increases and as the uncertainty in this pseudomeasurement is decreased. An analysis and possible cure for the energy-constraint bias problem is given in Section III.

E. *KALMAN FILTER APPLICATION*
    OF THE EXPLICIT JACOBIAN

Unlike the batch least squares process of Sections II.B and II.C, the Kalman filter is a recursive minimum-variance filter. Using two angle measurements at each observation point, this algorithm determines the current target state estimate such that the state covariance is minimized. Thus, the state estimate is conditioned on all measurements made up to that time. The Kalman filter is recursive in the sense that only the current measurement need be processed.

The minimum-variance estimate equations can be arrived at through the weighted least squares concept, so the minimum-variance estimate is also the weighted least squares estimate, and the Kalman filter can be said to be a recursive form of the batch filter. The Kalman filter uses the same linearization approximations as the batch filter of Section II.B for the measurement and transition matrices. However, due to its recursive nature, errors introduced through these approximations can build up resulting in a tracking performance that tends to deviate from the least squares batch filter results. The improved Kalman filter formulation presented here shall be shown to match the batch least squares results.

Using the notation developed in previous sections, the linearized model matrix equation for the recursive problem is

$$\Delta Y_n = A_n \, {}_n\phi_{n-1}\Delta X_{n-1} + \epsilon_n, \tag{60}$$

where ΔY_n is the deviation of the observations (y) from the nominal or calculated set ($\hat{y}$) at time t_n; A_n the matrix that transforms the state at time t_n to equivalent observation parameters; $\phi_{n,n-1}$ the transition matrix that transforms the state at time t_{n-1} to time t_n; ΔX_{n-1} the deviation from the estimate of the state at time t_{n-1}, which corresponds to ΔY_n; and ϵ_n the observation error.

The well-known Kalman filter solution to this problem is presented in Fig. 5. As discussed in Section II.B and repeated here for clarity, the transition matrix Φ in the usual approach is determined by solving the "linear" differential equation model

$$\frac{d}{dt}\Delta X(t) = F(X(t))\Delta X(t). \tag{61}$$

An "approximate" solution is given by

$$\Delta X(t + \Delta t) = \Delta X(t) + \Delta t F(X(t))\Delta X(t), \tag{62}$$

so that

$$\Phi(t + \Delta t, t) = I + \Delta t F(X(t)), \tag{63}$$

or

$$\phi_{n,n-1} = I + t F(X(t_n)), \tag{64}$$

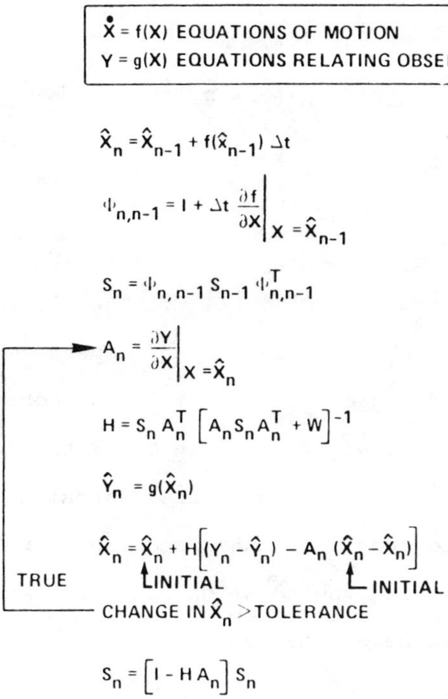

$$\dot{X} = f(X) \text{ EQUATIONS OF MOTION}$$
$$Y = g(X) \text{ EQUATIONS RELATING OBSERVATIONS TO STATES}$$

$$\hat{X}_n = \hat{X}_{n-1} + f(\hat{X}_{n-1})\,\Delta t$$

$$\Psi_{n,n-1} = I + \Delta t \left.\frac{\partial f}{\partial X}\right|_{X=\hat{X}_{n-1}}$$

$$S_n = \Psi_{n,n-1} S_{n-1} \Psi_{n,n-1}^T$$

$$A_n = \left.\frac{\partial Y}{\partial X}\right|_{X=\hat{X}_n}$$

$$H = S_n A_n^T \left[A_n S_n A_n^T + W \right]^{-1}$$

$$\hat{Y}_n = g(\hat{X}_n)$$

$$\hat{X}_n = \hat{X}_n + H\left[(Y_n - \hat{Y}_n) - A_n(\hat{X}_n - \hat{X}_n) \right]$$

TRUE   └ INITIAL   └ INITIAL

CHANGE IN $\hat{X}_n >$ TOLERANCE

$$S_n = \left[I - H A_n \right] S_n$$

Fig. 5.   Kalman filter algorithm.

where

$$F[(X(t_n))]_{i,j} = \partial f_i / \partial X_j \big|_{X=X(t_n)}$$

$$= \begin{bmatrix} \partial f_1/\partial x_1 & \cdots & \partial f_1/\partial x_6 \\ \vdots & & \vdots \\ \partial f_6/\partial x_1 & \cdots & \partial f_6/\partial x_6 \end{bmatrix}_{X=X(t_n)} , \tag{65}$$

$$f(X) = dX/dt. \tag{66}$$

As discussed in Section II.B, the accuracy of Φ can be improved
by breaking Δt up into equal fractions $h = \Delta t/m$ and using the
equation

$$\Phi(t + Kh, t) = [I + hF(t + (k - 1)h)]$$
$$\times \Phi(t + (k - 1)h, t), \quad k = 1, 2, 3, \ldots, m. \tag{67}$$

Additional accuracy can be achieved by adding another term to the approximate solution of Eq. (62):

$$\Delta X(t + \Delta t) = \Delta X(t) + \Delta t F(t) \Delta X(t) + \frac{1}{2} \Delta t^2$$

$$- (\dot{F}(t) = F^2(t)) \Delta X(t). \tag{68}$$

Another method for attacking the accuracy problem is to solve for the Jacobian matrix $[A_n \Phi_{n,n-1}]$ explicitly, as was done for the MLS algorithm in Section II.C. The partial derivatives of the measurement equations with respect to each state for the explicit Jacobian are identical to those derived in Section II.C for the MLS algorithm. In order to implement the explicit Jacobian into the Kalman filter formulation of Fig. 5, one need only to substitute the Jacobian wherever the pair $A_n \Phi_{n,n-1}$ occurs. Using the notation of Fig. 5, the Kalman gain matrix equation can be written as

$$H = \Phi_{n,n-1} S_{n-1} \Phi_{n,n-1}^T A_n^T$$

$$\times \left[A_n \Phi_{n,n-1} S_{n-1} \Phi_{n,n-1}^T A_n^T + W \right]^{-1}, \tag{69}$$

which would yield the following Jacobian formulation:

$$H = \Phi_{n,n-1} S_{n-1} J^T \left[J S_{n-1} J^T + W \right]^{-1}. \tag{70}$$

In similar fashion the updated state covariance matrix equation can be written as

$$S_n = \Phi_{n,n-1} S_{n-1} \Phi_{n,n-1}^T - H A_n \Phi_{n,n-1} S_{n-1} \Phi_{n,n-1}^T, \tag{71}$$

which would yield the following Jacobian formulation:

$$S_n = \Phi_{n,n-1} S_{n-1} \Phi_{n,n-1}^T - H J S_{n-1} \Phi_{n,n-1}^T. \tag{72}$$

The resulting Jacobian-Kalman formulation is presented in Fig. 6. This is only an intermediate formulation presented at this point for clarity.

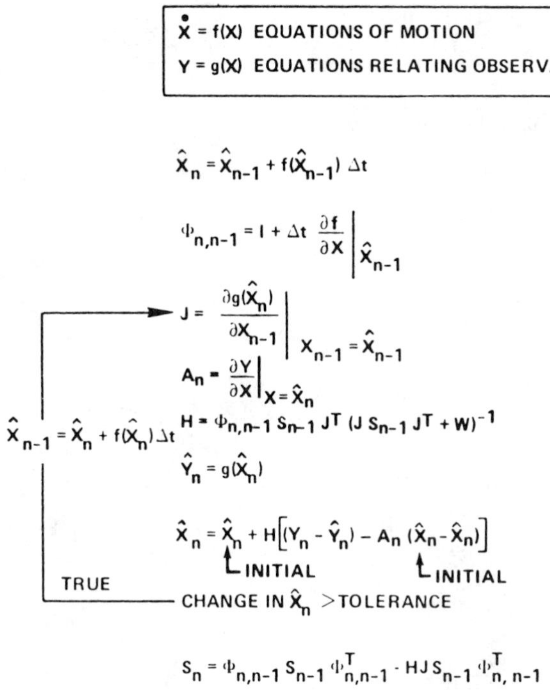

$$\dot{X} = f(X) \quad \text{EQUATIONS OF MOTION}$$

$$Y = g(X) \quad \text{EQUATIONS RELATING OBSERVATIONS TO STATES}$$

$$\hat{X}_n = \hat{X}_{n-1} + f(\hat{X}_{n-1})\,\Delta t$$

$$\Phi_{n,n-1} = I + \Delta t \left.\frac{\partial f}{\partial X}\right|_{\hat{X}_{n-1}}$$

$$J = \left.\frac{\partial g(\hat{X}_n)}{\partial X_{n-1}}\right|_{X_{n-1}=\hat{X}_{n-1}}$$

$$A_n = \left.\frac{\partial Y}{\partial X}\right|_{X=\hat{X}_n}$$

$$\hat{X}_{n-1} = \hat{X}_n + f(\hat{X}_n)\Delta t \qquad H = \Phi_{n,n-1}\, S_{n-1}\, J^T\, (J\, S_{n-1}\, J^T + W)^{-1}$$

$$\hat{Y}_n = g(\hat{X}_n)$$

$$\hat{X}_n = \hat{X}_n + H\left[(Y_n - \hat{Y}_n) - A_n\,(\hat{X}_n - \hat{X}_n)\right]$$

$$\overset{\llcorner \text{INITIAL}}{} \qquad \overset{\llcorner \text{INITIAL}}{}$$

TRUE

CHANGE IN $\hat{X}_n$ > TOLERANCE

$$S_n = \Phi_{n,n-1}\, S_{n-1}\, \Phi_{n,n-1}^T \cdot H\, J\, S_{n-1}\, \Phi_{n,\,n-1}^T$$

Fig. 6.   Jacobian-Kalman algorithm (intermediate).

From Fig. 6 the equation

$$\hat{X}_n = \hat{X}_n + H(Y_n - \hat{Y}_n), \tag{73}$$

can be written as

$$\Phi_{n,n-1}\hat{X}_{n-1} = \Phi_{n,n-1}\hat{X}_{n-1} + H(Y_n - \hat{Y}_n), \tag{74}$$

or

$$\hat{X}_{n-1} = \hat{X}_{n-1} + \Phi_{n,n-1}^{-1} H(Y_n - \hat{Y}_n). \tag{75}$$

From Fig. 6,

$$\Phi_{n,n-1}^{-1} H = S_{n-1} J^T \left(J S_{n-1} J^T + W\right)^{-1}$$

Letting $H' = \Phi_{n,n-1}^{-1} H$ and using Eqs. (75) and (76), the final
Jacobian-Kalman algorithm is presented in Fig. 7. This final
algorithm is very similar to the intermediate result of Fig. 6

$$\dot{X} = f(X) \quad \text{EQUATION OF MOTION}$$
$$Y = g(X) \quad \text{EQUATION RELATING OBSERVATIONS TO STATES}$$

$$\hat{X}_n = \hat{X}_{n-1} + f(\hat{X}_{n-1}) \, \Delta t$$

$$J = \left. \frac{\partial g(\hat{X}_n)}{\partial X_{n-1}} \right|_{X_{n-1} = \hat{X}_{n-1}}$$

$$H' = S_{n-1} J^T (J S_{n-1} J^T + W)^{-1}$$

$$A_n = \left. \frac{\partial Y}{\partial X} \right|_{X = \hat{X}_n}$$

$$\hat{Y}_n = g(\hat{X}_n)$$

$$\hat{X}_{n-1} = \hat{X}_{n-1} + H' \left[(Y_n - \hat{Y}_n) - A_n (\hat{X}_n - \hat{X}_n) \right]$$
$$\quad\quad\quad \text{INITIAL} \quad\quad\quad\quad\quad \text{INITIAL}$$

TRUE
CHANGE IN $\hat{X}_{n-1} >$ TOLERANCE

$$\hat{X}_n = \hat{X}_{n-1} + f(\hat{X}_{n-1}) \, \Delta t$$

$$\Phi_{n,n-1} = I + \Delta t \left. \frac{\partial f}{\partial X} \right|_{\hat{X}_{n-1}}$$

$$S_n = \Phi_{n,n-1} S_{n-1} \Phi_{n,n-1}^T - \Phi_{n,n-1} HJS_{n-1} \Phi_{n,n-1}^T$$

Fig. 7. Jacobian-Kalman algorithm (final).

but has the following advantages. In the final algorithm the transition matrix has been eliminated entirely from the equations used to determine the final state estimate. When the transition matrix is calculated for the state covariance update, it is evaluated with the final state estimate.

For the Kalman filter formulation Eqs. (37), (38), and (41) are evaluated with $\hat{X}_{n-1}$, whereas the R_x, R_y, and R_z terms of Eqs. (33) and (34) are evaluated with $\hat{X}_n$. Note that for this formulation the state estimate $\hat{X}_{n-1}$ must be projected forward to t_n each time an iteration is performed. This is accomplished with the f and g series [Eqs. (36) and (39)].

III.   PERFORMANCE EVALUATION

A. *TEST CASE DESCRIPTION*

Five test cases were used to demonstrate the performance of the tracking algorithms presented in this review.   These cases represent different degrees of difficulty in terms of relative engagement geometry.   In each case both the sensor and target are on ballistic trajectories.   For Cases 3, 4, and 5, the sensor states are identical.   The initial ECI state vectors for these test cases are given in Table I.   Some of the initial characteristic trajectory parameters which further describe these five test cases are given in Table II.

Case 1

In terms of tracking performance this case provides the best results of the five test cases.   The aspect angle in Table II confirms the favorable tracking geometry provided with this case.   The Marquardt least squares (MLS) results for this case are shown in Fig. 8 for both the constrained and unconstrained modes.   The results are for 50 Monte Carlo trials and assume an angle-measurement accuracy of 20 μrad (1σ).   A frame time of 10 sec was used so the initial results shown at 20-sec track time are for three measurements.   For the energy-constrained mode, three sets of results are shown for assumed 1σ energy uncertainties of 10, 20, and 40% [(σ_{energy}/energy) × 100]. Note that all constrained cases approach the unconstrained results.

The standard least squares batch filter results for this same case are shown in Fig. 9.   As expected, these agree with the MLS results of Fig. 8, since both algorithms provide the same least squares solution to the problem.   The somewhat

Table I. Test Cases: ECI Initial State Vectors

Test case	X (m)	Y (m)	Z (m)	$\dot{X}$ (m/sec)	$\dot{Y}$ (m/sec)	$\dot{Z}$ (m/sec)
Case 1						
Target	2,055,397.0	-2,486,767.2	6,353,203.9	-1837.40	-5320.04	-1810.57
Sensor	-369,437.1	-3,841,530.9	6,375,142.4	251.65	2616.77	-2004.02
Case 2						
Target	55,986.6	-1,977,835.6	8,149,027.5	171.90	-3871.66	-2046.71
Sensor	-409,131.8	-4,254,290.7	6,641,171.5	190.14	1977.13	-974.70
Case 3						
Target	-216,183.4	-2,011,906.4	7,467,052.0	-594.01	-4874.89	-2107.44
Sensor	-339,809.8	-4,448,893.3	5,091,404.1	-493.60	-1039.40	2544.72
Case 4						
Target	-255,729.4	-1,857,730.1	7,591,108.4	-566.59	-4828.72	-1896.57
Sensor	-339,809.8	-4,448,893.3	5,091,404.1	-493.60	-1039.40	2544.72
Case 5						
Target	-198,123.0	-1,731,424.1	7,661,489.0	-576.84	-4841.16	-1716.58
Sensor	-339,809.8	-4,448,893.3	5,091,404.1	-493.60	-1039.40	2544.72

Table II. Test Cases: Initial Characteristic Parameters

Case	Sensor altitude kft	Sensor altitude km	Target altitude kft	Target altitude km	Range nmi	Range km	Aspect angle (deg)
1	3573	1089	2506	764	1500	2778	44.8
2	5032	1534	6650	2027	1496	2770	10.7
3	1322	403	4520	1378	1839	3406	3.3
4	1322	403	4793	1461	1944	3601	3.8
5	1322	403	4917	1499	2021	3743	3.2

erratic behavior of the energy-constrained performance appears to be case dependent, since almost identical results were obtained for different sets of measurements simulated with different random-number sequences. Note that the standard least squares batch filter experienced convergence problems for the unconstrained mode when less than five measurements were utilized, whereas the MLS algorithm did not.

The bias error introduced through the a priori estimate of the target energy in the energy-constrained mode was investigated for this case. In this anlysis the effect of the error in the a priori estimated energy on the resulting bias was examined. The results of this analysis are shown in Fig. 10 for a track time of 150 sec and for assumed 1σ energy uncertainties of 10 and 20%. As an example, if the a priori energy measurement was 20% in error and a 1σ energy uncertainty of 10% was used, then the bias on the range estimate would be 30 km. One must therefore balance the benefits gained through the use of smaller energy uncertainties against the resulting increase in bias for incorrect a priori energy measurements.

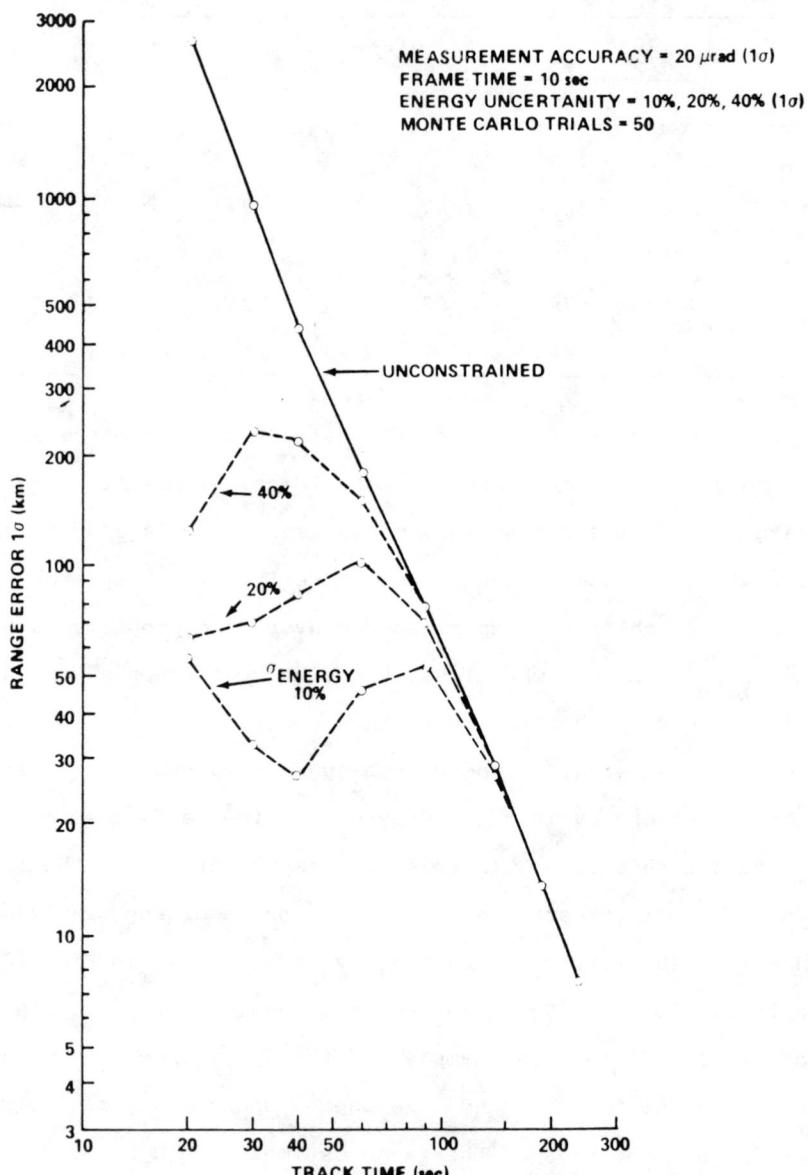

Fig. 8. Marquardt least squares for Case 1.

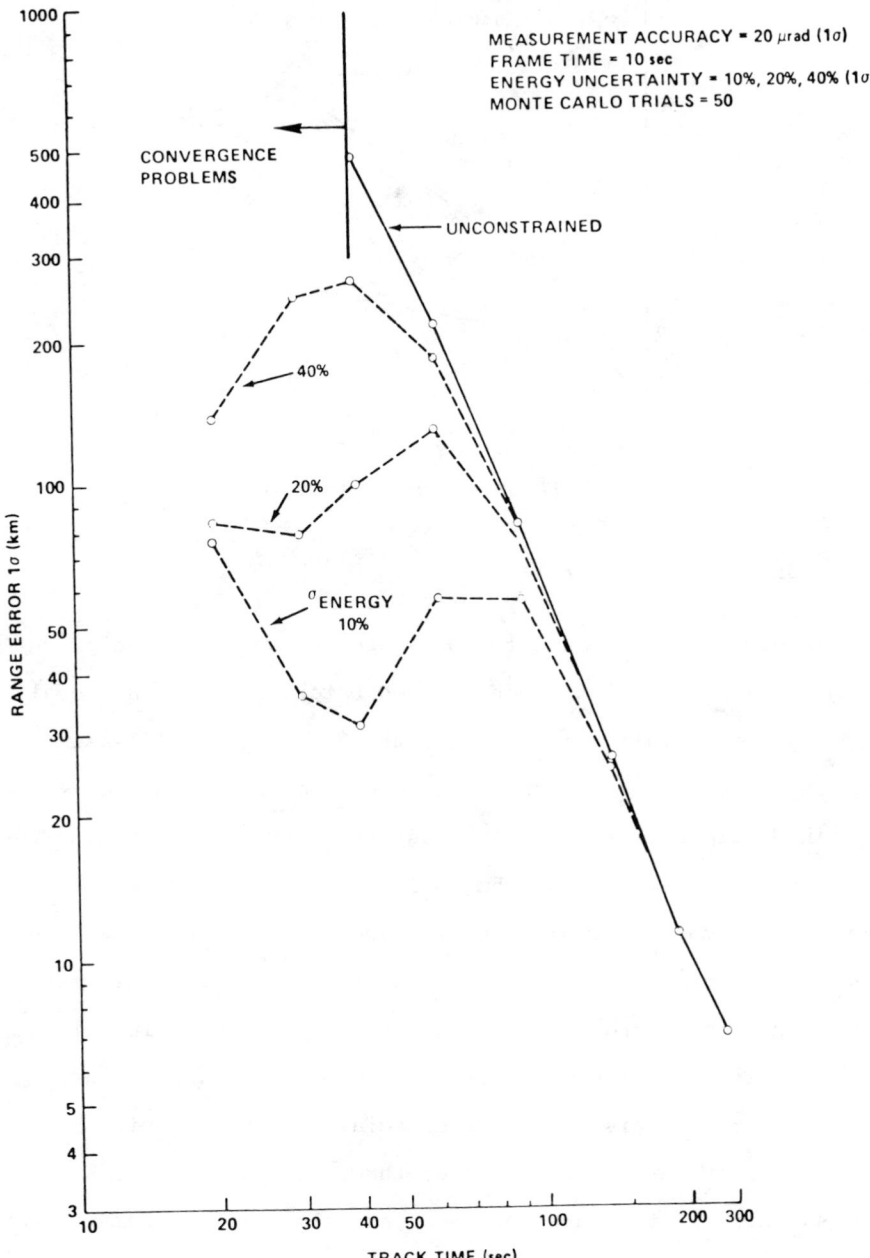

Fig. 9.   Standard least squares batch filter: Case 1.

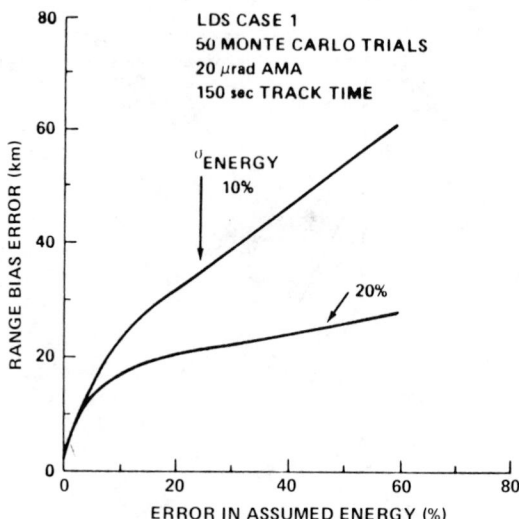

Fig. 10. Bias error due to incorrect assumed energy. Fifty Monte Carlo trials; AMA 20 μrad.

In order to eliminate the bias problem and the high processing requirements of the constrained batch process, the ideal approach is to initialize track with the constrained least squares batch algorithm and then to hand over to the unconstrained Kalman filter for the continuous-track function. The least squares batch algorithm would operate in the energy-constrained mode to improve performance and to guarantee convergence of the unconstrained Kalman filter. Using this method, any bias error introduced through the energy-constrained least squares batch process would be removed by the unconstrained Kalman filter. This method is especially attractive if the Kalman filter performance matches the least squares batch results, which is the case for Jacobian-Kalman filter formulation.

The Jacobian-Kalman filter results for Case 1 are presented in Fig. 11. A five-measurement-constrained MLS was used to initialize the Jacobian-Kalman filter. Three sets of results

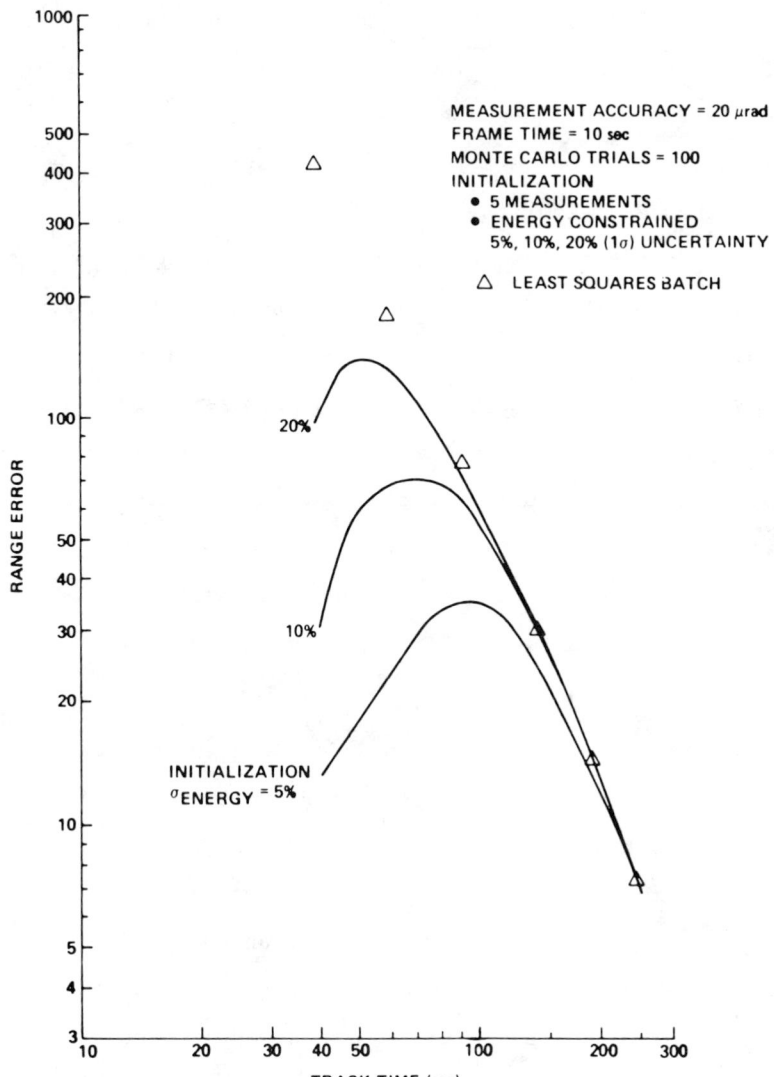

Fig. 11. Jacobian-Kalman: Case 1. One hundred Monte Carlo trials; frame time = 10 sec; measurement accuracy = 20 μrad. Δ = least squares batch.

are shown for a 5, 10, and 20% 1σ energy uncertainty for initialization. Also indicated are the MLS results from Fig. 8 to
show how the Jacobian-Kalman filter formulation matches the
least squares batch performance when initialized with an-energy
constrained MLS algorithm.

To demonstrate how the unconstrained Kalman filter removes
the bias error introduced through an incorrect a priori energy
estimate in the constrained least squares batch process, the
following test was made with Case 1. A bias error was introduced by assuming an a priori energy estimate for the five-
measurement least squares initialization algorithm which was
10% in error. The resulting bias error after initialization
(time, 40 sec) and after each subsequent measurement processed
by the Jacobian-Kalman filter is presented in Fig. 12.

Case 2

This is more difficult than Case 1 due to its near in-plane
geometry. The MLS results for Case 2 are shown in Fig. 13 for
both the constrained and unconstrained mode. The standard least
squares batch filter results for this same case are given in
Fig. 14 for comparison. Note that the unconstrained and 40%
energy uncertainty constrained cases were omitted for the standard least squares batch filter due to convergence problems.
Also note that in order to ensure convergence for this problem
beyond 100 sec of track, this algorithm must begin track at 10
measurements or less. These problems were not experienced with
the MLS algorithm.

The Jacobian-Kalman filter results for Case 2 are presented
in Fig. 15. A five-measurement MLS algorithm with energy constraints was used for initialization. Also indicated are the
MLS results from Fig. 13 for comparison.

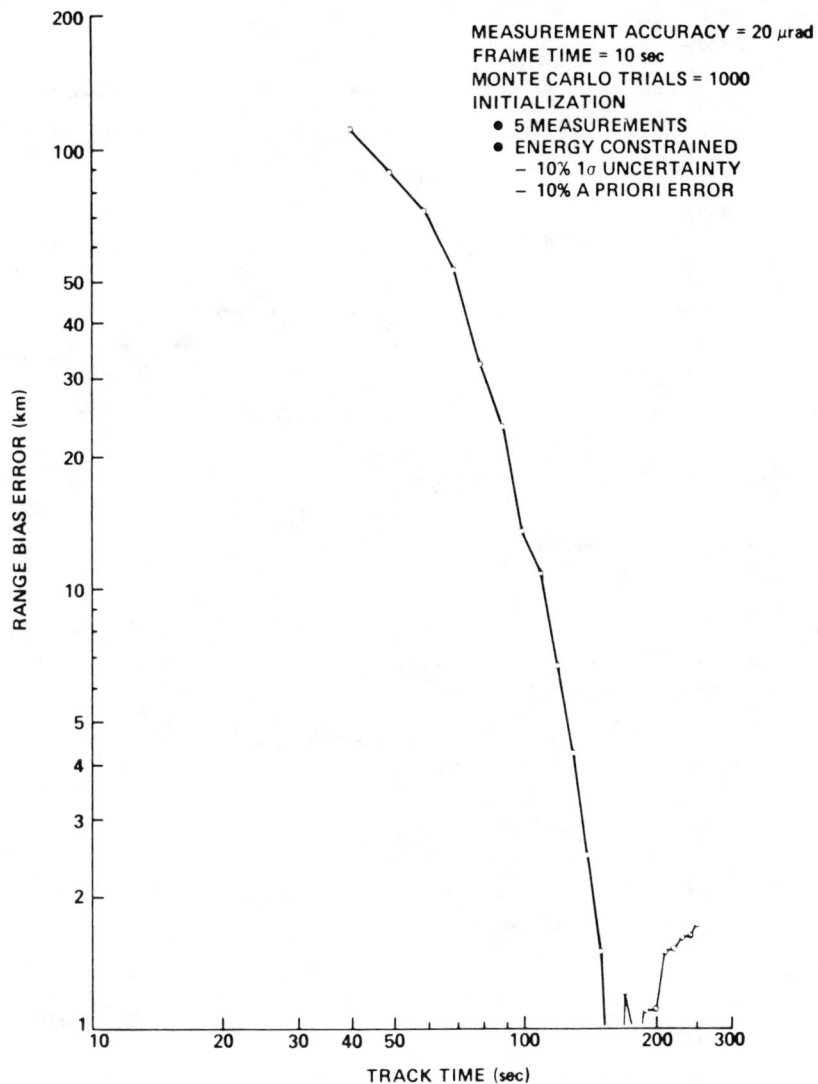

MEASUREMENT ACCURACY = 20 μrad
FRAME TIME = 10 sec
MONTE CARLO TRIALS = 1000
INITIALIZATION
   ● 5 MEASUREMENTS
   ● ENERGY CONSTRAINED
     − 10% 1σ UNCERTAINTY
     − 10% A PRIORI ERROR

Fig. 12.   Jacobian-Kalman: Case 1.

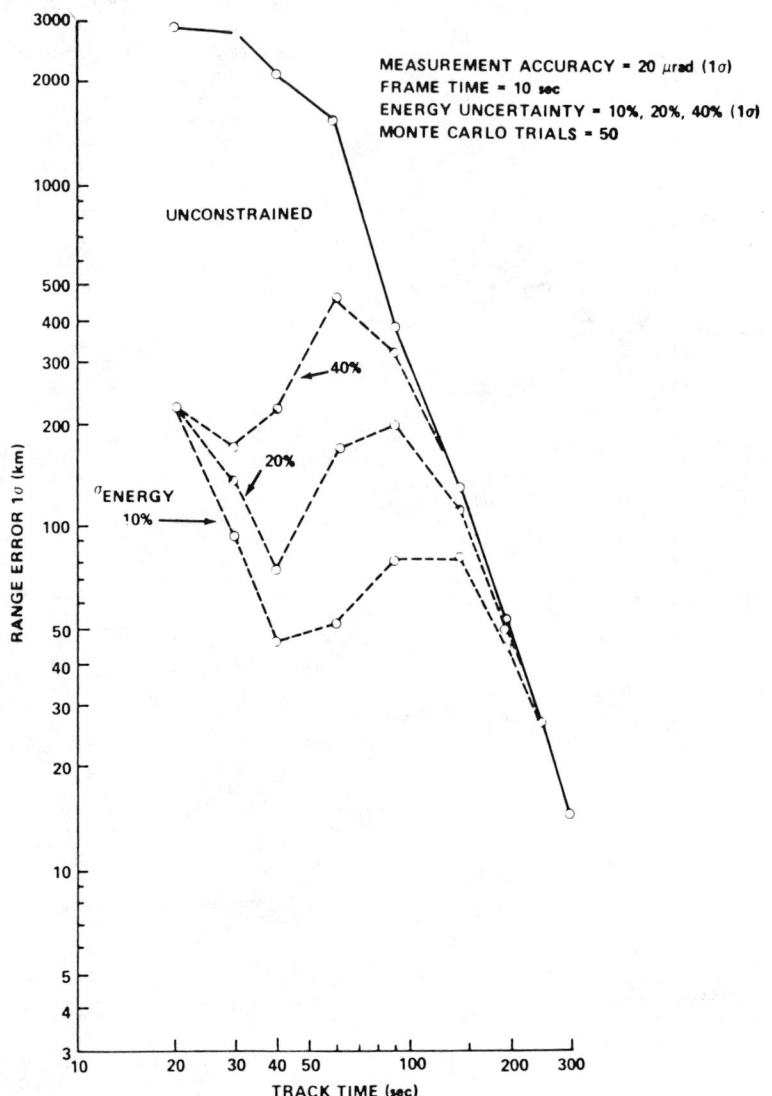

Fig. 13. Marquardt least squares: Case 2.

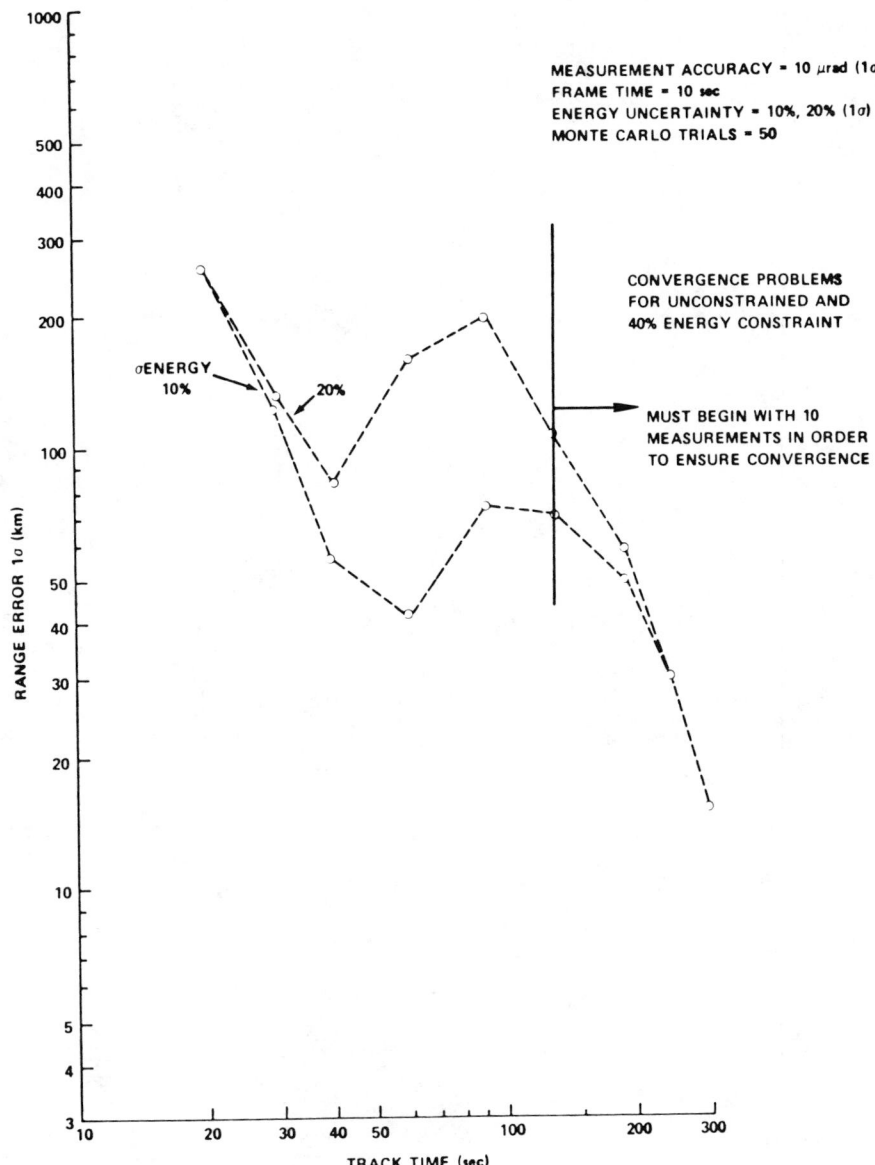

Fig. 14.   Standard least squares batch filter: Case 2.

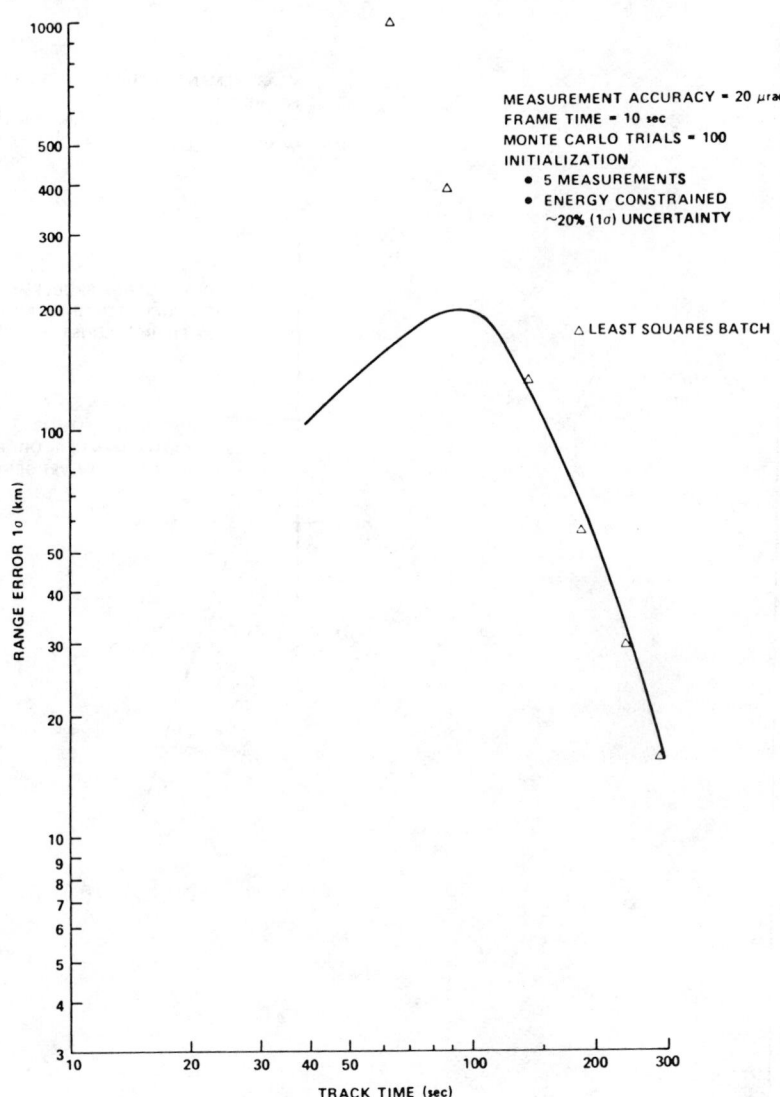

MEASUREMENT ACCURACY = 20 μrad
FRAME TIME = 10 sec
MONTE CARLO TRIALS = 100
INITIALIZATION
  • 5 MEASUREMENTS
  • ENERGY CONSTRAINED
    ~20% (1σ) UNCERTAINTY

△ LEAST SQUARES BATCH

Fig. 15. Jacobian-Kalman: Case 2.

Cases 3, 4, and 5

These three cases, which are all very similar in geometry, proved to be the most difficult. The aspect angles in Table II indicate a relative geometry very near to in-plane. The unconstrained MLS results for Cases 3, 4, and 5 are given in Fig. 16. Although no results are presented for the standard least squares batch filter, convergence problems were again experienced for both the constrained and unconstrained modes when using this algorithm.

The Jacobian-Kalman filter results for Cases 3, 4, and 5 are presented in Fig. 17. A five-measurement MLS algorithm with energy constraints was used for initialization. These results match the MLS performance shown in Fig. 16.

IV.   CONCLUSIONS

Based on the performance results of Section III, the MLS algorithm has been shown to be a superior algorithm in terms of stability and tracking performance when compared to existing least squares batch algorithms that use both a measurement and a transition matrix. This algorithm has never experienced convergence difficulties against even the most difficult of engagement geometries. In addition, this algorithm has been shown to converge for a very wide range of initial state guesses (±50% range guess). The use of the explicit Jacobian method enables higher accuracy by eliminating the need for the transition matrix used in the usual approach. This improved accuracy, along with the technique of Marquardt matrix conditioning, contributes to the enhanced stability of the algorithm. When used in an energy-constrained mode this algorithm serves as an ideal initialization technique for the Jacobian-Kalman filter algorithm.

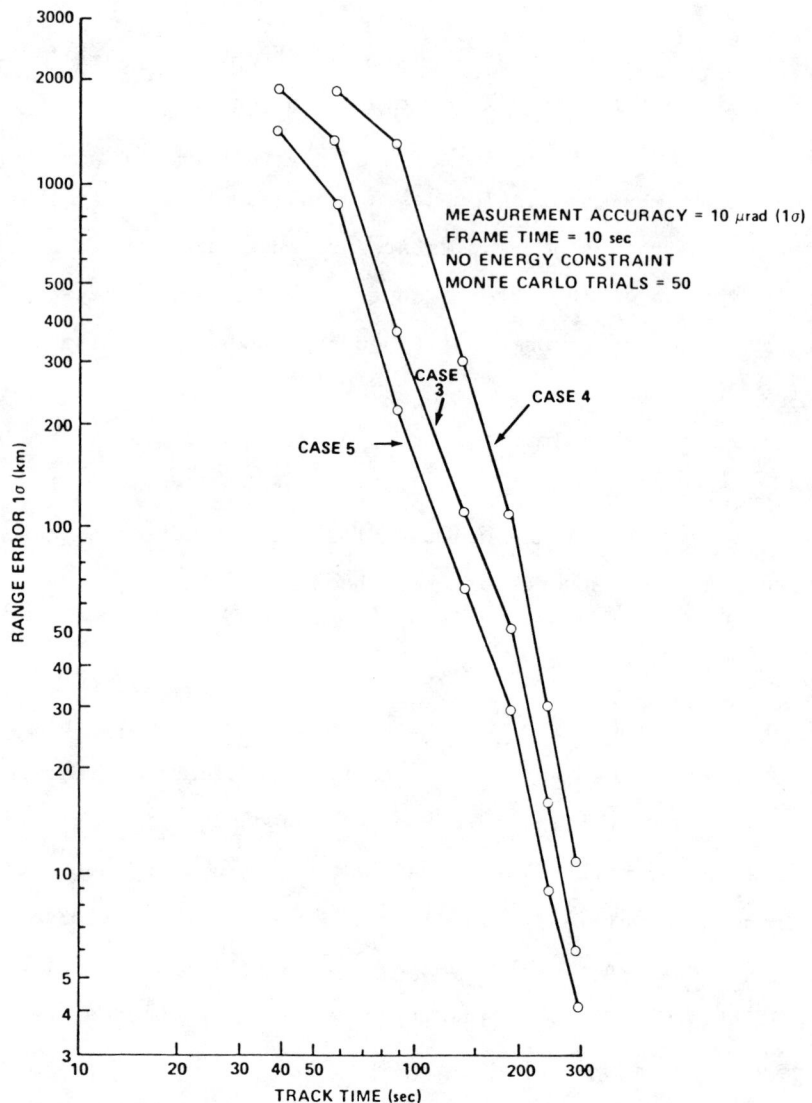

Fig. 16.   Marquardt least squares: Cases 3, 4, and 5.

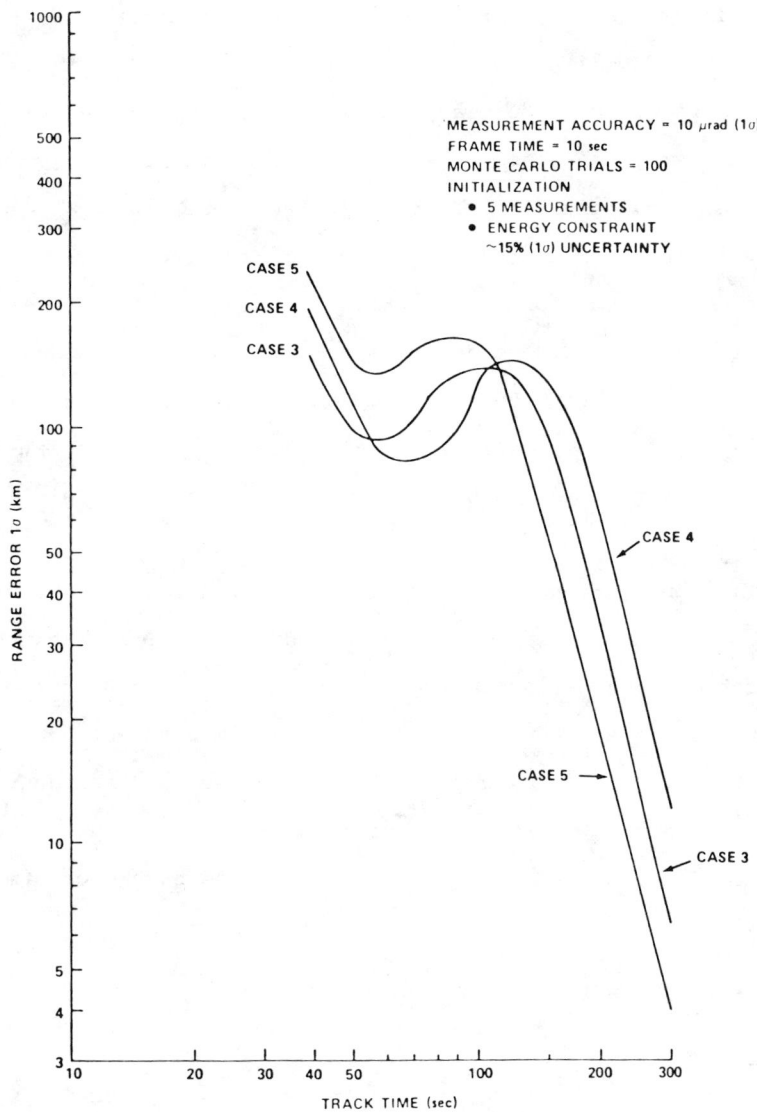

Fig. 17. Jacobian-Kalman: Cases 3, 4, and 5.

Bias errors are introduced in the energy-constrained mode of the MLS algorithm when the a priori estimated energy measurement is in error. The bias becomes greater as the uncertainty in the energy measurement is decreased. This problem can be eliminated by initializing track in the constrained mode and then handing over to the unconstrained Jacobian-Kalman filter algorithm for the continuous-track function. By this method, any bias error introduced in the energy-constrained initialization process is removed in the unconstrained Kalman filter. This approach is especially attractive since the Jacobian-Kalman filter performance matches the least squares batch results.

The application of the explicit Jacobian technique to the Kalman filter formulation has produced a recursive-tracking algorithm whose performance has been shown to match that of the least squares batch process. This is an improvement over existing Kalman filter algorithms that require both a measurement and a transition matrix that results in a tracking performance that tends to deviate from the least squares batch performance bound.

APPENDIX A.   TRANSITION EQUATION

STATE VECTOR

$$\underline{X} = \begin{bmatrix} x_1 \\ x_2 \\ x_3 \\ x_4 \\ x_5 \\ x_6 \end{bmatrix}.$$

EQUATIONS OF MOTION

$$\underline{\dot{X}} = \begin{bmatrix} \dot{x}_1 \\ \dot{x}_2 \\ \dot{x}_3 \\ \dot{x}_4 \\ \dot{x}_5 \\ \dot{x}_6 \end{bmatrix} = \underline{f}(X) = \begin{bmatrix} f_1 \\ f_2 \\ f_3 \\ f_4 \\ f_5 \\ f_6 \end{bmatrix} = \begin{bmatrix} x_1 \\ x_2 \\ x_3 \\ -GMX_1/R^3 \\ -GMX_2/R^3 \\ -GMX_3/R^3 \end{bmatrix},$$

where $R = \left(x_1^2 + x_2^2 + x_3^2\right)^{1/2}$, G is the gravitational constant, and M is the mass of the earth.

TRANSITION EQUATION

$$\Phi_{n-1,n} = I + F(X(t_n))(t_{n-1} - t_n).$$

F MATRIX

$$[F(X(t_n))]_{i,j} = \frac{\partial x_i}{\partial x_j}\bigg|_{X=X(t_n)} =$$

$$\begin{bmatrix} 0 & 0 & 0 & 1 & 0 & 0 \\ 0 & 0 & 0 & 0 & 1 & 0 \\ 0 & 0 & 0 & 0 & 0 & 1 \\ \dfrac{GM\left(2x_1^2-x_2^2-x_3^2\right)}{R^5} & \dfrac{3GMx_1x_2}{R^5} & \dfrac{3GMx_1x_2}{R^5} & 0 & 0 & 0 \\ \dfrac{3GMx_1x_2}{R_5} & \dfrac{GM\left(2x_1^2-x_1^2-x_3^2\right)}{R^5} & \dfrac{3GMx_2x_3}{R^5} & 0 & 0 & 0 \\ \dfrac{3GMx_1x_3}{R^5} & \dfrac{3GMx_2x_3}{R^5} & \dfrac{GM\left(2x_3^2-x_1^2-x_2^2\right)}{R^5} & 0 & 0 & 0 \end{bmatrix}_{X=X(t_n)}$$

APPENDIX B.   MEASUREMENT EQUATION

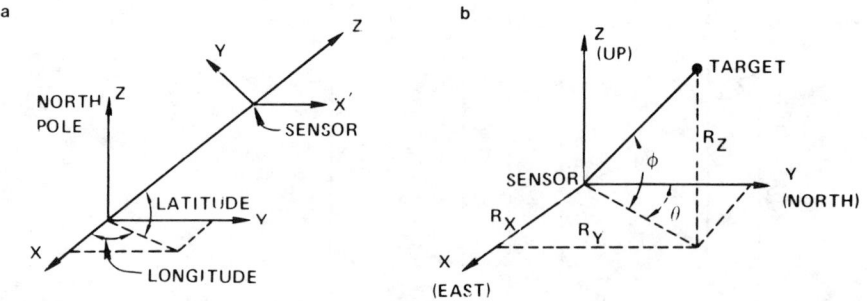

Fig. B.1.   (a) Earth-centered inertial (ECI) Coordinate
System; (b) sensor-centered topographic (SCT) Coordinate System.

RANGE VECTOR

$$\underline{R} = \underline{T} - \underline{S} \quad \text{or} \quad \begin{bmatrix} R_x \\ R_y \\ R_z \end{bmatrix} = \begin{bmatrix} T_x \\ T_y \\ T_z \end{bmatrix} - \begin{bmatrix} S_x \\ S_y \\ S_z \end{bmatrix}.$$

TRANSFORMATION EQUATION

$$\underline{R}(\text{SCT}) = T\underline{R}(\text{ECI}),$$

$$T = \begin{bmatrix} \sin(\text{long}) & \cos(\text{long}) & 0 \\ -\cos(\text{long})\sin(\text{lat}) & -\sin(\text{long})\sin(\text{lat}) & \cos(\text{lat}) \\ \cos(\text{long})\cos(\text{lat}) & \sin(\text{long})\cos(\text{lat}) & \sin(\text{lat}) \end{bmatrix}$$

MEASUREMENT VECTOR

$$\underline{Y} = \begin{bmatrix} \theta \\ \phi \end{bmatrix} = g(X) = \begin{bmatrix} g_1 \\ g_2 \end{bmatrix} = \begin{bmatrix} \tan^{-1}(R_x/R_y) \\ \tan^{-1}\left[R_z / \left(R_x^2 + R_y^2\right)^{1/2}\right] \end{bmatrix}.$$

MEASUREMENT EQUATION

$$\Delta Y_n = A_n \Delta X_n + \epsilon_n.$$

APPENDIX C.   DERIVATION OF INITIAL
              STATE GUESS ALGORITHM

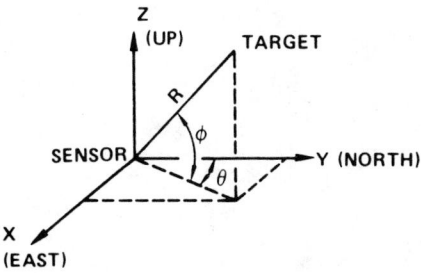

Fig. C.1.   Sensor-centered topographic (SCT) Coordinate
System.

RELATIVE STATE EQUATIONS

$$x = R \sin \theta \cos \phi, \quad y = R \cos \theta \cos \phi, \quad z = R \sin \phi,$$

$$\dot{x} = \dot{R} \sin \theta \cos \phi - R\dot{\phi} \sin \theta \sin \phi + R\dot{\theta} \cos \theta \cos \phi, \quad (C.1)$$

$$\dot{y} = \dot{R} \cos \theta \cos \phi - R\dot{\phi} \cos \theta \sin \phi - R\dot{\theta} \cos \theta \sin \phi,$$

Dividing $\dot{x}$, $\dot{y}$, *and* $\dot{z}$ *[Eq. (C.1)] by R,*

$$a_1 = \dot{x}/R = (\dot{R}/R) \sin \theta \cos \phi - \dot{\phi} \sin \theta \sin \phi$$

$$+ \dot{\theta} \cos \theta \cos \phi,$$

$$a_2 = \dot{y}/R = (\dot{R}/R) \cos \theta \cos \phi - \dot{\phi} \cos \theta \sin \phi \quad (C.2)$$

$$- \theta \cos \dot{\theta} \sin \phi,$$

$$a_3 = \dot{z}/R = (\dot{R}/R) \sin \phi + \dot{\phi} \cos \phi.$$

VELOCITY-VECTOR DEFINITIONS

$$\underline{V}_T = \underline{V}_S + \underline{V}_R, \quad \underline{V}_T = \begin{bmatrix} x_S \\ y_S \\ z_S \end{bmatrix} + \begin{bmatrix} x_R \\ y_R \\ z_R \end{bmatrix} = \begin{bmatrix} \dot{x}_S + a_1 R \\ \dot{y}_S + a_2 R \\ \dot{z}_S + a_3 R \end{bmatrix}. \quad (C.3)$$

where $\underline{V}_T$ is the target-velocity vector in sensor-centered topographic (SCT) system, $\underline{V}_S$ is the sensor-velocity vector in SCT
system, and $\underline{V}_R$ is the relative velocity vector (target-sensor).

Target-Velocity Magnitude
[Eq. (C.3)] Squared

$$|\underline{V}_T|^2 = (\dot{x}_S + a_1R)^2 + (\dot{y}_S + a_2R)^2 + (\dot{z}_S + a_3R)^2,$$

$$|\underline{V}_T|^2 = R^2\left(a_1^2 + a_2^2 + a_3^2\right) + R(2\dot{x}_S a_1 + 2\dot{y}_S a_2 + 2\dot{z}_S a_3) \quad \text{(C.4)}$$

$$+ \left(\dot{x}_S^2 + \dot{y}_S^2 + \dot{z}_S^2\right).$$

ENERGY EQUATION PER UNIT MASS

$$E_T = E_{kinetic} + E_{potential} = \frac{1}{2}|\underline{V}_T|^2 - \mu/R_T \quad \text{(C.5)}$$

where R_T is the distance from earth center to target, and
$\mu = G \times M$.

COMBINING EQS. (C.4) AND (C.5),

$$R^2\underbrace{\left(a_1^2 + a_2^2 + a_3^2\right)}_{a} + \underbrace{R(2\dot{x}_S a_1 + 2\dot{y}_S a_2 + 2\dot{z}_S a_3)}_{b}$$

$$+ \underbrace{\left(\dot{x}_S^2 + \dot{y}_S^2 + \dot{z}_S^2\right) - 2E_T - (2\mu/R_T) = 0.}_{c} \quad \text{(C.6)}$$

Quadratic Solution of R [Eq. (C.6)]

$$R = \left[-b + (b^2 - 4ac)^{1/2}\right]\Big/ 2a. \quad \text{(C.7)}$$

FROM GEOMETRY,

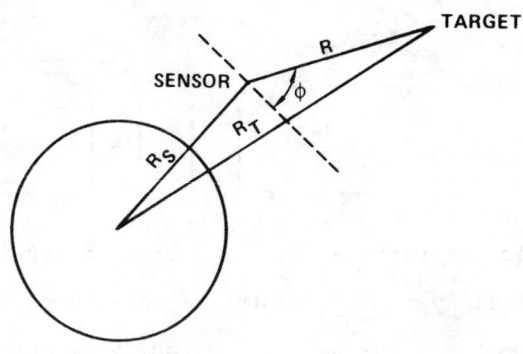

Fig. C.2.   Relative geometry definitions.

$$R_T = \left(R^2 + R_S^2 + 2RR_S \sin \phi \right)^{1/2} \tag{C.8}$$

FOR A FREE-FALLING BODY,

$$\dot{R}/R = \dot{\phi} \tan \phi - (\ddot{\theta}/2\dot{\theta}). \tag{C.9}$$

1. Calculate $\dot{R}/R$ [Eq. (C.9)].

2. Calculate a_1, a_2, a_3 [Eq. (C.2)].

3. Calculate a and b [Eq. (C.6)].

4. Calculate R_T [Eq. (C.8)].

5. Calculate c [Eq. (C.6)].

6. Calculate R [Eq. (C.7)].

7. Repeat steps 4, 5, and 6 until R converges.

8. Calculate $\dot{R}$ [Eq. (C.9)].

9. Calculate x, y, z, $\dot{x}$, $\dot{y}$, and $\dot{z}$ [Eq. (C.1).

REFERENCES

1. D. W. MARQUARDT, *J. Soc. Ind. Appl. Math. 5*, 32–38 (1957).

2. C. B. CHANG and K. P. DUNN, *MIT Lincoln Lab. Project Rep.* No. RMP-183 (1979).

3. P. R. ESCOBAL, "Methods of Orbit Determination," Wiley, New York, 1965.

Information Theoretic Smoothing Algorithms for Dynamic Systems with or without Interference

KERIM DEMIRBAS

School of Engineering and Applied Science
University of California
Los Angeles, California

I. INTRODUCTION

In target tracking, a motion model describing the motion of the target tracked as accurately as possible and an observation model are first obtained. In a clear environment (i.e., no interference such as jamming or clutter exists), these models are generally discrete and linear with respect to the disturbance and observation noises. Moreover, the observation noise is an additive Gaussian noise where the Gaussian assumption is due to the central limit theorem. Next, by using one of the estimation algorithms already developed in the literature [7-11] [e.g., the (extended)[1] Kalman filter algorithm], the target states are estimated.

If either the motion or the observation model is nonlinear (in general, this is the case), then the optimum solution to this estimation problem cannot be given due to nonlinear functions in the models. However, a suboptimal solution is given by using a nonlinear estimation algorithm. By a nonlinear estimation algorithm (e.g., the extended Kalman filter algorithm), the states are estimated as follows. First, using a Taylor series expansion, the nonlinear functions are linearized around some points so that the nonlinear estimation problem is reduced to a linear one. Then, using a (usually) known

---

[1]By convention, by the (extended) Kalman filter algorithm, either the Kalman filter algorithm or the extended Kalman filter algorithm is expressed.

solution to this linear estimation problem, the states are estimated. Hence, if the nonlinear functions are not smooth enough for a Taylor series expansion, then the nonlinear estimation algorithm may produce estimates which are much different from the actual values of the states. Therefore, some estimation algorithms for an estimation problem with nonlinear functions which are not smooth enough for a Taylor series expansion are needed.

If both the motion and the observation models are linear, we have a problem of state estimation in a linear discrete time system. The solution to this problem has been treated extensively in the literature [7-11].

In the tracking of a target in the presence of interference (such as jamming or clutter), which is, in general, not Gaussian noise [1], an observation model with only additive observation noise may not be used since it does not account for the possibility of utilizing measurements originating simultaneously from the interference source and the target. Still, if a classical estimation algorithm were used to track the target by using an observation model with only observation noise, the estimates of the target states may diverge from the actual values. Hence, some estimation algorithms are needed for discrete models with arbitrary random interference as well as an observation noise.

This chapter treats the problem of state estimation for discrete models with or without interference. As a result, three new smoothing algorithms are presented that satisfy the need just mentioned. The main idea for these smoothing algorithms is that of quantizing the states of the models to a finite set of states. This approach reduces the smoothing problem to a multiple (composite) hypothesis-testing problem. Further, three decoding techniques of information theory are used to develop the smoothing algorithms. The first smoothing algorithm is referred to as the optimum decoding-based smoothing algorithm, which uses the Viterbi decoding algorithm. The second smoothing algorithm is referred to as the stack sequential decoding-based smoothing algorithm, which uses a stack sequential decoding algorithm. The third is referred to as the suboptimum decoding-based smoothing algorithm, which uses a suboptimum decoding algorithm.

II. SMOOTHING ALGORITHMS

A. *MODELS AND ASSUMPTIONS*

Throughout this chapter, we deal with the following discrete models:

Motion model, $x(k + 1) = f(k, x(k), u(k), w(k))$,

Observation model, $z(k) = g(k, x(k), v(k))$,   (1)

for target tracking in a clear environment. In the presence of interference, we deal with the following models:

Motion model, $x(k + 1) = f(k, x(k), u(k), w(k))$,

Observation model, $z(k) = g(k, x(k), I(k), v(k))$,   (2)

where $x(0)$ is an $n \times 1$ initial (target) state random vector (which determines the considered target location at time 0); $x(k)$ is an $n \times 1$ (target) state vector at time k (which determines the considered target location at time k); $u(k)$ is a $q \times 1$ known pilot-command vector at time k; $w(k)$ is a $p \times 1$ disturbance-noise vector at time k with zero mean and known statistics; $v(k)$ is an $l \times 1$ observation-noise vector at time k with zero mean and known statistics; $z(k)$ is an $r \times 1$

observation vector at time k; and I(k) is an m × 1 interference vector with known statistics. Time k is time $t_0 + kT_0$ where t_0 and T_0 are the initial time and the observation interval, respectively; f(k, x(k), u(k), w(k)), g(k, x(k), I(k), v(k)), and g(k, x(k), v(k)) are linear or nonlinear vectors with appropriate dimensions. Furthermore, the random vectors x(0), w(j), w(k), v(l), v(m), I(n), and I(p) are assumed to be independent for all j, k, l, m, n, p. The goal is to estimate the state sequence {x(0), x(1), ..., x(L)} by using the observation sequence {z(1), z(2), ..., z(L)}, where L is a chosen integer. This is a nonlinear estimation problem. The estimation algorithms presented in this chapter are applicable to any type of estimation problem, not necessarily target tracking, as long as the estimation problem under consideration can be modeled in either Eq. (1) or Eq. (2).

B. QUANTIZATION OF STATES AND TRANSITION PROBABILITIES

Section II.B describes a type of quantization for target states and some difficulties in calculating transition probabilities between quantization levels. Let us consider the state x(k). It is a random vector whose range is in the space R^n (n-dimensional Euclidean space). Let us divide R^n into nonoverlapping subspaces R_i^n and assign a unique value x_{qi} to each subspace R_i^n, where the subscript q is quantization.

Definition 1. A function $x_q(\cdot) \triangleq Q\{x(\cdot)\}$ is a quantizer for the state x(·) if the following hold:

(1) $x_q(\cdot) \triangleq Q\{x(\cdot)\} = x_{qi}$ whenever $x(\cdot) \in R_i^n$; and

(2) x_{qi} is unique for each R_i^n.

Definition 2.   The function $x_q(\cdot)$ is the quantized state vector at time $(\cdot)$, and its possible values are sometimes called the quantization levels of the state $x(\cdot)$.

Definition 3.   Subspace R_i^n is sometimes called gate (or cell) R_i^n.

Definition 4.   The value x_{qi} is called the quantization level for the gate (cell) R_i^n.

Quantization means that whenever a random state vector $x(\cdot)$ falls within a given subspace, say R_i^n, the state $x(\cdot)$ is quantized to the unique value x_{qi} (Fig. 1).   Let us now define the transition probabilities which govern the target motion within the gates.

Definition 5.   The transition probability $\pi_{jm}(k)$ is the probability that the state $x(k + 1)$ will lie in the gate R_m^n when the state $x(k)$ is in the gate R_j^n; i.e.,

$$\pi_{jm}(k) \triangleq \text{Prob}\left\{x(k + 1) \in R_m^n | x(k) \in R_j^n\right\}. \tag{3}$$

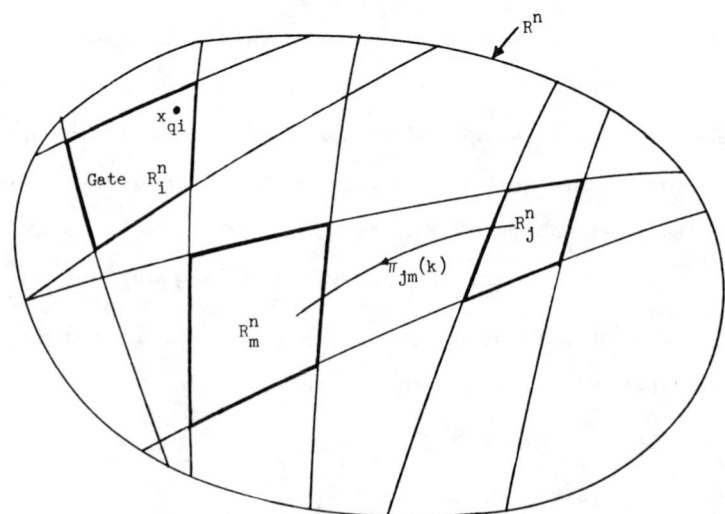

Fig. 1.   Quantization and transition probabilities.

The transition probability $\pi_{jm}(k)$ is a conditional probability. Hence it can be rewritten as

$$
\begin{aligned}
\pi_{jm}(k) &= \frac{\text{Prob}\left\{x(k+1) \in R_m^n, \; x(k) \in R_j^n\right\}}{\text{Prob}\left\{x(k) \in R_j^n\right\}} \\[2mm]
&= \left[\int_{R_j^n} p(x(k))\,dx(k)\right]^{-1} \\[2mm]
&\quad \times \left[\int_{R_j^n}\int_{R_m^n} p(x(k+1),\,x(k))\,dx(k+1)\,dx(k)\right] \\[2mm]
&= \left[\int_{R_j^n} p(x(k))\,dx(k)\right]^{-1} \\[2mm]
&\quad \times \left\{\int_{R_j^n}\left[\int_{R_m^n} p(x(k+1)\,|\,x(k))\,dx(k+1)\right] p(x(k))\,dx(k)\right\},
\end{aligned}
$$

(4)

where $p(x(k+1),\,x(k))$ is the joint probability density function of $x(k+1)$ and $x(k)$; $p(x(k))$ is the probability density function of $x(k)$; and $p(x(k+1)\,|\,x(k))$ is the conditional probability density function of $x(k+1)$ given $x(k)$. It is not usually easy to evaluate the transition probability $\pi_{jm}(k)$ analytically. The difficulties are due to the shapes of the gates R_j^n and R_m^n and the statistics of the disturbance-noise vectors $w(\cdot)$ and the initial state vector $x(0)$. In order to see this, consider the following linear motion example

$$x(k+1) = Ax(k) + w(k), \tag{5}$$

where $x(0)$ is an $n \times 1$ Gaussian initial state vector; $x(k)$ is an $n \times 1$ state vector at time k; $w(k)$ is an $n \times 1$ Gaussian disturbance vector at time k; and A is a constant transition matrix with appropriate dimension. Moreover, the random vectors $x(0)$, $w(k)$, and $w(l)$ are assumed to be statistically independent for all k, l. Hence $x(k+1)$ and $x(k)$ are linear

transformations of the Gaussian random vectors $x(0)$, $w(1)$,
..., and $w(k)$. Thus, $p(x(k))$ and $p(x(k + 1)|x(k))$ are
normal density functions. Therefore the evaluation of the
probability $p\{x(k + 1) \in R_m^n | x(k) \in R_j^n\}$ is not analytically
possible. The problem is more difficult if the motion model is
nonlinear. If the transition probability $\pi_{jm}(k)$ needs to be
calculated, it should be performed numerically. Even this may
be difficult. In other words, the evaluation of the exact
transition probabilities between gates is not practical. Therefore, Section II.C discusses an approximate target motion model
called the finite state model which is obtained by approximating
the disturbance noise vector $w(k)$ and the initial state vector
$x(0)$ by discrete random vectors (see Appendix B), and by quantizing the state $x(k)$ as previously described for all $k = 1$,
2, ... For this finite-state model, the transition probabilities can be easily calculated.

C. A FINITE-STATE MODEL
FOR THE TARGET MOTION

Throughout Section II.C, gates are assumed to be generalized rectangles such that the zero vector 0 (origin) is located
in the center of a generalized rectangle, say R_0^n (Fig. 2). Let
the lengths of the sides of a generalized rectangle, say R_i^n, be
g_{i1}, g_{i2}, ..., g_{in}. These lengths are said to be the sizes of
gate R_i^n. Moreover, the quantization levels for gates are assumed to be the center points of the gates, namely,

$$x_q(\cdot) \triangleq Q\{x(\cdot)\} = x_{qi} \qquad \text{if} \quad x(\cdot) \in R_i^n, \tag{6}$$

where x_{qi} is the center of the generalized rectangle (gate) R_i^n.
Let us now define the finite-state model which approximates
the target-motion model. The flowchart of this finite-state
model is in Fig. 3. For each k, the disturbance noise vector

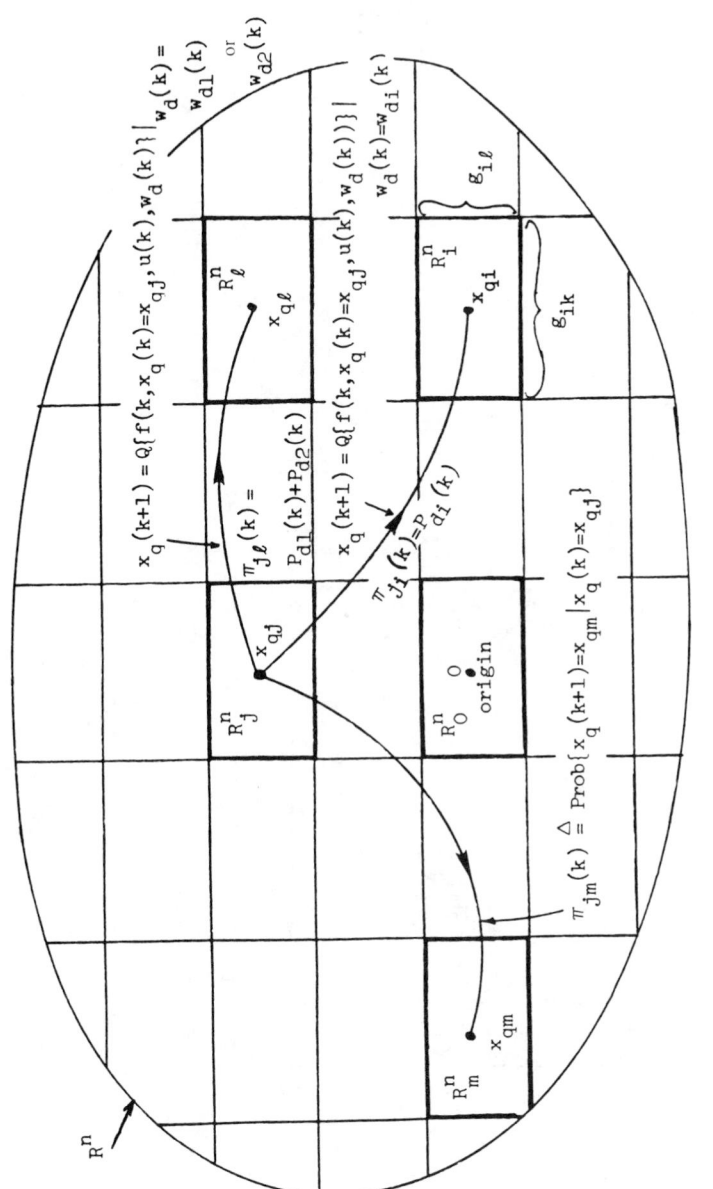

Fig. 2. Quantization with generalized rectangles.

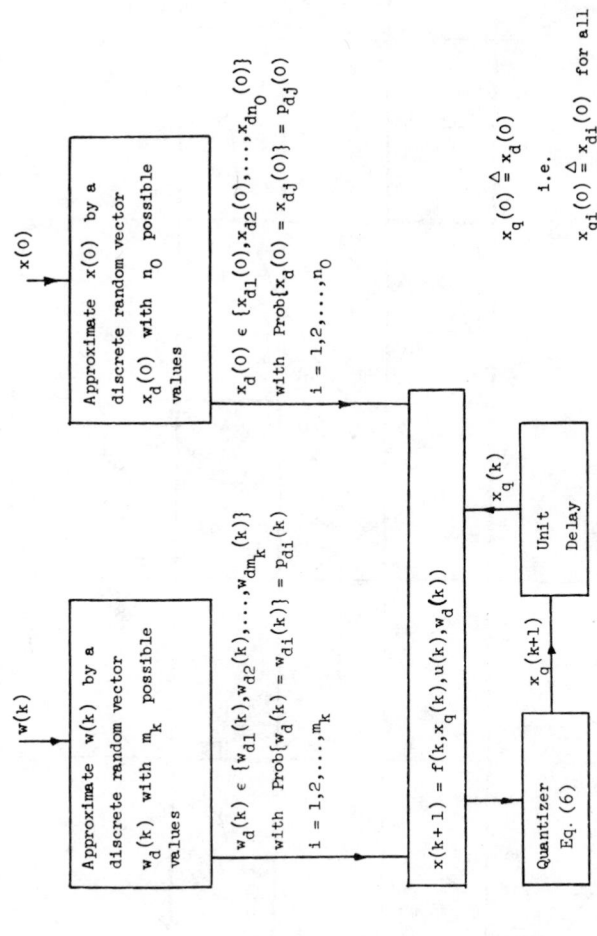

Fig. 3. The flowchart of the finite-state model.

$w(k)$ is approximated by a discrete random vector $w_d(k)$ whose

possible values are $w_{d1}(k)$, $w_{d2}(k)$, ..., $w_{dm_k}(k)$; the corre-

sponding probabilities are $p_{d1}(k)$, $p_{d2}(k)$, ..., $P_{dm_k}(k)$, i.e.,

$\text{Prob}\{w_d(k) = w_{di}(k)\} = p_{di}(k)$, where $i = 1, 2, ..., m_k$ and m_k

is a positive integer (see Appendix B). Also, the initial state

vector $x(0)$ is approximated by a discrete random vector $x_d(0)$

whose possible values are $x_{d1}(0)$, $x_{d2}(0)$, ..., $x_{dn_0}(0)$; the

corresponding probabilities are $p_{d1}(0)$, $p_{d2}(0)$, ..., $P_{dn_0}(0)$,

i.e., $\text{Prob}\{x_d(0) = x_{di}(0)\} = p_{di}(0)$, where $i = 1, 2, ..., n_0$

and n_0 is a positive integer. The positive integers n_0 and m_k

are chosen so that the random vectors $x(0)$ and $w(k)$ are satis-

factorily approximated by the discrete vectors $x_d(0)$ and $w_d(k)$

for the considered estimation problem respectively. Further,

in the target-motion model, by replacing the disturbance noise

vector $w(k)$ and the initial state vector $x(0)$ with the discrete

random vectors $w_d(k)$ and $x_d(0)$, respectively, and then quan-

tizing the states by Eq. (6), the target-motion model is reduced

to the finite-state model

$$x_q(k + 1) = Q\{f(k, x_q(k), u(k), w_d(k))\}, \tag{7}$$

where $Q\{\cdot\}$ is the quantizer [Eq. (6)]; $x_q(k)$ is the quantized

state vector at time k and its possible values, i.e., the quan-

tization levels of the state vector $(x(k))$ are $x_{q1}(k)$, $x_{q2}(k)$,

..., $x_{qn_k}(k)$ where n_k is the number of possible quantization

levels of the state vector $x(k)$; $x_q(0) \triangleq x_d(0)$ [by definition,

$x_{qi}(0) \triangleq x_{di}(0)$, $i = 1, 2, ..., n_0$; in other words, the quan-

tization levels of $x(0)$ are assumed to equal the possible values

of the discrete random vector $x(0)$]. Throughout this chapter

whenever the target-motion model (or the state equations or the

target motion) is mentioned we refer to Eq. (7); i.e., it is
assumed that the target motion is described by the finite-state
model.

The transition probability $\pi_{jl}(k)$, which is defined by the
conditional probability that the quantized state vector
$x_q(k + 1)$ will be equal to the quantization level x_{ql} for gate
R_l^n, given that the quantized state vector $x_q(k)$ is equal to the
quantization level x_{qj} for gate R_j^n, namely,

$$\pi_{jl}(k) = \text{Prob}\{x_q(k + 1) = x_{ql} | x_q(k) = x_{qj}\} \tag{8}$$

is determined as follows (Fig. 2).

Let us assume that the quantized state vector $x_q(k)$ is
equal to the quantization level x_{qj} for gate R_j^n (i.e., the target is in R_j^n at time k). The transitions from this quantization level to the others are determined by the discrete random
vector $w_d(k)$ and the function $Q\{f(k, x_q(k) = x_{qj}, u(k), w_d(k))\}$.
The discrete random vector $w_d(k)$ can take any value in the set
$\{w_{d1}(k), w_{d2}(k), \ldots, w_{dm_k}(k)\}$ with corresponding probabilities
$P_{d1}(k), P_{d2}(k), \ldots, P_{dm_k}(k)$. Thus, the quantized state vector
$x_q(k + 1)$ can be equal to at most m_k various quantization levels.
If the function $f(k, x_q(k) = x_{qj}, u(k), w_d(k))$ maps x_{qj} into
another gate, say R_i^n for only one possible value, say $w_{di}(k)$,
of the discrete random vector $w_d(k)$, then the transition probability $\pi_{ji}(k)$ from gate R_j^n to gate R_i^n is the probability that
the possible value $w_{di}(k)$ of $w_d(k)$ occurs, i.e., $\pi_{ji}(k) = P_{di}(k)$.
However, if the function $f(k, x_q(k) = x_{qj}, u(k), w_d(k))$ maps
x_{qj} into another gate, say R_l^n, for more than one possible value,
say $w_{d1}(k)$ and $w_{d2}(k))$ of $w_d(k)$, the transition probability
$\pi_{jl}(k)$ from gate R_j^n to gate R_l^n is the probability that the discrete random vector $w_d(k)$ is equal to either of the possible

values $w_{d1}(k)$ or $w_{d2}(k)$, i.e., $\pi_{j\ell}(k) = \Sigma_n \, p_{dn}(k) = p_{d1}(k)$
$+ \, p_{d2}(k)$, where the summation is over all n such that $Q\{f(k,$
$x_q(k) = x_{qj}, \, u(k), \, w_{dn}(k))\} = x_{q\ell}$. Having determined the finite-state model, we can represent the target motion by a
trellis diagram.

D.   A TRELLIS DIAGRAM
     FOR THE TARGET MOTION

Let us assume that the quantized state vector $x_q(k)$ has n_k
possible values, say $x_{q1}(k)$, $x_{q2}(k)$, ..., $x_{qn_k}(k)$ where n_k is
a positive integer. To represent the target motion by a graph,
we adopt the following conventions

1. Each possible value of $x_q(k)$ is represented on the kth
    column by a point (sometimes called node) with the
    corresponding quantization level so that the kth column
    contains the possible quantization levels of x(k) (in
    other words, the possible gates in which the target can
    lie at time (k)) where k = 0, 1, 2, ...

2. The transition from one quantization level to another
    is represented by a line having a direction indicating
    the direction of the target motion.

Hence, the target motion from time zero to time L can be represented by a directed graph shown in Fig. 4, which is called the
trellis diagram for the target motion from time zero to time L.

Definition 6. A path in the trellis diagram is any sequence
of directed lines where the final vertex of one is the initial
vertex of the next.

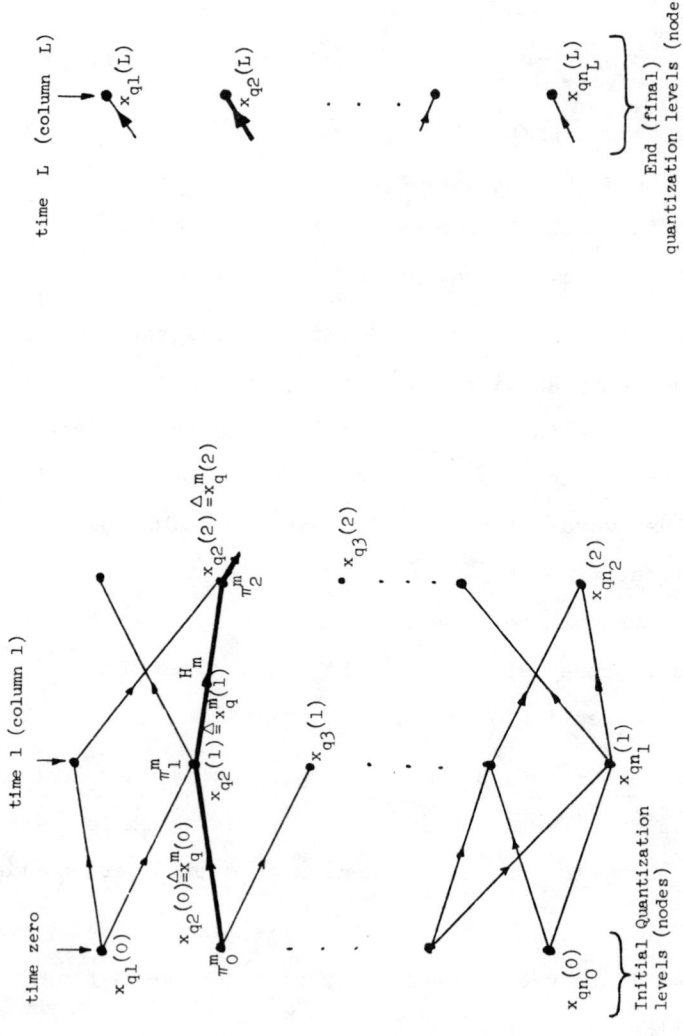

Fig. 4. The trellis diagram for the target motion.

E. APPROXIMATE OBSERVATION MODELS

So far the target-motion model has been reduced to a finite-state model which uses the quantized state vector $x_q(\cdot)$. However, the observation models in Eqs. (1) and (2) use the target state vector $x(\cdot)$. Thus, in the observation models in Eqs. (1) and (2), by replacing the state vector $x(k)$ with the quantized state vector $x_q(k)$, the following approximate observation models are obtained:

$$z(k) = \begin{cases} g(k, x_q(k), v(k)), & \text{in clear environments,} \\ g(k, x_q(k), I(k), v(k)), & \text{in the presence of} \\ & \text{interference.} \end{cases} \qquad (9)$$

From now on, when we discuss the observation model (or the measurement model equation) we refer to the models in Eq. (9), which are used in the following analyses.

Let us consider the trellis diagram in Fig. 4 where it is assumed that, without loss of generality, the target will be tracked from time zero up to and including time L. Therefore, the trellis diagram is drawn from time zero to time L. Time zero refers to the initial state. Let us now define the following symbols which are used throughout our further analyses:

n_i Number of quantization levels for the gates in which the target may lie at time i; in other words, the number of possible values of the quantized state vector $x_q(i)$ where $i = 0, 1, 2, \ldots, L$

$\underset{\sim}{x}(i)$ Set of all the quantization levels for the gates in which the target may lie at time i, namely, $\underset{\sim}{x}(i) \triangleq \{x_{q1}(i), x_{q2}(i), \ldots, x_{qn_i}(i)\}$, where $i = 0, 1, 2, \ldots, L$

M   Number of possible paths through the trellis diagram; this number is equal to or less than

$$\prod_{j=0}^{L} n_j$$

H_m The mth path through the trellis diagram, indicated by a bold line in Fig. 4

$x_q^m(i)$ Quantization level for the gate in which the target lies at time i when it follows path H_m. In other words, the possible value of the quantized state vector $x_q(i)$ through which the mth path passes. For example, in trellis diagram of Fig. 4, $x_q^m(0) = x_{q2}(0); \; x_q^m(1) = x_{q2}(1); \; x_q^2(2) = x_q^m(2), \; \ldots$

π_0^m Probability that the possible value of the initial state vector $x_d(0)$ from which the mth path starts occurs, namely, $\pi_0^m = \text{Prob}\left\{ x_q(0) = x_q^m(0) \right\}$. For example, in trellis diagram of Fig. 4, $\pi_0^m = \text{Prob}\{ x_q(0) = x_{q2}(0) \}$

π_i^m Transition probability from the (i − 1)th gate for the mth path (i.e., the gate from which the target passes at time i − 1 when it follows the path H_m) to the ith gate for the mth path. In other words, it is the transition probability that the target will be at the ith quantization level (node) of path H_m at time i when it is at the (i − 1)th quantization level node) of H_m at time i − 1; that is $\pi_i^m \triangleq \text{Prob}\left\{ x_q(i) = x_q^m(i) \,\middle|\, x_q(i-1) = x_q^m(i-1) \right\}$. For example, in trellis diagram of Fig. 4,

$$\pi_1^m = \text{Prob}\left\{ x_q(1) = x_q^m(1) \,\middle|\, x_q(0) = x_q^m(0) \right\}$$

$$\triangleq \text{Prob}\{ x_q(1) = x_{q2}(1) \,|\, x_q(0) = x_{q2}(0) \},$$

$$\pi_2^m = \text{Prob}\left\{ x_q(2) = x_q^m(2) \,\middle|\, x_q(1) = x_q^m(1) \right\}$$

$$\triangleq \text{Prob}\{ x_q(2) = x_{q2}(2) \,|\, x_q(1) = x_{q2}(1) \}$$

π_0^{max} Maximum of the probabilities that the quantization levels at time zero occur, i.e.,

$$\pi_0^{max} \triangleq \max_{a \in \underline{x}(0)} \text{Prob}\{ x_q(0) = a \}$$

π_i^{max} Maximum of the transition probabilities from the
quantization levels at time i - 1 to the quantization levels at time i (where i = 1, 2, ..., L);
that is,

$$\pi_i^{max} = \max_{\substack{a \in \underset{\sim}{x}(i) \\ b \in \underset{\sim}{x}(i-1)}} \text{Prob}\{x_q(i) = a | x_q(i - 1) = b\}$$

π_0^{min} Minimum of the probabilities that the quantization
levels at time zero occur, i.e.,

$$\pi_0^{min} = \min_{a \in \underset{\sim}{x}(0)} \text{Prob}\{x_q(0) = a\}$$

π_i^{min} Minimum of the transition probabilities from the
quantization levels at time i - 1 to the quantization levels at time i (where i = 1, 2, ..., L),
namely,

$$\pi_i^{min} = \min_{\substack{a \in \underset{\sim}{x}(i) \\ b \in \underset{\sim}{x}(i-1)}} \text{Prob}\{x_q(i) = a | x_q(i - 1) = b\}$$

$\underset{\sim}{x}_L^m \triangleq \left\{ x_q^m(0), x_q^m(1), \ldots, x_q^m(L) \right\}$

Sequence of the quantization levels (nodes) which
the mth path passes through; obviously,
$x_q^m(i) \in \underset{\sim}{x}(i)$, where i = 0, 1, 2, ..., L

$z^L = \{z(1), z(2), \ldots, z(L)\}$
Observation sequence from time 1 to time L

$I^L \triangleq \{I(1), I(2), \ldots, I(L)\}$
Interference sequence from time 1 to time L

Obviously, the target motion occurs along one of the possible
paths through the trellis diagram. Hence, our aim is to decide
upon a path through the trellis diagram which is most likely
(probably) followed by the target by using the observation sequence z^L. Because of randomness in the models, our approach
must be statistical, i.e., a statistical optimization problem.
Based on the observations, we shall guess which path was (most
likely) followed by the target. Hence, a criterion is needed.

For a tracking problem, a suitable criterion may be the minimum
error probability criterion, which is a special case of Bayes'
criterion in detection theory [12]. Using this criterion reduces the problem of finding the path most likely followed by
the target to a multiple-(composite) hypothesis-testing problem.

F.   MINIMUM ERROR
PROBABILITY CRITERION

In Section II.E, we labeled the M possible paths through
the trellis diagram H_1, H_2, ..., H_M. Sometimes these paths are
referred to as hypotheses. Hence, using the minimum error
probability criterion and the observation sequence, we would
like to decide which hypothesis is true (in other words, we
would like to find the path most likely followed by the target). To accomplish this we develop a decision rule that as-
signs each point in the observation space D to one of the hypotheses. Therefore, the decision rule divides the whole observation space D into M subspaces D_1, D_2, ..., and D_M (Fig. 5).
If the observations fall in the subspace D_i, we decide that the
target (most likely) followed path H_i (i.e., H_i is true).
Subspace D_i is called the decision region for hypothesis H_i.
We must therefore choose the decision regions D_1, D_2, ..., D_M
in such a way that the overall error probability is minimized.

The overall error probability, sometimes called the Bayes
risk R, is defined by

$$R \triangleq \sum_{j=1}^{M} \sum_{\substack{i=1 \\ i \neq j}}^{M} \left\{ \int_{z^L \in D_i} p(H_j) p'\left(z^L | H_j\right) dz^L \right\}, \tag{10}$$

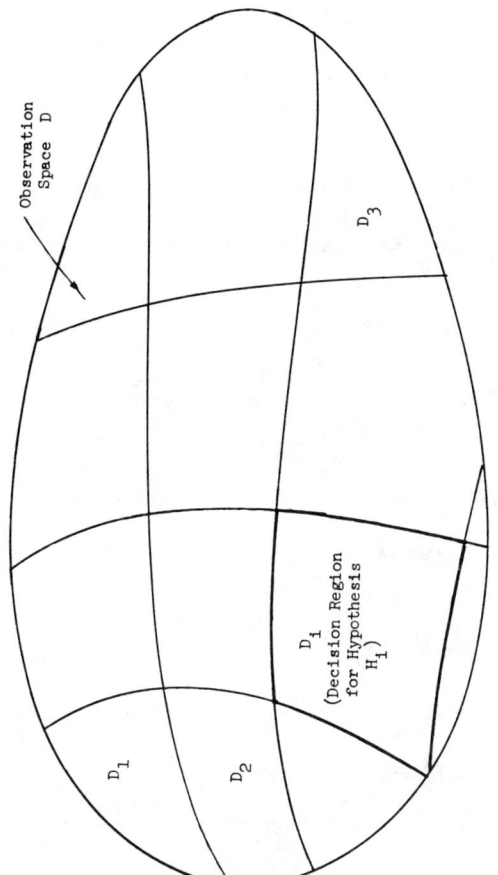

Observation
Space D

D_3

D_1
(Decision Region
for Hypothesis
H_1)

D_1

D_2

Fig. 5. Observation space and decision regions.

where

$$p'\left(z^L|H_j\right) = \begin{cases} p\left(z^L|H_j\right) & \text{in clear environments} \\ \displaystyle\iint_{I^L} p\left(z^L|H_j,\ I^L\right)p(I^L)dI^L & \text{in the presence} \\ & \text{of interference,} \end{cases} \qquad (11)$$

$p(H_j)$ Probability that the hypothesis H_j (path H_j) is true. This is called the a priori probability of hypothesis H_j

$p\left(z^L|H_j\right)$ Conditional probability of the observation sequence z^L in clear environments given that hypothesis H_j is true (i.e., the target followed the path H_j)

$p\left(z^L|H_j,\ I^L\right)$ Conditional probability of the observation sequence z^L in the presence of interference given hypothesis H_j (i.e., hypothesis H_j is true) and the interference sequence I^L

$p(I^L)$ Joint density function of the interference sequence I^L

In order to find the optimal decision rule, we vary the decision regions D_1, D_2, ..., D_M so that the risk R is minimized. It is well known that the optimum decision rule [12,18] is

choose H_i if $p(H_i)p'\left(z^L|H_i\right) > p(H_j)p'\left(z^L|H_j\right)$

$$\text{for all} \quad j \neq i. \qquad (12)$$

For a given observation sequence z^L, if the inequality in Eq. (12) becomes an equality for one or more hypotheses H_j, any one of these H_j and H_i can be chosen as the decision. This does not change the average error probability. Convention: throughout this chapter for all observation sequences z^L for which the inequality in Eq. (12) becomes an equality, the decision is made at random (e.g., by the flip of a coin) among the hypotheses satisfying the equality. Hence the optimum decision region D_i

for hypothesis H_i becomes

$$D_i \triangleq \left\{ z^L : p(H_i) p'\left(z^L | H_i\right) > p(H_j) p'\left(z^L | H_j\right) \text{ for all } j \neq i \right.$$

and all the observation sequences z^L for which the inequality

in Eq. (12) becomes an equality, and then the hypothesis H_i is

$\left. \text{chosen} \right\}.$ (13)

Note that the decision regions are nonoverlapping, namely,

$D_i \cap D_j = \phi$ for $i \neq j$, where $\cap$ and ϕ stand for the inter-

section and the empty set, respectively. However, the union

of all the decision regions covers the whole observation space

D, that is, $D = \cup_{i=1}^{M} D_i$. Hence, the optimum decision rule may

be interpreted as follows: If the observation sequence z^L falls

within the optimum decision region D_i, then choose hypothesis

(path) H_i as the decision; i.e.,

$$\text{choose } H_i \quad \text{if } z^L \in D_i. \tag{14}$$

Having determined the optimum decision rule [Eq. (12) or Eq.

(14)] with respect to the minimum error probability criterion,

we apply it to tracking problems in Section II.G, next.

G. OPTIMUM DECISION RULE FOR THE TARGET PATHS

Let us consider the motion model in Eq. (7) and the obser-

vation models in Eq. (9). The a priori probability of hypo-

thesis H_i can be rewritten as

$$p(H_i) = \prod_{k=0}^{L} \pi_k^i, \tag{15}$$

since the disturbance noise vector $w(k)$ is assumed to be inde-

pendent of $w(j)$ and $x(0)$ for all $j \neq k$, where π_k^i is as defined

in Section II.E. Further, using the assumption that the

interference vector $I(k)$ is independent of $I(j)$ for all $k \neq j$,
we can rewrite the joint density function of the interference
sequence I^L as

$$p(I^L) = \prod_{k=1}^{L} p(I(k)), \tag{16}$$

where $p(I(k))$ is the probability density function of the interference vector $I(k)$. Moreover, recognizing that the sequence
$\underset{\sim}{x}_L^i$ defined in Section II.E describes hypothesis H_i completely
and using Eq. (16) and the assumption that the observation
noise is independent from sample to sample, the function
$p'\left(z^L|H_i\right)$ in Eq. (12) can be rewritten as

$$p'\left(z^L|H_i\right) = p'\left(z^L|\underset{\sim}{x}_L^i\right) = \prod_{k=1}^{L} p'\left(z(k)|x_q^i(k)\right), \tag{17}$$

where

$$p'\left(z(k)|x_q^i(k)\right) = \begin{cases} p\left(z(k)|x_q^i(k)\right), & \text{in clear environments,} \\ \int_{I(k)} p\left(z(k)|x_q^i(k),\ I(k)\right) \\ \qquad \times\ p(I(k))dI(k), & \text{in the presence} \\ & \text{of interference;} \end{cases}$$

and $p\left(z(k)|x_q^i(k)\right)$ is the conditional probability of the obser-
vation $z(k)$ in clear environments in Eq. (9) given that $x_q(k) =$
$x_q^i(k)$, i.e., $p\left(z(k)|x_q^i(k)\right) \triangleq p\left(z(k)|x_q(k) = x_q^i(k)\right)$; and
$p\left(z(k)|x_q^i(k),\ I(k)\right)$ is the conditional probability of the observation $z(k)$ in the presence of interference in Eq. (9) given
that $x_q(k) = x_q^i(k)$ and $I(k)$, i.e., $p\left(z(k)|x_q^i(k),\ I(k)\right) =$
$p\left(z(k)|x_q(k) = x_q^i(k),\ I(k)\right)$. Let us now consider the function

$p'\left(z(k)\,|\,x_q^i(k)\right)$ in the presence of interference; that is,

$$p'\left(z(k)\,|\,x_q^i(k)\right)$$

$$= \int_{I(k)} p\left(z(k)\,|\,x_q^i(k),\ I(k)\right) p(I(k))\,dI(k). \tag{19}$$

Whether or not this integral can be evaluated in a closed form depends on the function $g(k,\ x_q(k) = x_q^i(k),\ I(k),\ v(k))$ and the statistics of the interference $I(k)$ and the observation noise $v(k)$. In some cases, a numerical integration might be used to evaluate the integral in Eq. (19). However, throughout this chapter, approximating the interference vector $I(k)$ by a discrete random vector $I_d(k)$ whose possible values are $I_{d1}(k)$, $I_{d2}(k)$, ..., $I_{dr_k}(k)$, with corresponding probabilities $p(I_{d1}(k))$, $p(I_{d2}(k))$, ..., and $p(I_{dr_k}(k))$ [i.e., $\text{Prob}\{I_d(k) = I_{di}(k)\} = p(I_{di}(k))$], the integral in Eq. (19) is reduced to a summation

$$\int_{I(k)} p\left(z(k)\,|\,x_q^i(k),\ I(k)\right) p(I(k))\,dI(k)$$

$$\approx \sum_{l=1}^{r_k} p\left(z(k)\,|\,x_q^i(k),\ I_{dl}(k)\right) p(I_{dl}(k)), \tag{20}$$

where r_k is the number of possible values of the approximating discrete vector $I_d(k)$. In other words, by changing the interference $I(k)$ to $I_d(k)$, we make another approximation for the observation model in the presence of interference in Eq. (9). The observation model becomes

$$z(k) = g(k,\ x_q(k),\ I(k) = I_d(k),\ v(k))$$

$$\triangleq g(k,\ x_q(k),\ I_d(k),\ v(k)). \tag{21}$$

From now on, throughout the chapter, if it is not easy to calculate the integral in Eq. (19), the integral will be approximated by Eq. (20). In other words, when the integral in

Eq. (19) cannot be easily evaluated, the observation model of Eq. (21) rather than that of Eq. (9) will be used for further analyses.

Substituting Eqs. (15) and (17) into the optimum decision rule of Eq. (12), we obtain the following:

$$\text{Choose} \quad H_i \quad \text{if} \quad \pi_0^i \prod_{k=1}^{L} \pi_k^i p' \Big(z(k) \,|\, x_q^i(k) \Big)$$

$$> \pi_0^j \prod_{k=1}^{L} \pi_k^j p' \Big(z(k) \,|\, x_q^j(k) \Big)$$

$$\text{for all} \quad j \neq i. \tag{22}$$

Since it is frequently more convenient to perform summations that multiplications, and the natural logarithm function is monotonically increasing, taking the natural logarithms of both sides of the inequalities in Eq. (22), we get the following:

$$\text{Choose} \quad H_i \quad \text{if} \quad \ln \pi_0^i + \sum_{k=1}^{L} \Big\{ \ln \pi_k^i + \ln p' \Big(z(k) \,|\, x_q^i(k) \Big) \Big\}$$

$$> \ln \pi_0^j + \sum_{k=1}^{L} \Big\{ \ln \pi_k^j + \ln p' \Big(z(k) \,|\, x_q^j(k) \Big) \Big\}$$

$$\text{for all} \quad j \neq i, \tag{23}$$

where in clear environments,

$$p' \Big(z(k) \,|\, x_q^i(k) \Big) = p \Big(z(k) \,|\, x_q^i(k) \Big); \tag{24a}$$

in the presence of interference,

$$p' \Big(z(k) \,|\, x_q^i(k) \Big) = \begin{cases} \displaystyle\int p \Big(z(k) \,|\, x_q^i(k), \ I(k) \Big) p(I(k)) \, dI(k), \\ \qquad\qquad \text{using Eq. (9);} \\[4pt] \displaystyle\sum_{\ell=1}^{r_k} p \Big(z(k) \,|\, x_q^i(k), \ I_{d\ell}(k) \Big) p(I_{d\ell}(k)), \\ \qquad\qquad \text{using Eq. (21);} \end{cases} \tag{24b}$$

where

$$p\Big(z(k)\,|\,x_q^i(k),\ I_{d\ell}(k)\Big)\ =\ p\Big(z(k)\,|\,x_q(k)\ =\ x_q^i(k),\ I_d(k)\ =\ I_{d\ell}(k)\Big),$$

which is the conditional probability of $z(k)$ in Eq. (21) given

that $x_q(k) = x_q^i(k)$ and $I_d(k) = I_{d\ell}(k)$. Either one of the ex-

pressions in Eqs. (22) and (23) with the convention in Section

II.F is the optimum decision rule for deciding the path most

probably followed by the target.

Now we are going to verify the following equalities for the

observation models in the presence of interference in Eqs. (9)

and (21):

$$p'\Big(z(k)\,|\,x_q^i(k)\Big)\ =\ p\Big(z(k)\,|\,x_q^i(k)\Big),$$

$$(25)$$

$$p\Big(z^L\,|\,H_i\Big)\ =\ \prod_{k=1}^{L}\ p\Big(z(k)\,|\,x_q^i(k)\Big)\ =\ p'\Big(z^L\,|\,H_i\Big),$$

where $p\Big(z(k)\,|\,x_q^i(k)\Big)$ is the conditional probability of the ob-

servation $z(k)$ in the presence of interference given that

$x_q(k) = x_q^i(k)$, and $p\Big(z^L\,|\,H_i\Big)$ is the conditional probability of

the observation sequence z^L in the presence of interference

given that hypothesis H_i is true (i.e., the target followed

the path H_i). Let us prove the equalities in Eq. (25) for the

observation model in the presence of interference in Eq. (9).

From Eqs. (24a) and (24b), we have

$$p'\Big(z(k)\,|\,x_q^i(k)\Big)$$

$$=\int_{I(k)}\ p\Big(z(k)\,|\,x_q^i(k),\ I(k)\Big)p(I(k))\,dI(k).\qquad(26)$$

Using the Baye rule and the assumption that the interference

$I(k)$ is independent of the initial state vector $x(0)$ and of the disturbance-noise vector $w(j)$ for all j, k, we obtain

$$p\left(z(k) \mid x_q^i(k), I(k)\right) = \frac{p\left(z(k), x_q^i(k), I(k)\right)}{p(x_q^i(k))p(I(k))}, \qquad (27)$$

Hence, substituting Eq. (27) into Eq. (26) and recalling that
the integration is taken over all the sample space of $I(k)$, the
first equality in Eq. (25) is obtained. The second equality
in Eq. (25) follows from the assumption that the interference
and the observation noise are independent from sample to sample.
The equalities in Eq. (25) can also be verified for the observation model of Eq. (21). Hence for all the observation models
already considered, the function $p'(\cdot \mid \cdot)$ is a conditional probability density function.

Let us now present some definitions to be used later in
the chapter:

Definition 7. An initial node is a quantization level at
time zero. The metric, denoted by $MN(x_{qi}(0))$, of the initial
node $x_{qi}(0)$ is defined by

$$MN(x_{qi}(0)) = \ln[\text{Prob}\{x_q(0) = x_{qi}(0)\}]. \qquad (28)$$

Consequently, $MN\left(x_q^m(0)\right) = \ln \pi_0^m$.

Definition 8. The metric, denoted by $M(x_{qj}(k-1) \to x_{qi}(k))$,
of the branch which connects the quantization level (node)
$x_{qj}(k-1)$ to the quantization level $x_{qi}(k)$ is defined by

$$M(x_{qj}(k-1) \to x_{qi}(k))$$

$$\underline{\underline{\Delta}} \ \ln[\text{Prob}\{x_q(k) = x_{qi}(k) \mid x_q(k-1) = x_{qj}(k-1)\}]$$

$$+ \ln p'(z(k) \mid x_{qi}(k)). \qquad (29)$$

Definition 9. The metric of a path from time zero to time
i is the summation of the metric of the initial node from which
the path starts and the metrics of the branches of which the

path consists. For example, the metric, denoted by $M\left(x_q^m(i)\right)$, of the portion between the nodes $x_q^m(0)$ and $x_q^m(i)$ of the path (hypothesis) H_m is

$$M\left(x_q^m(i)\right) = \ln \pi_0^m$$

$$+ \sum_{k=1}^{i} \left[\ln \pi_k^m + \ln p'\left(z(k)|x_q^m(k)\right)\right]. \tag{30}$$

Consequently, the metric, sometimes denoted by $M(H_m)$, of the path H_m (through the trellis) is

$$M(H_m) \triangleq M\left(x_q^m(L)\right) = \ln\left[p(H_m)p'\left(z^L|H_m\right)\right], \tag{31}$$

where $x_q^m(L)$ is the end node of the path H_m, and $p(H_m)$ and $p'\left(z^L|H_m\right)$ are given by Eqs. (15), (17), (24a) and (24b).

   Definition 10. The error probability of a path, say H_m, through a trellis diagram T with M possible paths H_1, H_2, ..., H_M is the probability of deciding that a path which is different from H_m is the one most probably followed by the target when the target actually followed the path H_m. This error probability is denoted by either $P_{E_m}(H_1, H_2, ..., H_M)$ or $P_{E_m}(T)$ where subscripts E and m are the error and the mth path, respectively. Hence

$$P_{E_m}(H_1, ..., H_M) \triangleq P_{E_m}(T)$$

$$\triangleq \text{Prob}\left\{z^L \in \bar{D}_m|H_m\right\} = \int_{z^L \in \bar{D}_m} p\left(z^L|H_m\right)dz^L,$$

$$\tag{32}$$

where $\bar{D}_m$ is the complement of the decision region D_m for the path H_m, and $p\left(z^L|H_m\right)$ is the probability density function of the observation sequence z^L when the target actually followed the path H_m. Hence from Eq. (10), the overall error probability

for the detection of the path most likely followed by the target [denoted by R, P_E, $P_E(H_1, \ldots, H_M)$ or $P_E(T)$] can be expressed in terms of the path error probabilities as follows:

$$P_E = \sum_{m=1}^{M} p(H_m) P_{E_m}(H_1, H_2, \ldots, H_M), \qquad (33)$$

where $p(H_m)$ is given by Eq. (15). By convention, the overall
error probability P_E is assigned to the trellis diagram T, in
other words, it is sometimes said that the trellis diagram T
has the overall error probability P_E.

Definition 11. The density function of the observation
sequence z^L when the target actually followed the path H_m
$\left[\text{i.e., } p\left(z^L | H_m\right)\right]$ is referred to as the likelihood function for
the path (hypothesis) H_m.

It follows from Def. 9 and Eq. (23) that the optimum decision rule (for deciding the path most likely followed by the
target from time zero to time L) is to choose the path (from
time zero to time L) with the largest metric through the trellis diagram as the decision [i.e., as the path (most likely)
followed by the target]. This can be handled by using the
Viterbi decoding algorithm (VDA) [13,14], which is the optimum
decoding algorithm. The algorithm which obtains a trellis diagram for the target motion (model), as described before, and
which finds the path most likely followed by the target by using
the VDA is referred to as the optimum decoding-based smoothing
algorithm (ODSA).

*H. OPTIMUM DECODING-BASED
SMOOTHING ALGORITHM*

Preliminary step. Reduce the target-motion model to a finite-state model, as described before, and obtain a trellis
diagram for the target motion (model) from time zero to the

time, say L, until which the target will be tracked.   Then
assign to each initial node its metric.

Step 1.   For each node at time 1; use the observation z(1)
and evaluate the metrics of the branches connecting the initial
nodes to the node at time 1.   Add these metrics to the metrics
of the initial nodes from which the branches start, find the
metrics of the paths merging at the node at time 1, label the
path with the largest metric (which is called the best path for
the node at time 1), and then discard the other paths.   Finally,
assign the largest metric to the node at time 1 (which is called
the metric of the node at time 1).

Step k.   For each node at time k; use the observation at
time k, [i.e., z(k)], and calculate the metrics of the branches
connecting the nodes at time k - 1 to the node at time k.   Add
these metrics to the metrics of the nodes at time k - 1 from
which the branches start, find the metrics of the paths merging
at the node at time k, label the path with the largest metric
(which is called the best path for the node at time k), and
then discard the other paths.   Finally, assign the largest
metric to the node at time k (which is called the metric of
the node at time k).

At the end of step L, stop and choose from among the nodes
at time L that with the largest metric.   Then decide that the
best path for this node is the path followed by the target.

1.   *An Example of the ODSA*

Let us consider a target whose motion from time zero to
time 2 is described by Fig. 6a.   Using the ODSA, we would like
to find the path in the trellis diagram which was most likely
followed by the target from time zero to time 2.

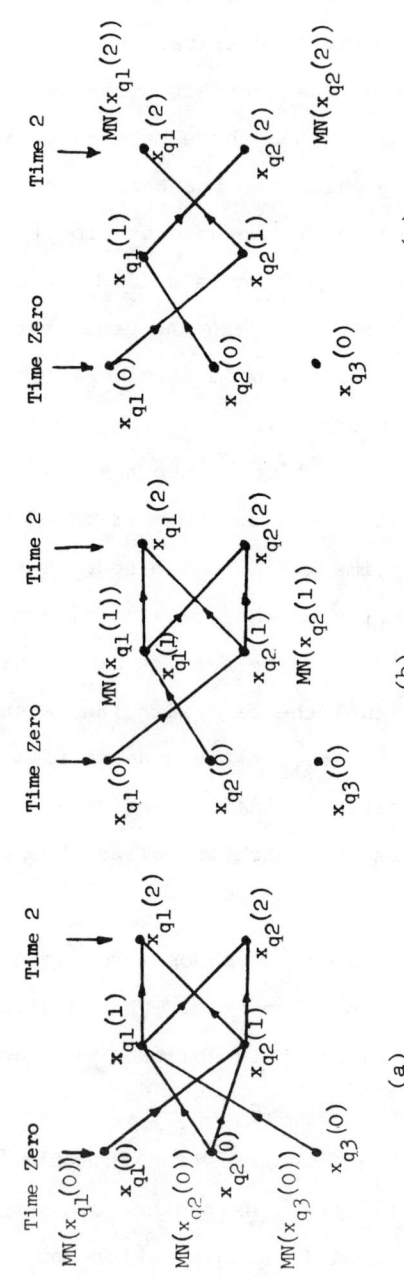

Fig. 6. Diagrams for the example of the ODSA: (a) trellis diagram for target motion from time zero to time 2; (b) diagram at end of first step; (c) diagram at end of second step.

Preliminary step. To each intial node, assign its metric, i.e., $MN(x_{qi}(0)) = Prob\{x_q(0) = x_{qi}(0)\}$, where $i = 1, 2, 3$. From now on, the metric of the node $x_{qi}(k)$ is represented by $MN(x_{qi}(k))$.

Step 1. Consider the node $x_{q1}(1)$. The branches $x_{q2}(0)x_{q1}(1)$ and $x_{q3}(0)x_{q1}(1)$ are the only ones connecting the nodes at time zero to the node $x_{q1}(1)$. Hence calculate the metrics of these branches, then add these metrics to the metrics of the nodes $x_{q2}(0)$ and $x_{q3}(0)$ and obtain the following:

$$A_{11} \triangleq M(x_{q2}(0) \rightarrow x_{q1}(1)) + MN(x_{q2}(0)),$$

$$A_{12} \triangleq M(x_{q3}(0) \rightarrow x_{q1}(1)) + MN(x_{q3}(0)).$$

Further, assuming that $A_{11} \geq A_{12}$, the path $x_{q2}(0)x_{q1}(1)$ is chosen as the best path for the node $x_{q1}(1)$, and A_{11} is assigned to the node $x_{q1}(1)$ as its metric, i.e., $MN(x_{q1}(1)) = A_{11}$. The path $x_{q3}(0)x_{q1}(1)$ is then discarded. Let us now assume that the following are similarly found for the node $x_{q2}(1)$: $x_{q1}(0)x_{q2}(1)$ is the best path for $x_{q2}(1)$, and $MN(x_{q2}(1)) = M(x_{q1}(0) \rightarrow x_{q2}(1)) + MN(x_{q1}(0))$. Hence, we have Fig. 6b at the end of step 1.

Step 2. Consider the node $x_{q1}(2)$. The branches $x_{q1}(1)x_{q1}(2)$ and $x_{q2}(1)x_{q1}(2)$ are those connecting the nodes at time 1 to the node $x_{q1}(2)$. Hence calculating the metrics of these branches and adding these metrics to the metrics of the nodes $x_{q1}(1)$ and $x_{q2}(1)$, we obtain the following:

$$A_{21} \triangleq M(x_{q1}(1) \rightarrow x_{q1}(2)) + MN(x_{q1}(1)),$$

$$A_{22} \triangleq M(x_{q2}(1) \rightarrow x_{q1}(2)) + MN(x_{q2}(1)).$$

Further, assuming that $A_{22} \geq A_{21}$, the path $x_{q1}(0)x_{q2}(1)x_{q1}(2)$ is chosen as the best path for the node $x_{q1}(2)$, and A_{22} is assigned to the node $x_{q1}(2)$ as its metric, that is, $MN(x_{q1}(2))$ = A_{22}. The path $x_{q2}(0)x_{q1}(1)x_{q1}(2)$ is then discarded. Let us now assume that the following are similarly found for the node $x_{q2}(2)$: $x_{q2}(0)x_{q1}(1)x_{q2}(2)$ is the best path for $x_{q2}(2)$, and $MN(x_{q2}(2)) = M(x_{q1}(1) \rightarrow x_{q2}(2)) + MN(x_{q1}(1))$. Hence, we have Fig. 6c at the end of step 2. In addition, assuming that $MN(x_{q2}(2)) \geq MN(x_{q1}(2))$, the path $x_{q2}(0)x_{q1}(1)x_{q2}(2)$ is chosen as the path followed by the target from time zero to time 2.

Having defined the ODSA, its performance should be determined. This is discussed in Sections II.H.2 and II.H.3, next.

2. An Upper Bound for the Overall Error Probability

Let us consider a target whose motion from time zero to time L is described by a trellis diagram with M possible paths H_1, H_2, ..., H_m from time zero to time L. The evaluation of the overall error probability P_E for the detection of the path followed by the target from time 0 to time L is conceptually easy; however, it is in general computationally impractical since it contains multidimensional integrals. On the other hand, upper bounds on P_E are available [1,14,15] which in some cases approximate P_E quite well. One of these bounds is now presented.

Let Γ_i be a subset of the observation space D such that

$$\Gamma_i \triangleq \left\{ z^L: \ M(H_j) \geq M(H_i) \quad \text{for some} \ j \neq i \right\}, \tag{34}$$

where $M(H_i)$ is the metric of the path (hypothesis) H_i and is given by Eq. (31). The set Γ_i contains the complement $\overline{D}_i$ of the optimum decision region D_i since the observation sequences z^L for which the inequality in Eq. (13) becomes an equality are

resolved at random into the decision regions satisfying the
equality. It then follows from Eq. (32) that the error probability of the path H_i can be upper bounded by

$$P_{E_i}(H_1, H_2, \ldots, H_M) \leq \int_{z^L \in \Gamma_i} p\left(z^L | H_i\right) dz^L$$

$$= \int_{z^L \in \Gamma_i} p\left(z^L | H_i\right) \phi(z^L) dz^L, \tag{35}$$

where the function $\phi(z^L)$ is defined by

$$\phi(z^L) \triangleq \begin{cases} 1 & \text{if } z^L \in \Gamma_i, \\ 0 & \text{elsewhere.} \end{cases} \tag{36}$$

Furthermore, $\phi(z^L)$ can be upper bounded as follows.

If $z^L \in \Gamma_i$, it then follows from Eq. (34) that for some
$j \neq i$, $M(H_j) - M(H_i) \geq 0$. Hence $\exp \alpha[M(H_j) - M(H_i)] \geq 1$ for
some $j \neq i$ and any nonnegative number α. Thus, for any nonnegative numbers α and ρ, we have

$$\left(\sum_{j \neq i} \exp \alpha[M(H_j) - M(H_i)]\right)^\rho \geq 1 \qquad \text{for any } \alpha, \rho \geq 0. \tag{37}$$

On the other hand, if $z^L \notin \Gamma_i$, then the expression on the left-
hand side of the inequality in Eq. (37) is at least a nonnegative number. Therefore for all z^L, we obtain

$$\phi(z^L) \leq \left(\sum_{j \neq i} \exp \alpha[M(H_j) - M(H_i)]\right)^\rho \qquad \text{for any } \alpha, \rho \geq 0. \tag{38}$$

Further, substituting Eq. (38) into Eq. (35) yields

$$P_{E_i}(H_1, \ldots, H_M) \leq \int_{z^L \in \Gamma_i} p\left(z^L | H_i\right)$$

$$\times \left(\sum_{j \neq i} \exp \alpha [M(H_j) - M(H_i)]\right)^{\rho} dz^L. \quad (39)$$

The integrand in Eq. (39) is obviously nonnegative. Hence en-
larging the domain of the integration in Eq. (39) makes the value
of the integral larger. Therefore the error probability of H_i
can be further upper bounded by taking the integration over the
whole observation space

$$P_{E_i}(H_1, \ldots, H_M) \leq \int_{z^L} p\left(z^L | H_i\right) \exp - \alpha\rho M(H_i)$$

$$\times \left[\sum_{j \neq i} \exp \alpha M(H_j)\right]^{\rho} dz^L$$

$$\text{for any } \alpha, \rho \geq 0. \quad (40)$$

Since α and ρ are any arbitrary nonnegative numbers, the inequality in Eq. (40) holds for $\alpha = 1/(1 + \rho)$. Hence taking
$\alpha = 1/(1 + \rho)$ and, further, using Eqs. (25), (31), and the
equality $a^b = e^{b \ln a}$ [for any number b and any positive number
a], the error probability of the path H_i can be bounded by

$$P_{E_i}(H_1, H_2, \ldots, H_M) \leq \int_{z^L} \left(\prod_{k=0}^{L} \frac{1}{\pi_k^i}\right)^{\rho/(1+\rho)}$$

$$\times \left[\prod_{k=1}^{L} p'\left(z(k) | x_q^i(k)\right)\right]^{1/(1+\rho)}$$

(Eq. (41) continues)

$$\times \left\{ \sum_{\substack{j \neq i}} \left(\prod_{k=0}^{L} \pi_k^j \right)^{1/(1+\rho)} \right.$$

$$\left. \times \left[\prod_{k=1}^{L} p'\left(z(k) \,|\, x_q^j(k) \right) \right]^{1/(1+\rho)} \right\}^{\rho} dz^L,$$

for any $\rho \geq 0,$ (41)

where $p'\left(z(k) \,|\, x_q^i(k) \right)$ is given by Eqs. (24a) and (24b) and π_k^i is defined in Section II.E. The bound in Eq. (41) is that of the Gallager type. Substituting this bound for the error probability of the path H_i in Eq. (33) yields an upper bound on the overall error probability for the detection of the path followed by the target.

3. *An Ensemble Upper Bound*
   for the Overall Error Probability

Let us consider a target whose motion is described by a trellis diagram T with M possible paths H_1, H_2, ..., H_M from time zero to time L (Fig. 4), and let H_i pass through the quantization levels $x_q^i(0)$, $x_q^i(1)$, ..., $x_q^i(L)$. In order to derive an ensemble bound, let us start defining the following symbols, which are used in the subsequent analyses:

Xe Set of all possible quantization levels from time 1 to
        time L, namely, $X^e \triangleq \{$all possible values of $x_q(k)$
        for $k = 1, 2, ..., L\}$

N^e Number of elements in X^e

$\underset{\sim}{H}^e$ Set of all L-tuples of X^e

E Ensemble (or set) of all M-tuples of $\underset{\sim}{H}^e$; hence E con-
        tains $(N^e)^{LM}$ elements in it

H_i^e Set of all quantization levels which the path H_i passes
        through from time 1 to time L in T, i.e., $H_i^e \triangleq \{x_q^i(1),$
        $x_q^i(2), ..., x_q^i(L)\} \in \underset{\sim}{H}^e$, which is an L-tuple of X^e

$$T^e \quad \triangleq \quad \left\{ H^e_1, \ H^e_2, \ \ldots, \ H^e_M \right\} \in E$$

E_M Ensemble of all possible trellis diagrams with M pos-
          sible paths from time zero to time L, which are ob-
          tained from the trellis diagram T by replacing only
          T^e by elements of E; hence this ensemble contains
          $(N^e)^{LM}$ elements in it; E_M is referred to as the en-
          semble of each motion (or trellis diagram) in itself.
          Obviously, E_M is the ensemble of T too (since $T \in E_M$).

The exact expressions for both the overall error probability and the upper bound given in Eq. (33) and Section II.H.2
contain multidimensional integrals which are generally very
complex to evaluate. Therefore, instead of evaluating these,
we consider an average overall error probability bound over
the ensemble E. Averaging an error probability over an ensemble is referred to as "random coding," which is the central
technique of information theory [14,15]. An (upper) bound (on
the overall error probability) averaged over an ensemble is
called an ensemble (upper) bound (for the overall error probability) which may turn out to be quite simple to evaluate. Obviously, at least one trellis diagram in E_M must have an (overall) error probability as small as this ensemble bound. In
other words, an ensemble upper bound will give us an upper
bound on the (overall) error probability for the best trellis
diagram in E_M (i.e., the trellis diagram with minimum (overall)
error probability in E_M).

In order to derive an ensemble (overall) error probability
(or an ensemble bound), first a probability density function
$Q^e(\cdot)$ is defined on the ensemble E such that

$$Q^e(T^e) \triangleq \prod_{i=1}^{M} Q\left(H^e_i\right), \qquad Q\left(H^e_i\right) \triangleq \prod_{k=1}^{L} q\left(x^i_q(k)\right), \qquad (42)$$

where $q(\cdot)$ is an arbitrary probability density function on X^e.
Hence $Q(\cdot)$ is a probability density function on the set $\underline{H}^e$.
Then the (overall) error probability (or an error probability
bound) is averaged with respect to $Q^e(\cdot)$ over the ensemble.

The ensemble (overall) error probability, denoted by either
$\overline{P}_E$ or $\overline{P_E(T)}$ for the detection of the path (through a trellis
diagram T in E_M) most likely followed by the target is defined
by

$$\overline{P}_E = \sum_{T^e \in E} Q^e(T^e) P_E(T) = \sum_{T \in E_M} Q^e(T^e) P_E(T) , \qquad (43)$$

where $P_E(T)$ is the overall error probability for the detection
of the path (through T) most likely followed by the target.
Substituting Eq. (33) into Eq. (43) and changing the order of
summations, the ensemble overall error probability can be
rewritten in terms of the path ensemble error probabilities
as

$$\overline{P}_E = \sum_{i=1}^{M} p(H_i) \overline{P_{E_i}(T)} , \qquad (44)$$

where

$$\overline{P_{E_i}(T)} = \sum_{T^e \in E} Q^e(T^e) P_{E_i}(T) , \qquad (45)$$

where $P_{E_i}(T)$ is the error probability of the path H_i, and
$\overline{P_{E_i}(T)}$ is referred to as the ensemble error probability of the
path H_i: an overbar denotes the ensemble average of a particular
quantity. An ensemble upper bound for the detection of the
path most probably followed by the target can be obtained by
averaging upper bounds for the path error probabilities over

the ensemble. Let $P_{EB_i}(T)$ be an upper bound for the error probability of the path H_i, i.e.,

$$P_{E_i}(T) \leq P_{EB_i}(T). \tag{46}$$

Substituting this bound for the error probability of H_i in Eq. (45) yields the following bound for the ensemble error probability of the path H_i:

$$\overline{P_{E_i}(T)} \leq \sum_{T^e \in E} Q^e(T^e) P_{EB_i}(T) \triangleq \overline{P_{EB_i}(T)}, \tag{47}$$

where $\overline{P_{EB_i}(T)}$ is referred to as an ensemble upper bound for the error probability of the path H_i. Further, substituting the bound in Eq. (47) for the ensemble error probability of the path H_i in Eq. (44) yields the following bound for the ensemble overall error probability

$$\overline{P_E} \leq \sum_{i=1}^{M} p(H_i) \overline{P_{EB_i}(T)}. \tag{48}$$

Let us now derive an ensemble bound for the overall error probability by using the bound in Eq. (41). Substituting the bound in Eqs. (41) and (42) into Eq. (47), the ensemble error probability of the path H_i can be upper bounded as

$$\overline{P_{E_i}(T)} \leq \sum_{H_1^e \in \underset{\sim}{H}^e} \cdots \sum_{H_M^e \in \underset{\sim}{H}^e} Q\left(H_1^e\right) Q\left(H_2^e\right) \cdots Q\left(H_M^e\right)$$

$$\times \int_{z^L} b_i^{-\rho/(1+\rho)} \left[p'\left(z^L | H_i\right)\right]^{1/(1+\rho)}$$

$$\times \left\{\sum_{j \neq i} b_j^{1/(1+\rho)} \left[p'\left(z^L | H_j\right)\right]^{1/(1+\rho)}\right\}^{\rho} dz^L,$$

$$\text{for any } \rho \geq 0, \tag{49}$$

where $p'\left(z^L|H_i\right)$ is defined by Eq. (17), and

$$b_i \triangleq \prod_{k=0}^{L} \pi_k^i. \tag{50}$$

Changing the order of summations and integration, Eq. (49) can be rewritten as

$$\overline{P_{E_i}(T)} \leq \int_{z^L} b_i^{-\rho/(1+\rho)} \left\{ \sum_{H_i^e \in \underset{\sim}{H}^e} Q\left(H_i^e\right)\left(p'\left(z^L|H_i\right)\right)^{1/(1+\rho)} \right\}$$

$$\times \left\{ \sum_{\underset{H_j^e \in \underset{\sim}{H}^e}{j\neq i}} \cdots \sum \left(\prod_{j\neq i} Q\left(H_j^e\right)\right)\left[\sum_{j\neq i} b_j^{1/(1+\rho)} \right. \tag{51}$$

$$\left. \times \left(p'\left(z^L|H_j\right)\right)^{1/(1+\rho)}\right]^{\rho} \right\},$$

where ρ is an arbitrary nonnegative number. If we restrict the parameter ρ to lie in the interval [0, 1], then the term in the last braces can be further upper bounded by using Jensen's inequality. Let $f(R)$ be the term in the last brackets, namely,

$$f(R) \triangleq R^{\rho} \triangleq \left[\sum_{j\neq i} b_j^{1/(1+\rho)}\left(p'\left(z^L|H_j\right)\right)^{1/(1+\rho)}\right]^{\rho}, \tag{52}$$

where

$$R \triangleq \sum_{j\neq i} b_j^{1/(1+\rho)}\left(p'\left(z^L|H_j\right)\right)^{1/(1+\rho)}. \tag{53}$$

The function $f(R)$ is a convex $\cap$ function for any $\rho \in [0, 1]$, since R^{ρ} is a convex $\cap$ function for any $\rho \in [0, 1]$. Furthermore, the term in the last braces in Eq. (51) is the expectation of $f(R)$ with respect to the following probability density function:

$$\sum_{\underset{H_j^e \in \underset{\sim}{H}^e}{j\neq i}} \cdots \sum \left(\prod_{j\neq i} Q\left(H_j^e\right)\right). \tag{54}$$

Therefore, using Jensen's inequality [17] (which states that if R is a random variable, $f(R)$ is a convex $\cap$ function of R, and $E\{R\}$ is finite, then

$$E\{f(R)\} \leq f(E\{R\}), \tag{55}$$

where E is the expectation) and recognizing that H_j^e is summed over the same space H^e for all j and that

$$\sum_{H_j^e \in \underset{\sim}{H}^e} Q\left(H_j^e\right) = 1, \tag{56}$$

we obtain the following bound for the term in the last braces in Eq. (51):

$$\sum_{\substack{H_j^e \in \underset{\sim}{H}^e}} \cdots_{j \neq i} \sum_{j \neq i} \prod_{j \neq i} Q\left(H_j^e\right) \left[\sum_{j \neq i} b_j^{1/(1+\rho)}\left(p'\left(z^L | H_j\right)\right)^{1/(1+\rho)}\right]^\rho$$

$$\leq \left[\sum_{j \neq i} b_j^{1/(1+\rho)}\right]^\rho \left[\sum_{H_i^e \in \underset{\sim}{H}^e} Q\left(H_i^e\right)\left(p'\left(z^L | H_i\right)\right)^{1/(1+\rho)}\right]^\rho,$$

$$\text{for any } \rho \in [0, 1]. \tag{57}$$

Substituting this bound into Eq. (51), we get

$$\overline{P_{E_i}(T)} \leq b_i^{-1/(1+\rho)} \left[\sum_{j \neq i} b_j^{1/(1+\rho)}\right]^\rho$$

$$\times \int_{z^L} \left[\sum_{H_i^e \in \underset{\sim}{H}^e} Q\left(H_i^e\right)\left(p'\left(z^L | H_i\right)\right)^{1/(1+\rho)}\right]^{\rho+1} dz^L. \tag{58}$$

Further, using the following inequalities:

$$\prod_{k=0}^{L} \pi_k^{min} \leq b_i \leq \prod_{k=0}^{L} \pi_k^{max} \quad \text{for all } i, \tag{59}$$

the term outside the integral in Eq. (58) can be further upper
bounded by

$$b_i^{-\rho/(1+\rho)}\left[\sum_{j\neq i} b_j^{1/(1+\rho)}\right]^{\rho} \leq (M-1)^{\rho} \prod_{k=0}^{L}\left(\frac{\pi_k^{max}}{\pi_k^{min}}\right)^{\rho/(1+\rho)}$$

$$\text{for all } i. \tag{60}$$

Let us now consider the term in the last brackets in Eq. (58).
Substituting Eqs. (17) and (42) into this term and changing
the order of summations and multiplications, we obtain

$$\sum_{H_i^e \in \underset{\sim}{H}^e} Q\left(H_i^e\right)\left(p'\left(z^L|H_i\right)\right)^{1/(1+\rho)}$$

$$= \sum_{H_i^e \in \underset{\sim}{H}^e} \prod_{k=1}^{L} q\left(x_q^i(k)\right)\left(p'\left(z(k)|x_q^i(k)\right)\right)^{1/(1+\rho)}$$

$$= \prod_{k=1}^{L}\left\{\sum_{x_q^i(k)\in X^e} q\left(x_q^i(k)\right)\left(p'\left(z(k)|x_q^i(k)\right)\right)^{1/(1+\rho)}\right\}$$

$$= \prod_{k=1}^{L}\left\{\sum_{x\in X^e} q(x)(p'(z(k)|x))^{1/(1+\rho)}\right\}, \tag{61}$$

where the last equality follows from the fact that for all i
and k, $x_q^i(k)$ is summed over the same space X^e. Substituting
Eqs. (60) and (61) into Eq. (58) yields the following ensemble
upper bound for the error probability of H_i.

$$\overline{P_{E_i}(T)} \leq (M-1)^{\rho}\left(\prod_{k=0}^{L}\frac{\pi_k^{max}}{\pi_k^{min}}\right)^{\rho/(1+\rho)}$$

$$\times \prod_{k=1}^{L}\left\{\int_{z(k)}\left[\sum_{x\in X^e} q(x)(p'(z(k)|x))^{1/(1+\rho)}\right]^{\rho+1} dz(k)\right\}$$

$$\triangleq \overline{B(T)} \quad \text{for all } i \text{ and any } \rho \in [0, 1], \tag{62}$$

where $p'(z(k)|x)$ is given by Eqs. (24a) and (24b), π_k^{max} and π_k^{min} are defined in Section II.E, and $q(\cdot)$ is an arbitrary density function on X^e. Substituting the bound in Eq. (62) into Eq. (48) and recognizing that this bound does not depend on i (that is, the paths), we obtain

$$\overline{P}_E \leq \sum_{i=1}^{M} p(H_i)\overline{B(T)} = \overline{B(T)} \sum_{i=1}^{M} p(H_i) = \overline{B(T)}. \tag{63}$$

Hence, the bound in Eq. (62) is also an upper bound for the ensemble overall error probability for the detection of the path most likely followed by the target.

If the function $g(k, ., ., .)$ in the observation model being considered and the statistics of the observation noise $v(k)$ and the interference $I(k)$ (in the presence of interference) are time invariant, then the term in braces in Eq. (62) is time invariant. Hence in this case the ensemble upper bound in Eq. (62) becomes

$$\overline{P}_E \leq (M - 1)^\rho \left(\prod_{k=0}^{L} \frac{\pi_k^{max}}{\pi_k^{min}} \right)^{\rho/(1+\rho)}$$

$$\times \left\{ \int_{z(k)} \left[\sum_{x \in X^e} q(x)(p'(z(k)|x))^{1/(1+\rho)} \right]^{\rho+1} dz(k) \right\}^{L},$$

$$\text{for any } \rho \in [0, 1]. \tag{64}$$

Using the relation that $\exp[\ln a] = a$ for any $a > 0$, the bound in Eq. (64) can be rewritten as

$$\overline{P}_E \leq \exp -L\left\{ E_0(\rho, \underset{\sim}{q}) - \rho \frac{\ln(M-1)}{L} - \frac{\rho}{1+\rho} \frac{\ln G}{L} \right\},$$

$$\text{for any } \rho \in [0, 1], \tag{65}$$

where

$$E_0(\rho, \underset{\sim}{q}) \triangleq -\ell n \left\{ \int_{z(k)} \left[\sum_{x \in X^e} q(x) (p'(z(k)|x))^{1/(1+\rho)} \right]^{\rho+1} \right\},$$

(66)

$$G \triangleq \prod_{k=0}^{L} \left(\pi_k^{max} / \pi_1^{min} \right),$$

and $E_0(\rho, \underset{\sim}{q})$ is referred to as the Gallager function, since it was first defined by Gallager [15]. Recalling that ρ is any arbitrary number in [0, 1] and $q(\cdot)$ is an arbitrary probability density function on X^e leads us to obtain the tightest bound on $\overline{P}_E$ by minimizing the right-hand side of Eq. (65) over ρ and $\underset{\sim}{q}$. This gives us the following bound:

$$\overline{P}_E \leq exp -LE(M, G, L),$$

(67)

where

$$E(M, G, L) \triangleq \max_{\underset{\sim}{q}} \max_{\rho \in [0,1]}$$

$$\times \left[E_0(\rho, \underset{\sim}{q}) - \rho \frac{\ell n(M-1)}{L} - \frac{\rho}{1+\rho} \frac{\ell n\, G}{L} \right],$$

$$= \max_{\rho \in [0,1]} \left[\max_{\underset{\sim}{q}} E_0(\rho, \underset{\sim}{q}) \right.$$

(68)

$$\left. - \rho \frac{\ell n(M-1)}{L} - \frac{\rho}{1+\rho} \frac{\ell n\, G}{L} \right].$$

As has been noted, the maximization is taken over all $\rho \in [0, 1]$ and the set of all possible probability density functions on X^e. In order to evaluate $E(M, G, L)$, it is necessary to analyze $E_0(\rho, \underset{\sim}{q})$ as a function of ρ. The important properties of this function are stated in the following Theorem 1. The proof of this theorem is presented in Ref. [15].

Theorem 1. Assume that the average mutual information $I(\underset{\sim}{q})$ which is defined by

$$I(\underset{\sim}{q}) \triangleq \int_{z(k)} \sum_{x \in X^e} q(x)p'(z(k)|x)$$

$$\times \ln\left[\frac{p'(z(k)|x)}{\sum_{x \in X^e} q(x)p'(z(k)|x)}\right] dz(k), \tag{69}$$

is nonzero [in fact, $I(\underset{\sim}{q})$ is always nonnegative]. Then $E_0(\rho, \underset{\sim}{q})$ has the following properties [14,15]:

$$E_0(\rho, \underset{\sim}{q}) = 0, \quad \text{for} \quad \rho = 0, \tag{70}$$

$$E_0(\rho, \underset{\sim}{q}) > 0, \quad \text{for} \quad \rho > 0, \tag{71}$$

$$\partial E_0(\rho, \underset{\sim}{q})/\partial\rho > 0, \quad \text{for} \quad \rho > 0, \tag{72}$$

$$\partial E_0(\rho, \underset{\sim}{q})/\partial\rho\big|_{\rho=0} = I(q), \tag{73}$$

$$\partial^2 E_0(\rho, \underset{\sim}{q})/\partial\rho^2 \leq 0, \tag{74}$$

with equality in Eq. (74) if and only if

$$\ln\left[\frac{p'(z(k)|x)}{\sum_{x \in X^e} q(x)p'(z(k)|x)}\right] = I(\underset{\sim}{q}), \tag{75}$$

for all $x \in X^e$ and all $z(k)$ in the space of all possible observations at time k such that $q(x)p'(z(k)|x) > 0$. Therefore for a given $\underset{\sim}{q}$, $E_0(\rho, \underset{\sim}{q})$ is a positive increasing convex $\cap$ function of $\rho \in [0, \infty)$ with a slope at the origin equal to $I(\underset{\sim}{q})$. Also, the function

$$\rho \frac{\ln(M-1)}{L} + \frac{\rho}{1+\rho} \frac{\ln G}{L}$$

is a convex $\cap$ function of $\rho \in [0, \infty)$. Hence we can easily perform the maximization in Eq. (68) over $\rho \in [0, 1]$ for a given $q(\cdot)$ so that $E(M, G, L)$ can be expressed parametrically as follows:

$$E(M, G, L) = \begin{cases} 0, & \text{if } R > C, \\[2em] \begin{aligned} &\max_{\underset{\sim}{q}} E_0(\rho, \underset{\sim}{q}) \\ &- \rho \frac{\ln(M-1)}{L} \\ &- \rho \frac{\ln G}{(1+\rho)L}, \end{aligned} & \text{if } \frac{\partial}{\partial \rho}[\max_{\underset{\sim}{q}} E_0(\rho, \underset{\sim}{q})]\Big|_{\rho=1} \leq R \leq C, \\[1em] \text{where } \rho \text{ is determined} \\ \text{from the equation} \\[1em] \frac{\partial}{\partial \rho}[\max_{\underset{\sim}{q}} E_0(\rho, \underset{\sim}{q})] \\[1em] = \frac{\ln(M-1)}{L} + \frac{\ln G}{(1+\rho)^2 L}, \\[2em] \begin{aligned} &\max_{\underset{\sim}{q}} E_0(1, \underset{\sim}{q}) \\ &- \frac{\ln(M-1)}{L} \\ &- \frac{\ln G}{2L} \end{aligned} & \text{if } R \leq \frac{\partial}{\partial \rho}[\max_{\underset{\sim}{q}} E_0(\rho, \underset{\sim}{q})]\Big|_{\rho=1}, \end{cases} \tag{76}$$

where

$$R \triangleq [\ln(M-1)/L] + (\ln G/4L), \qquad C \triangleq \max_{\underset{\sim}{q}} I(\underset{\sim}{q}). \tag{77}$$

The maximization of the Gallager function $E_0(\rho, \underset{\sim}{q})$ and the average mutual information $I(\underset{\sim}{q})$ over the space of all possible probability density functions on X^e has been treated in the literature. Two theorems related to this maximization are stated. Their proof can be found in Refs. [14,15].

Theorem 2. A probability density function $q_0(\cdot)$ on X^e maximizes the function $E_0(\rho, \underset{\sim}{q}_0)$ for a given $\rho \geq 0$ if and only if the following holds:

$$\int_{z(k)} [p'(z(k)|x)]^{1/(1+\rho)} [\alpha(z(k), \underset{\sim}{q}_0)]^\rho$$

$$\geq \int_{z(k)} [\alpha(z(k), \underset{\sim}{q}_0)]^{1+\rho}, \qquad \text{for all } x \in X^e, \tag{78}$$

with equality for all $x \in X^e$ for which $q_0(x) > 0$ where

$$\alpha(z(k), \underset{\sim}{q}_0) \triangleq \sum_{x \in X^e} q_0(x) [p'(z(k)|x)]^{1/(1+\rho)}. \tag{79}$$

Theorem 3. A probability density function $q_0(\cdot)$ on X^e maximizes the average mutual information $I(\underset{\sim}{q}_0)$ if and only if the following holds:

$$\int_{z(k)} p'(z(k)|x) \ \ln\left[\frac{p'(z(k)|x)}{\sum_{x \in X^e} q_0(x)p'(z(k)|x)}\right]$$

$$\leq I(\underset{\sim}{q}_0), \qquad \text{for all} \quad x \in X^e, \tag{80}$$

with equality for all x for which $q_0(x) > 0$.

It then follows that neither Theorem 2 nor Theorem 3 is very useful in finding the maximum of the Gallager function or the average mutual information. But both are useful in verifying if a hypothesized solution indeed maximizes these functions. For example, using Theorems 2 and 3, it can be verified that the uniform distribution on X^e, that is,

$$q(x) = 1/N^e, \qquad \text{for all} \quad x \in X^e, \tag{81}$$

is not the optimum (maximizing) distribution for either $E_0(\rho, \underset{\sim}{q})$ or $I(\underset{\sim}{q})$. In general, the maximization over the probability density functions on X^e must be performed numerically. Even if the optimum (maximizing) probability density function is known, the evaluation of the Gallager function or the average mutual information is, in general, not at all easy since the related expressions contain multidimensional integrals; hence the evaluation of them must be performed numerically. Throughout this chapter, as the ensemble upper bound (which is used as the performance measure of the ODSA), the bound using $\rho = 1$ and the uniform distribution for $q(\cdot)$ in Eq. (62), is used by virtue of a nice feature arising from the uniform distribution (which is stated next in Theorem 4) and the fact that the simplest function to calculate among $E_0(\rho, \underset{\sim}{q})$ is $E_0(1, \underset{\sim}{q})$. Obviously, this bound is in general not the tightest bound for all $\rho \in [0, 1]$ and all probability density functions $q(\cdot)$ on X^m. Substituting $\rho = 1$ and $(1/N^e)$ for $q(x)$ in Eq. (62), we obtain the following ensemble upper bound for the overall error probability:

$$\overline{P}_E \leq D \prod_{k=1}^{L} \left\{ \int_{z(k)} \left[\sum_{x \in X^e} (p'(z(k)|x))^{1/2} \right]^2 \right\} \triangleq B^e, \tag{82}$$

where

$$D \triangleq (M - 1) \left[\prod_{k=0}^{L} \left(\frac{\pi_k^{max}}{\pi_k^{min}} \right)^{1/2} \right] \left(\frac{1}{N^e} \right)^{2L}. \tag{83}$$

If the function $g(k, ., ., .)$ in the observation model being considered and the statistics of the observation noise $v(k)$ and the interference $I(k)$ (in the presence of interference) are time invariant, then the bound in Eq. (82) becomes

$$\overline{P}_E \leq D \left\{ \int_{z(k)} \left[\sum_{x \in X^e} (p'(z(k)|x))^{1/2} \right]^2 \right\}^L. \tag{84}$$

Let us now prove Theorem 4, which gives us the reason that a uniformly weighted ensemble bound, such as Eq. (82), is used as the performance measure. A uniformly weighted ensemble bound is an ensemble bound obtained by using the uniform density function $q(\cdot)$ on X^m, i.e.,

$$q(x) = 1/N^e, \qquad \text{for all} \quad x \in X^e;$$

$$\tag{85}$$

$$\text{hence,} \qquad Q^e(T^e) = 1/(N^e)^{LM}, \qquad \text{for all} \quad T^e \in E.$$

Theorem 4. For a given uniformly weighed ensemble upper bound B^e for the overall error probability, there exists a subset E_{M_s} of the ensemble E_M such that E_{M_s} contains at least half the elements in E_M, and every element (trellis diagram) in E_{M_s} must have an overall error probability which is less than or equal to two times B^e, i.e.,

$$P_E(T) \leq 2B^e, \qquad \text{for all} \quad T \in E_{M_s}, \tag{86}$$

where $P_E(T)$ is the overall error probability of T (i.e., the overall error probability for the detection of the path through T most likely followed by the target).

Proof. Let $\overline{E_{M_s}}$ be the subset of E_M such that

$$P_E(T) > 2B^e, \qquad \text{for all} \quad T \in \overline{E_{M_s}}, \tag{87}$$

and assume that the number K of elements in $\overline{E_{M_s}}$ is greater than half the number of elements in E_M (otherwise there is nothing to prove), i.e.,

$$K > (N^e)^{LM}/2. \tag{88}$$

Since both $Q^e(T^e)$ and $P_E(T)$ are positive for all T in E_M, and $\overline{E_{M_s}}$ is contained in E_M, the ensemble error probability $\overline{P}_E$ defined by Eq. (43) can be lower bounded by

$$\overline{P}_E \geq \sum_{T \in \overline{E_{M_s}}} Q^e(T^e) P_E(T). \tag{89}$$

Substituting the bound in Eq. (87) for the overall error probability of T, the ensemble error probability can be further lower bounded as

$$\overline{P}_E > 2B^e \sum_{T \in \overline{E_{M_s}}} Q^e(T^e). \tag{90}$$

Substituting Eq. (85) into Eq. (90) (since the uniformly weighted ensemble bound is considered), we get

$$\overline{P}_E > 2B^e \sum_{T \in \overline{E}_{M_s}} \frac{1}{(N^e)^{LM}} = 2B^e K \frac{1}{(N^e)^{LM}}. \tag{91}$$

Further, substituting the lower bound in Eq. (88) for K in Eq. (91), the bound in Eq. (91) can again be lower bounded so that we have $\overline{P}_E > B^e$, This contradicts the assumption that B^e is an ensemble upper bound. This completes the proof.

If the trellis diagram under consideration is a member of E_{M_s}, then the overall error probability of this trellis is upper bounded by $2B^e$; otherwise, we do not have any idea about the overall error probability. Since E_{M_s} contains at least half the elements in the ensemble E_M, there is a good chance that the trellis diagram being considered belongs to E_{M_s}. However, for nonuniformly weighted ensembles, a large subset (of E_M), every element of which has an overall error probability bounded by a constant bound, may not easily be obtained since the arguments must include the effect of nonuniform weighting. That is why uniformly weighted ensemble bounds are used as the performance measure.

I. *STACK SEQUENTIAL DECODING-BASED SMOOTHING ALGORITHM*

In tracking a target from time zero to time L by using the optimum decoding-based smoothing algorithm (ODSA), deciding the path most likely followed by the target [multiple- (composite) hypothesis testing or decoding problem] is accomplished by finding the path with the largest metric through a trellis diagram from time zero to time L. The ODSA does this by using the Viterbi decoding algorithm, which systematically examines

(searches) all possible paths in the trellis diagram. Hence,
if the number of possible paths in the trellis diagram is very
large, the ODSA requires a huge amount of memory and computation.

If there were a way to guess the correct path without calculating the metric of every path in the trellis diagram, most
of the computation and memory requirement in ODSA could be
avoided. One way to do so is to use a smoothing algorithm that
uses a stack sequential decoding algorithm (SSDA), which is
suboptimum (i.e., it does not minimize the overall error probability) [13,14]. Such a smoothing algorithm, which at any time
(step) stores a "stack" of previously searched paths of varying
length ordered according to their metrics, is now presented.
This algorithm is referred to as the stack sequential decoding-
based smoothing algorithm (SSDSA) [14].

Preliminary step. Reduce the target-motion model to a
finite-state model as described before and obtain a trellis
diagram for the target-motion model from time zero to time, say
L, until which the target will be tracked. Calculate the
metrics of all initial paths (by convention, an initial node
and its metric are referred to as an initial path and the metric
of this initial path, respectively). Then store these paths
and their metrics and order them according to their metrics.

Recursive step. Compute the metrics of the paths which are
the single-branch continuations of the best path in the stack
(see Definition 12, next) and replace the best path and its metric by these paths and their metrics. If any of the newly added
paths merges with a path already in the stack, discard the one
with smaller metric. Then reorder the remaining paths. If
the best path in the stack terminates in a final node of the

trellis diagram, stop and choose the best path as the path most probably followed by the target; otherwise repeat the process (i.e., continue to search by extending the best path in the stack).

Definition 12. The best path in a stack of already searched paths of varying length is that with the largest metric. If there is more than one path with the identical largest metric, then the best path is that with the longest length. If there is more than one path with the identical largest metric and length, then the best path is only one of these paths (this is chosen at random).

Example 1 illustrates the SSDSA.

Example 1. Let us consider a target whose motion from time zero to time 2 is described by Fig. 7. Using the SSDSA, we would like to find the path (through the trellis diagram) most likely followed by the target from time zero to time 2. Let us first adopt the following conventions:

(1)   In a stack a searched path is represented by the node sequence that the path passes through and its metric following the sequence. The sequence and the metric are separated by a comma. The metric of the path is denoted by MS($\cdot$) where the term in parantheses is the final node of the path. Further, various searched paths are separated by semicolons.

(2)   A stack is ordered in such a way that the best path is placed at the end of the stack, and the path with the second largest metric is placed before the best path, etc.

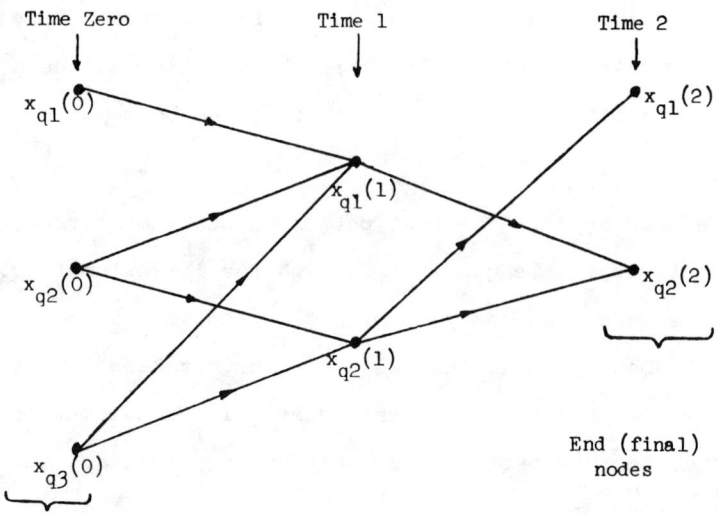

$x_{q1}(0)$

$x_{q1}(1)$

$x_{q2}(0)$

$x_{q2}(1)$

$x_{q1}(2)$

$x_{q2}(2)$

End (final) nodes

$x_{q3}(0)$

Initial nodes

Fig. 7.   The trellis diagram for the example of the SSDSA.

Then the path most likely followed by the target is obtained as follows:

Preliminary step. Assuming that the metrics of the initial paths (i.e., initial nodes) $x_{q1}(0)$, $x_{q2}(0)$, and $x_{q3}(0)$ are such that

$$MS(x_{q2}(0)) \geq MS(x_{q3}(0)) \geq MS(x_{q1}(0)), \tag{92}$$

where $MS(\cdot) \triangleq MN(\ )$ is defined by Eq. (28), the following stack is obtained:

$$x_{q1}(0), \ MS(x_{q1}(0)); \ x_{q3}(0), \ MS(x_{q3}(0));$$

$$x_{q2}(0), \ MS(x_{q2}(0)). \tag{93}$$

Step 1. The paths $X_{q2}(0)x_{q1}(1)$ and $x_{q2}(0)x_{q2}(1)$ are the single-branch continuations of the best path $x_{q2}(0)$ in Stack (93). Hence calculating the metric of these paths, that is,

$$MS(x_{q1}(1)) = M(x_{q2}(0) \rightarrow x_{q1}(1)) + MS(x_{q2}(0)),$$

$$MS(x_{q2}(1)) = M(x_{q2}(0) \rightarrow x_{q2}(1)) + MS(x_{q2}(0)), \tag{94}$$

where $M(x \to y)$ is defined by Eq. (29), and replacing the best path and its metric [i.e., $x_{q2}(0)$, $MS(x_{q2}(0))$] by these paths and their metrics, we get

$$x_{q1}(0), \ MS(x_{q1}(0)); \ x_{q3}(0), \ MS(x_{q3}(0));$$
$$x_{q2}(0)x_{q1}(1), \ MS(x_{q1}(1)); \ x_{q2}(0)x_{q2}(1), \ MS(x_{q2}(1)). \tag{95}$$

Now assuming that

$$MS(x_{q3}(0)) \geq MS(x_{q2}(1)) \geq MS(x_{q1}(1)) \geq MS(x_{q1}(0)), \tag{96}$$

and then reordering the paths according to their metrics, we obtain the stack

$$x_{q1}(0), \ MS(x_{q1}(0)); \ x_{q2}(0)x_{q1}(1), \ MS(x_{q1}(1));$$
$$X_{q2}(0)x_{q2}(1), \ MS(x_{q2}(1)); \ x_{q3}(0), \ MS(x_{q3}(0)). \tag{97}$$

Hence, the best path $x_{q3}(0)$ does not terminate in a final node in Fig. 7. We shall therefore continue to search by similarly extending the best path $x_{q3}(0)$.

Step 2. The paths $x_{q3}(0)x_{q1}(1)$ and $x_{q3}(0)x_{q2}(1)$ are the single-branch continuations of the best path $x_{q3}(0)$ in Stack (97). Thus calculating the metrics of these paths and then replacing the best path and its metric in Stack (97) by these paths and their metrics yields

$$x_{q1}(0), \ MS(x_{q1}(0)); \ x_{q2}(0)x_{q1}(1), \ MS(x_{q1}(1));$$
$$x_{q2}(0)x_{q2}(1), \ MS(x_{q2}(1)); \ x_{q3}(0)x_{q1}(1), \ MS(x_{q1}(1)); \tag{98}$$
$$x_{q3}(0)x_{q2}(1), \ MS(x_{q2}(1)).$$

Hence the newly added paths $x_{q3}(0)x_{q1}(1)$ and $x_{q3}(0)x_{q2}(1)$ merge with the paths $x_{q2}(0)x_{q1}(1)$ and $x_{q2}(0)x_{q2}(1)$ (which are already in the stack), respectively. Assuming that

$$MS(x_{q1}(1)) \ \text{ of } \ x_{q2}(0)x_{q1}(1) \geq MS(x_{q1}(1)) \ \text{ of } \ x_{q3}(0)x_{q1}(1),$$

and

$$MS(x_{q2}(1)) \ \text{ of } \ x_{q2}(0)x_{q2}(1) \geq MS(x_{q2}(1)) \ \text{ of } \ x_{q3}(0)x_{q2}(1),$$

the paths $x_{q3}(0)x_{q1}(1)$ and $x_{q3}(0)x_{q2}(1)$ and their metrics are discarded. Hence we have the stack

$$x_{q1}(0), \ MS(x_{q1}(0)); \ x_{q2}(0)x_{q1}(1), \ MS(x_{q1}(1));$$

$$x_{q2}(0)x_{q2}(1), \ MS(x_{q2}(1)). \qquad (99)$$

Still, the best path $x_{q2}(0)x_{q2}(1)$ in Stack (99) does not terminate in one of the end nodes of the trellis diagram. Hence the best path $x_{q2}(0)x_{q2}(1)$ is extended.

Step 3. Computing the metrics of the paths $x_{q2}(0)x_{q2}(1)x_{q1}(2)$ and $x_{q2}(0)x_{q2}(1)x_{q2}(2)$, namely,

$$MS(x_{q1}(2)) = M(x_{q2}(1) \rightarrow x_{q1}(2)) + MS(x_{q2}(1)),$$
$$MS(x_{q2}(2)) = M(x_{q2}(1) \rightarrow x_{q2}(2)) + MS(x_{q2}(1)), \qquad (100)$$

replacing the best path and its metric in Stack (99) by these paths and their metrics, and then reordering all the paths according to their metrics, let us assume that the following stack is obtained:

$$x_{q1}(0), \ MS(x_{q1}(0)); \ x_{q2}(0)x_{q1}(1), \ MS(x_{q1}(1));$$

$$x_{q2}(0)x_{q2}(1)x_{q2}(2), \ MS(x_{q2}(2)); \qquad (101)$$

$$x_{q2}(0)x_{q2}(1)x_{q1}(2), \ MS(x_{q1}(2)).$$

Since the best path $x_{q2}(0)x_{q2}(1)x_{q1}(2)$ terminates in the final node $x_{q1}(2)$ in the trellis diagram, it is decided that the path $x_{q2}(0)x_{q2}(1)x_{q1}(2)$ was most probably followed by the target.

Having defined the stack sequential decoding based smoothing algorithm, its performance is going to be discussed in the following two sections.

2. *An Upper Bound for the Overall*
   Error Probability

Let us consider a target whose motion from time zero to time L is described by a trellis diagram T with M possible paths H_1, H_2, ..., and H_M from time zero to time L. Let H_m

and H_m, be two paths through the trellis diagram T such that H_m is the correct path (i.e., it is the path actually followed by the target) and H_m, is an incorrect path (i.e., a path which was not followed by the target) (Fig. 8). It can easily be verified that since at each step the SSDSA extends only the best path in the stack by only one branch; the path H_m, cannot be chosen as that most likely followed by the target (as the decision) [14]:

$$\text{if} \quad M\left(x_q^m(i)\right) > M\left(x_q^{m'}(j)\right) \quad \text{for all} \quad i \in S$$
$$\text{and some} \quad j \in S, \qquad (102)$$

or

$$\text{if} \quad \gamma_m > M\left(x_q^{m'}(j)\right) \quad \text{for some} \quad j \in S;$$

however, the path H_m, may be chosen as the decision

$$\text{if} \quad M\left(x_q^{m'}(j)\right) \geq \gamma_m \quad \text{for all} \quad j \in S, \qquad (103)$$

where $M\left(x_q^m(i)\right)$ is the metric of the portion between the nodes $x_q^m(0)$ and $x_q^m(i)$ of the path H_m, which is defined by Eq. (30), and

$$S \triangleq \{0, 1, 2, \ldots, L\}, \qquad \gamma_m \triangleq \min_{i \in S} M\left(x_q^m(i)\right). \qquad (104)$$

Let Γ_m be a subset of the observation space D such that

$$\Gamma_m \triangleq \left\{ z^L : \quad M\left(x_q^{m'}(j)\right) \geq \gamma_m \quad \text{for all} \quad j \in S \right.$$
$$\left. \text{and all} \quad m' \neq m \right\}. \qquad (105)$$

Since any incorrect path H_m, can be chosen as the decision only if Eq. (103) is valid, the set Γ_m contains the complement $\overline{D}_m$ of the decision region D_m for the path H_m (the decision region D_m is by definition the subset of the observation space such that whenever the observation sequence z^L falls within this subset, the SSDSA decides that the path H_m is that most likely followed

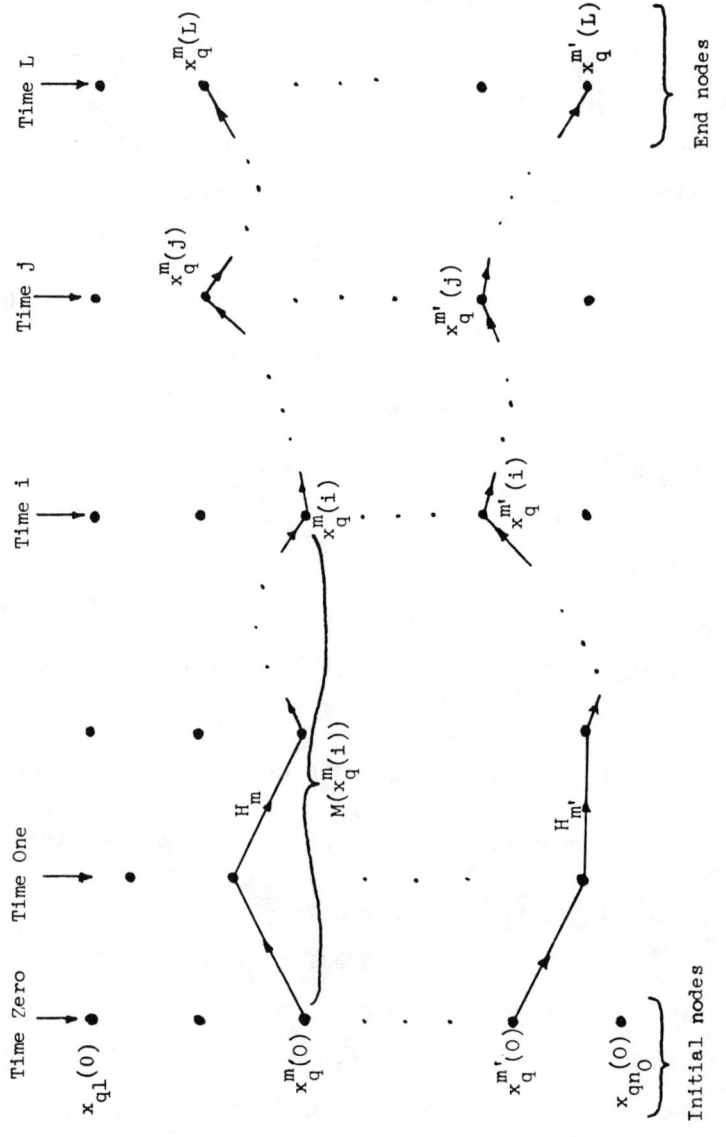

Fig. 8. The trellis diagram for the performance analysis of the SSDSA.

by the target). Let $\tilde{\Gamma}_m$ be another subset of the observation space D such that

$$\tilde{\Gamma}_m = \left\{ z^L: \quad M\left(x_q^{m'}(L)\right) \geq \gamma_m \quad \text{for all} \quad m' \neq m \right\}. \tag{106}$$

Since the inequality in Eq. (105) implies the inequality in Eq. (106) (the converse is not true), Γ_m is contained in $\tilde{\Gamma}_m$. Hence $\bar{D}_m$ is a subset of $\tilde{\Gamma}_m$. Therefore the error probability of the path H_m [see Eq. (32)] can be upper bounded as

$$P_{E_m}(H_1, H_2, \ldots, H_M) \leq \int_{z^L \in \tilde{\Gamma}_m} p\left(z^L | H_m\right) dz^L$$

$$\triangleq \int_{z^L} p\left(z^L | H_m\right) \phi(z^L) dz^L, \tag{107}$$

where

$$\phi(z^L) \triangleq \begin{cases} 1 & \text{if} \quad z^L \in \tilde{\Gamma}_m, \\ 0 & \text{elsewhere.} \end{cases} \tag{108}$$

Moreover, the function $\phi(z^L)$ can be upper bounded by

$$\phi(z^L) \leq \left[\sum_{m' \neq m} \exp \alpha\left(M\left(x_q^{m'}(L)\right) - \gamma_m\right) \right]^\rho$$

$$\text{for any} \quad \alpha, \rho \geq 0. \tag{109}$$

The reason that Eq. (109) is valid is as follows: if $z^L \in \tilde{\Gamma}_m$, it follows from Eq. (106) that there exists at least an $m' \neq m$ such that $M\left(x_q^{m'}(L)\right) - \gamma_m \geq 0$. Consequently, $\exp \alpha\left(M(x_q^{m'}(L)) - \gamma_m\right) \geq 1$ for any $\alpha \geq 0$. Therefore the summation in brackets in Eq. (109) is ≥ 1; obviously any nonnegative power, say ρ, of it is ≥ 1. On the other hand if $z^L \notin \Gamma_m$, the term in brackets in Eq. (109) is at least nonnegative. Hence the inequality in Eq. (109) is valid. Further, substituting Eq. (109) into

Eq. (107) yields

$$P_{E_m}(H_1, H_2, \ldots, H_M) \leq \int_{z^L} p\left(z^L \mid H_m\right) (\exp -\alpha\rho\gamma_m)$$

$$\times \left[\sum_{m' \neq m} \exp \alpha M\left(x_q^{m'}(L)\right) \right]^{\rho} dz^L$$

for any $\alpha, \rho \geq 0.$ \hfill (110)

Also from Eq. (104), we have

$$\exp -\alpha\rho\gamma_m \leq \sum_{i=0}^{L} \exp -\left[\alpha\rho M\left(x_q^m(i)\right)\right],$$ \hfill (111)

since at least one term in the summation is equal to the left-hand side of the inequality, and the other terms in the summation are at least nonnegative. Substituting Eq. (111) into Eq. (110), we obtain

$$P_{E_m}(H_1, H_2, \ldots, H_M) \leq \int_{z^L} p\left(z^L \mid H_m\right) \left[\sum_{i=0}^{L} \exp -\alpha\rho M\left(x_q^m(i)\right) \right]$$

$$\times \left[\sum_{m' \neq m} \exp \alpha M\left(x_q^{m'}(L)\right) \right]^{\rho} dz^L$$

for any $\alpha, \rho \geq 0.$ \hfill (112)

Moreover, by using Eqs. (25), (30), and the equality $\exp(\ln a)$ = a for all $a > 0$, Eq. (112) can be rewritten as

$$P_{E_m}(H_1, \ldots, H_M)$$

$$\leq \int_{z^L} \left[\left(\pi_0^m\right)^{-\alpha\rho} \prod_{k=1}^{L} p'\left(z(k) \mid x_q^m(k)\right) + \sum_{i=1}^{L} \left(\pi_0^m\right)^{-\alpha\rho} \prod_{k=1}^{i} \left(\pi_k^m\right)^{-\alpha\rho} \right.$$

$$\times \left. \left(p'\left(z(k) \mid x_q^m(k)\right)\right)^{1-\alpha\rho} \prod_{j=i+1}^{L} p'\left(z(j) \mid x_q^m(j)\right) \right]$$

[Eq. (113) continues]

$$\times \left\{ \sum_{m' \neq m} \left[\pi_0^{m'} \prod_{k=1}^{L} \pi_k^{m'} p' \Big(z(k) \,|\, x_q^{m'}(k) \Big) \right]^{\alpha} \right\}^{\rho} dz^{L}$$

$$\text{for any } \alpha, \ \rho \geq 0. \tag{113}$$

Moreover, by using the inequalities

$$\pi_k^{min} \leq \pi_k^{m} \leq \pi_k^{max} \qquad \text{for all } m \text{ and } k, \tag{114}$$

the bound in Eq. (113) can be upper bounded further so that we obtain the following bound for the error probability of the path H_m:

$$P_{E_m}(H_1, H_2, \ldots, H_M)$$

$$\leq \int_{z^L} \left[\left(\pi_0^{min} \right)^{-\alpha\rho} \prod_{k=1}^{L} p' \Big(z(k) \,|\, x_q^{m}(k) \Big) + \sum_{i=1}^{L} \left(\pi_0^{min} \right)^{-\alpha\rho} \prod_{k=1}^{i} \left(\pi_k^{min} \right)^{-\alpha\rho} \right.$$

$$\left. \times \left(p' \Big(z(k) \,|\, x_q^{m}(k) \Big) \right)^{1-\alpha\rho} \prod_{j=i+1}^{L} p' \Big(z(j) \,|\, x_q^{m}(j) \Big) \right]$$

$$\times \left\{ \sum_{m' \neq m} \left[\pi_0^{max} \prod_{k=1}^{L} \pi_k^{max} p' \Big(z(k) \,|\, x_q^{m'}(k) \Big) \right]^{\alpha} \right\}^{\rho} dz^{L}$$

$$\text{for any } \alpha, \ \rho \geq 0, \tag{115}$$

where $p' \Big(z(k) \,|\, x_q^{m}(k) \Big)$ is given by Eqs. (24a) and (24b); π_k^{min} and π_k^{max} are defined in Section II.E. Moreover, substituting this bound into Eq. (33), we obtain an upper bound for the overall error probability. Since it contains multidimensional integrals as in Eq. (115), it cannot be easily evaluated. Therefore, an ensemble upper bound on the overall error probability is considered next.

3. An Ensemble Upper Bound for the Overall Error Probability

Setting α equal to $1/(1 + \rho)$, the bound for the error probability of the path H_m in Eq. (115) can be rewritten as

$$P_{E_m}(T) \leq \int_{z^L} A_m(z^L)$$

$$\times \left\{ \sum_{m' \neq m} B_{m'}(z^L) \right\}^{\rho} dz^L \qquad \text{for any} \quad \rho \geq 0, \qquad (116)$$

where

$$A_m(z^L) \triangleq \left(\pi_0^{\min}\right)^{-\rho/(1+\rho)} \prod_{k=1}^{L} p'\left(z(k)\,|\,x_q^m(k)\right)$$

$$+ \sum_{i=1}^{L} \left\{ \left(\pi_0^{\min}\right)^{-\rho/(1+\rho)} \prod_{k=1}^{i} \left(\pi_k^{\min}\right)^{-\rho/(1+\rho)} \right.$$

$$\hspace{4cm} (117)$$

$$\left. \times\, p'\left(z(k)\,|\,x_q^m(k)\right)^{1/(1+\rho)} \prod_{j=i+1}^{L} p'\left(z(j)\,|\,x_q^m(j)\right) \right\},$$

$$B_{m'}(z^L) \triangleq \left(\pi_0^{\max}\right)^{1/(1+\rho)} \prod_{k=1}^{L} \left(\pi_k^{\max}\right)^{1/(1+\rho)} p'\left(z(k)\,|\,x_q^{m'}(k)\right)^{1/(1+\rho)}.$$

Averaging this bound over the ensemble E as in Section II.H.3, we obtain the following bound for the ensemble error probability of the path H_m:

$$\overline{P_{E_m}(T)} \leq \left[\prod_{k=1}^{L} \sum_{H_k^e \in \underset{\sim}{H}^e} Q\left(H_k^e\right)\right] \int_{z^L} A_m(z^L) \left\{ \sum_{m' \neq m} B_{m'}(z^L) \right\}^{\rho} dz^L$$

$$\hspace{8cm} (118)$$

$$= \int_{z^L} \left[\sum_{H_m^e \in \underset{\sim}{H}^e} Q\left(H_m^e\right) A_m(z^L) \right] \left\{ \prod_{\substack{k=1 \\ k \neq m}}^{L} \left(\sum_{H_k^e \in \underset{\sim}{H}^e} Q\left(H_k^e\right) \left[\sum_{m' \neq m} B_{m'}(z^L) \right] \right)^{\rho} \right\} dz^L$$

where ρ is any nonnegative number. Hence, restricting ρ in the interval [0, 1], and using Jensen's inequality (as in Section II.H.3), the term in braces in Eq. (118) can be upper bounded so that we obtain

$$\overline{P_{E_m}(T)} \leq \int_{z^L} \left[\sum_{H_m^e} Q\left(H_m^e\right) A_m(z^L) \right] \left[\sum_{m' \neq m} \sum_{H_{m'}^e} Q\left(H_{m'}^e\right) B_{m'}(z^L) \right]^\rho dz^L$$

$$= (M - L)^\rho \int_{z^L} \left[\sum_{H_m^e} Q\left(H_m^e\right) A_m(z^L) \right] \tag{119}$$

$$\times \left[\sum_{H_{m'}^e} Q\left(H_{m'}^e\right) B_{m'}(z^L) \right]^\rho dz^L \qquad \text{for any} \quad \rho \in [0, 1].$$

The last equality follows from the fact that the summations are performed over the same space $\underline{H}^e$. Further, using Eq. (42), we can easily obtain the following equalities:

$$\sum_{H_m^e} Q\left(H_m^e\right) A_m(z^L) = \left(\pi_0^{min}\right)^{-\rho/(1+\rho)} \prod_{k=1}^{L} \overline{p'(z(k)|x)}$$

$$+ \sum_{i=1}^{L} \left(\pi_0^{min}\right)^{-\rho/(1+\rho)} \prod_{k=1}^{i} \left(\pi_k^{min}\right)^{-\rho/(1+\rho)}$$

$$\times \overline{p'(z(k)|x)}^{1/(1+\rho)} \prod_{j=i+1}^{L} \overline{p'(z(j)|x)}, \tag{120}$$

$$\sum_{H_{m'}^e} Q\left(H_{m'}^e\right) B_{m'} \cdot (z^L)$$

$$= \left[\prod_{k=0}^{L} \left(\pi_k^{max}\right)^{1/(1+\rho)}\right] \prod_{k=1}^{L} \frac{1}{\overline{p'(z(k)|x)}^{1/(1+\rho)}} \qquad \text{for any} \quad \rho \in [0, 1]$$

where

$$\overline{p'(z(k)|x)} \triangleq \sum_{x \in X^e} q(x) p'(z(k)|x),$$

$$\overline{p'(z(k)|x)^{1/(1+\rho)}} \triangleq \sum_{x \in X^e} q(x) p'(z(k)|x)^{1/(1+\rho)}, \qquad (121)$$

where $p'(z(k)|x)$ is given by Eqs. (24a) and (24b); $q(\cdot)$ is an arbitrary probability density function on the set X^e, and X^e is defined in Section II.H.3. Substituting these equalities into Eq. (119) and changing the order of integrations and multiplications, we obtain the following bound for the ensemble error probability of the path H_m:

$$\overline{P_{E_m}(T)} \leq F\left\{\prod_{k=1}^{L} C_k + \sum_{i=1}^{L} \prod_{k=1}^{i} \left(\pi_k^{min}\right)^{-\rho/(1+\rho)} D_k \prod_{j=i+1}^{L} C_j\right\},$$

$$\text{for any} \quad \rho \in [0, 1], \qquad (122)$$

where

$$C_k \triangleq \int_{z(k)} \overline{p'(z(k)|x)} \left[\overline{p'(z(k)|x)^{1/(1+\rho)}}\right]^{\rho} dz(k),$$

$$D_k \triangleq \int_{z(k)} \left[\overline{p'(z(k)|x)^{1/(1+\rho)}}\right]^{1+\rho} dz(k),$$

$$F \triangleq (M-1)^{\rho} \left(\frac{\pi_0^{max}}{\pi_0^{min}}\right)^{\rho/(1+\rho)} \left[\prod_{k=1}^{L} \left(\pi_k^{max}\right)^{\rho/(1+\rho)}\right], \qquad (123)$$

where π_k^{max}, π_k^{min}, and M are as defined in Section II.E.

If the function $g(k, ., ., .)$ in the observation model under consideration, and the statistics of the observation noise $v(k)$ and the interference $I(k)$ (in the presence of interference) are time-invariant, then C_k and D_k are time invariant. Hence in this case, the bound in Eq. (122) becomes

$$\overline{P_{E_m}}(T) \leq F\left\{C_k^L + \sum_{i=1}^{L}\left[\prod_{k=1}^{i}\left(\pi_k^{min}\right)^{-\rho/(1+\rho)}\right]D_k^i C_k^{L-i}\right\}$$

for any $\rho \in [0, 1]$. (124)

Since the bound in Eq. (122) does not depend on the path index m, it follows from Eq. (44) that this bound is also an upper bound for the ensemble error probability $\overline{P}_E$ for the detection of the path most likely followed by the target being considered. The integrals in Eq. (122) may not easily be evaluated for all $\rho \in [0, 1]$. The easiest bound to calculate is the one with $\rho = 1$. This fact, along with the nice feature of the uniformly weighted ensemble bound (Theorem 4), leads us to using the bound with $\rho = 1$ and the uniform density function $q(\cdot)$ [that is, $q(\cdot) = 1/N^e$ where N^e is defined in Section II.H.3] in Eq. (122) as the performance measure of the SSDSA.

J. SUBOPTIMUM DECODING-BASED SMOOTHING ALGORITHM

As described earlier, in order to decide the path most likely followed by a target from time zero to time L, by using the OSDSA or the SSDSA, we first obtain a trellis diagram (denoted by T) for the target-motion model from time zero to time L and then use the VDA or a stack sequential decoding algorithm respectively. The number of paths in the trellis diagram depends on L as well as on n_0, m_1, m_2, ..., m_L, and

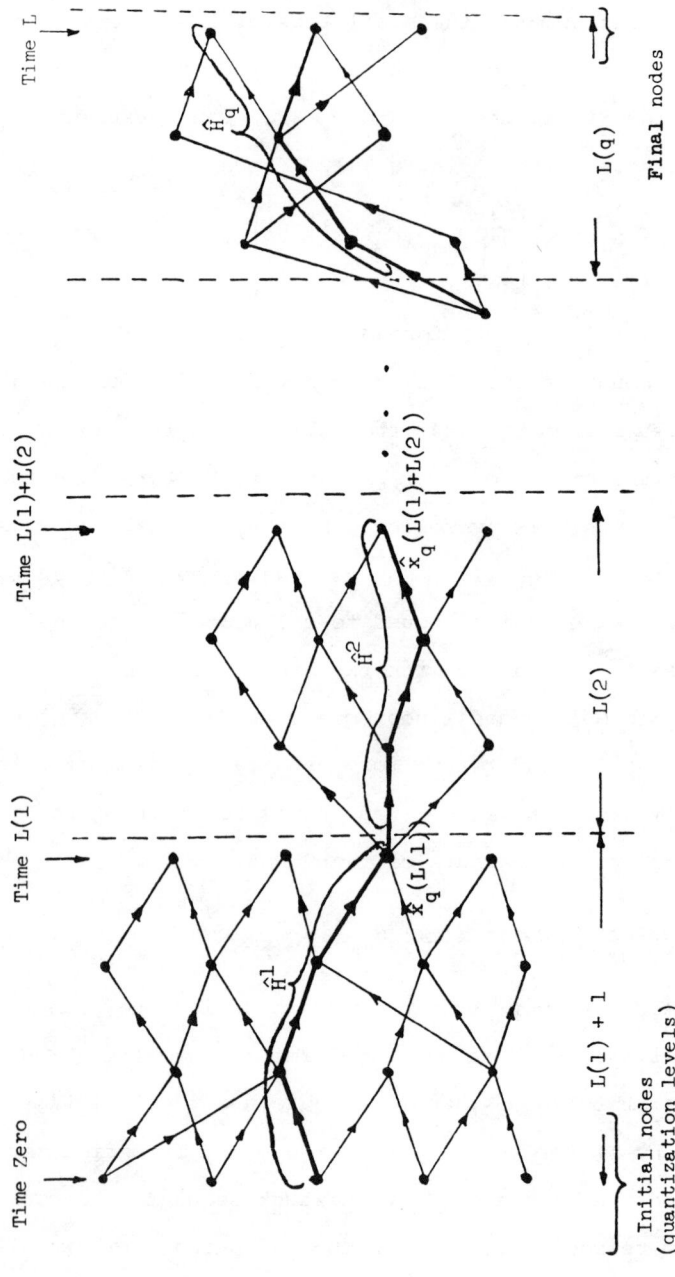

Fig. 9. The trellis diagram for the SDSA.

the gate sizes used to reduce the target-motion model to a
finite-state model where n_0 and m_k are the number of possible
values of the discrete random vectors $x_d(0)$ and $w_d(k)$ (Section
II.C). In particular, if L is very large (in other words, the
target needs to be tracked for a long time), the trellis diagram may contain a huge amount of paths. In such a case, the
SSDSA may require a very large memory for the storage of stacks
of searched paths and comparisons to reorder the paths in stacks
according to their metrics, whereas the ODSA requires a huge
memory and computation to compare the metrics of all paths in
the trellis diagram. Hence, in most target tracking problems,
both the ODSA and SSDSA becomes impractical. Therefore, a
smoothing algorithm which requires a constant memory to decide
the path most probably followed by the target from time zero to
any time L is needed. Such an algorithm is presented here. It
is based on a suboptimum decoding algorithm. Hence, it does not
minimize the overall error probability for the detection of the
path most likely followed by the target. This smoothing algorithm is called the suboptimum decoding-based smoothing algorithm (SDSA), which is as follows.

Preliminary step. After obtaining the first L(1) observations [i.e., the observation sequence from time 1 to time L(1)],
find the path most probably followed by the target from time
zero to time L(1) by using the ODSA. Let this path be $\hat{H}^1$
(Fig. 9).

Step 1. Obtain the next L(2) observations [i.e., the observation sequence from time L(1) + 1 to time L(1) + L(2)] and
assume that the target in fact followed the path $\hat{H}^1$ from time
zero to time L(1) [in other words, assume that the target was
at the end point, denoted by $\hat{x}_q(L(1))$, of $\hat{H}^1$ at time L(1) with

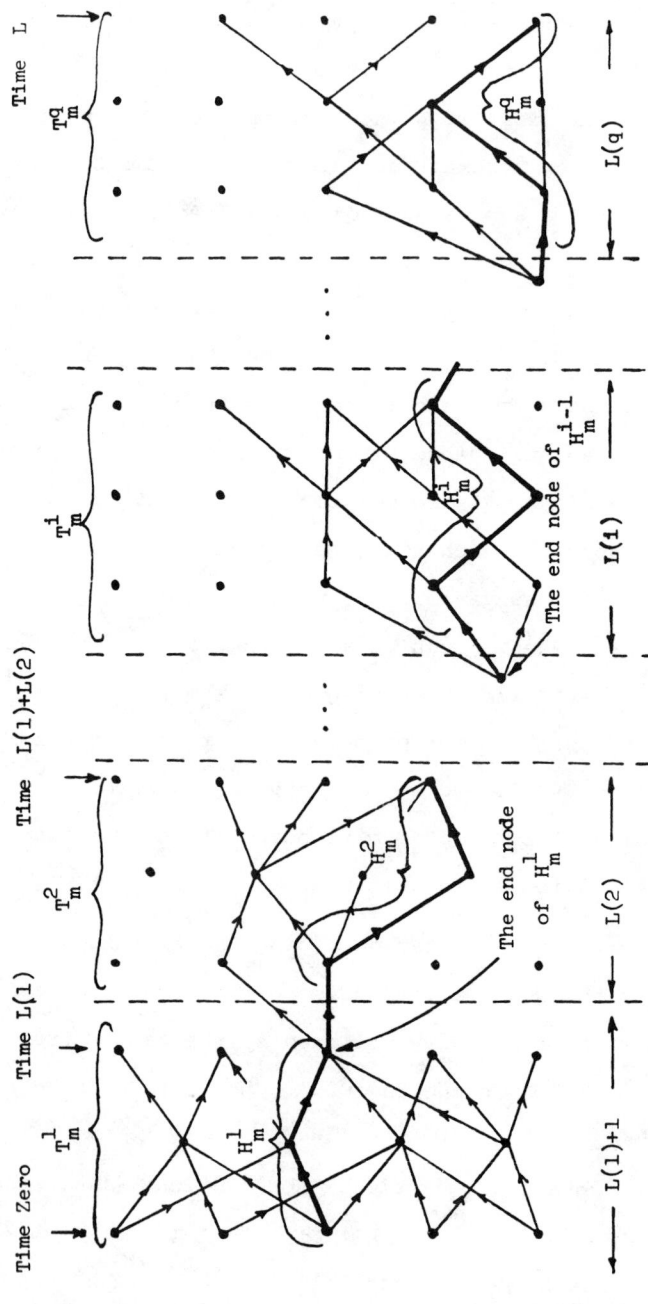

Fig. 10. The trellis diagram for the performance analysis of the SDSA.

probability 1]. Then, using the ODSA, find the path most likely followed by the target from time $L(1) + 1$ to $L(1) + L(2)$. Let this path be $\hat{H}^2$.

Step 2. Obtain the following $L(3)$ observations [i.e., the observation sequence from time $L(1) + L(2) + 1$ to time $L(1) + L(2) + L(3)$] and assume that the path $\hat{H}^1\hat{H}^2$ was actually followed by the target from time zero to time $L(1) + L(2)$ [in other words, assume that the target was at the end node, denoted by $\hat{x}_q(L(1) + L(2))$ of the path $\hat{H}^2$ with probability 1]. Then, using the ODSA, find the path most likely followed by the target from time $L(1) + L(2) + 1$ to time $L(1) + L(2) + L(3)$. Let this path be $\hat{H}^3$. The other steps similarly continue until $L = \sum_{k=1}^{q} L(k)$. At the end, decide that the path composed of the paths $\hat{H}^1$, $\hat{H}^2$, ..., $\hat{H}^q$ is the path ($\hat{H}$) most likely followed by the target from time zero to time L, i.e., $\hat{H} = \hat{H}^1\hat{H}^2 \cdots \hat{H}^q$, where q is the number of observation sequences considered from time zero to time L. The number $L(i)$ of observations in the ith observation sequence is chosen such that at the $(i - 1)$th step of the SDSA, the ODSA finds the path $\hat{H}^i$ without requiring a huge amount of memory and computation:

Let us divide the trellis diagram T into q parts such that the first part contains $L(1) + 1$ columns of quantization levels starting from time zero; the second part contains the next $L(2)$ columns of quantization levels; the third part contains the following $L(3)$ columns of quantization levels; and so on. Now we are going to define some symbols which are used in the following analyses (Fig. 10):

H_m The mth path through the trellis diagram T

H_m^i Portion of H_m in the ith part of T, where $i = 1$, 2, ..., q

$H_{\hat{m}}$ Path (through T) which the SDSA decides is most likely followed by the target when the target actually follows the path H_m

$H_{\hat{m}}^i$ Portion of $H_{\hat{m}}$ in the ith part of T, where $i = 1, 2, \ldots, q$

T_m^1 First part of the trellis diagram T (i.e., the part from time zero to time $L(1)$)

T_m^i Trellis diagram composed of the paths in the ith part of T which start at the end (final) node of H_m^{i-1}, where $i = 2, 3, \ldots, q$

T_m Trellis diagram composed of T_m^1, T_m^2, $\ldots$, T_m^q

M Number of possible paths (hypotheses) through T

M_m^i Number of possible paths through T_m^i

M_m Multiplication of M_m^1, M_m^2, $\ldots$, M_m^q

X_m^e Set of all quantization levels in T_m except the quantization levels at time zero

X_m^{le} Set of all quantization levels in T_m^1 except the quantization levels at time zero

X_m^{ie} Set of all quantization levels in T_m^i, where $i = 2, 3, \ldots, q$

$\underset{\sim}{H}_m^{ie}$ Set of all $L(i)$-tuples of X_m^{ie}, where $i = 1, 2, \ldots, q$

$\underset{\sim}{H}_m^e$ Cartesian product of the sets $\underset{\sim}{H}_m^{le}$, $\underset{\sim}{H}_m^{2e}$, $\ldots$, $\underset{\sim}{H}_m^{qe}$; that is, $\underset{\sim}{H}_m^e = \underset{\sim}{H}_m^{le} \times \underset{\sim}{H}_m^{2e} \times \cdots \times \underset{\sim}{H}_m^{qe}$

E_m^i Ensemble (or set) of all M_m^i-tuples of $\underset{\sim}{H}_m^{ie}$

E_m Ensemble (or set) of all M_m-tuples of $\underset{\sim}{H}_m^e$

$z^{L(i)}$ Observation sequence used at the $(i - 1)$ step of the SDSA (i.e., the observation sequence associated with the ith part of T): that is, $z^{L(i)} \triangleq \{z(n + 1), z(n + 2), \ldots, z(n + L(i))\}$

where $n \triangleq \sum_{k=1}^{i-1} L(k)$; by definition, $n = 0$ for $i = 1$. The per-formance of the SDSA is discussed in Sections II.J.1 and II.J.2.

1. An Upper Bound for the Overall Error Probability

Let us start calculating the correct detection probability of the path H_m, which is defined by the probability of choosing H_m as the path most likely followed by the target when the target in fact followed the path H_m. In other words, the correct detection probability $P_{C_m} (H_1, H_2, \ldots, H_M)$ is the probability of choosing $H_m \left(= H_m^1 \ H_m^2 \cdots H_m^q \right)$, given the observation sequence z^L and that the path H_m was followed by the target. Hence, we have

$$P_{C_m} (H_1, H_2, \ldots, H_M)$$

$$\triangleq \text{Prob} \left\{ H_{\hat{m}}^1 = H_m^1, \quad H_{\hat{m}}^2 = H_m^2, \quad \ldots, \quad H_{\hat{m}}^q = H_m^q \big| z^L, H_m \right\}$$

$$= \text{Prob} \left\{ H_{\hat{m}}^q = H_m^q \big| H_m, \ z^L, \ H_{\hat{m}}^1 = H_m^1, \ \ldots, \ H_{\hat{m}}^{q-1} = H_m^{q-1} \right\}$$

$$\times \text{Prob} \left\{ H_{\hat{m}}^{q-1} = H_m^{q-1} \big| H_m, \ z^L, \ H_{\hat{m}}^1 = H_m^1, \ \ldots, \ H_{\hat{m}}^{q-2} = H_m^{q-2} \right\}$$

$$\times \cdots \text{Prob} \left\{ H_{\hat{m}}^1 = H_m^1 \big| H_m, \ z^L \right\}. \tag{125}$$

On the other hand, for the SDSA, the probability of choosing H_m^i as the decision $\left(\text{i.e., } H_{\hat{m}}^i = H_m^i \right)$; given the observation sequence z^L, that the path H_m was followed by the target, and the correct detection of the paths $H_m^1, H_m^2, \ldots, H_m^{i-1}$; is the probability of choosing H_m^i given the observation sequence $z^{L(i)}$ and that the path H_m^i was followed by the target; that is,

$$\text{Prob} \left\{ H_{\hat{m}}^i = H_m^i \big| H_m, \ z^L, \ H_{\hat{m}}^1 = H_m^1, \ \ldots, \ H_{\hat{m}}^{i-1} = H_m^{i-1} \right\}$$

$$= \text{Prob} \left\{ H_{\hat{m}}^i = H_m^i \big| z^{L(i)}, \ H_m^i \right\}. \tag{126}$$

We also have

$$\text{Prob}\left\{H_{\hat{m}}^i = H_m^i \mid z^{L(i)}, \; H_m^i\right\} = 1 - P_{E_m}^i\left(T_m^i\right), \tag{127}$$

where

$$P_{E_m}^i(T_m^i) \triangleq \text{Prob}\left\{H_{\hat{m}}^i \neq H_m^i \mid z^{L(i)}, \; H_m^i\right\}, \tag{128}$$

which is the error probability of H_m^i when only the trellis diagram T_m^i is considered. First, substituting Eq. (127) into Eq.
(126) and then Eq. (126) into Eq. (125), we obtain the correct
detection probability of H_m:

$$P_{C_m}(H_1, H_2, \dots, H_M) = \prod_{i=1}^{q}\left[1 - P_{E_m}^i\left(T_m^i\right)\right]. \tag{129}$$

Moreover, from the definitions of the correct detection and
error probabilities of the path H_m (Definition 10), we have

$$P_{E_m}(H_1, H_2, \dots, H_M) \triangleq P_{E_m}(T)$$

$$= 1 - P_{C_m}(H_1, H_2, \dots, H_M). \tag{130}$$

Hence substituting Eq. (129) into Eq. (130), we get the error
probability of H_m as

$$P_{E_m}(H_1, H_2, \dots, H_M) = 1 - \prod_{i=1}^{q}\left[1 - P_{E_m}^i\left(T_m^i\right)\right]. \tag{131}$$

Furthermore, using a bound denoted by $B_m^i \, T_m^i$ for the error
probability of H_m^i, the error probability of H_m can be upper
bounded by

$$P_{E_m}(H_1, H_2, \dots, H_M) \leq 1 - \prod_{i=1}^{q}\left[1 - B_m^i\left(T_m^i\right)\right], \tag{132}$$

where, for example, $B_m^i\left(T_m^i\right)$ is a Gallager type bound [Eq. (41)]
when only the trellis diagram T_m^i is considered.

Therefore substituting the bound in Eq. (132) for the error probability of the path H_m in Eq. (33) yields a bound for the overall error probability.

2. *An Ensemble Upper Bound for the Overall Error Probability*

An ensemble bound for the overall error probability for the detection of the path most likely followed by the target being considered can be obtained as follows.

First, for every path, say H_m, through the trellis diagram T, a probability density function $Q_m^e(\cdot)$ is defined on the ensemble E_m such that

$$Q_m^e\left(T_m^e\right) = \prod_{i=1}^{q} Q_m^{ie}\left(T_m^{ie}\right) \qquad \text{for all} \quad T_m^e \in E_m, \tag{133}$$

where $Q_m^{ie}(\cdot)$ is a probability density function on the ensemble E_m^i, and $T_m^{ie} \in E_m^i$. Then, averaging the error probability of the path H_i over the ensemble E_m with respect to the probability density function in Eq. (133), the following ensemble error probability, denoted by $\overline{P_m(T)}$, of the path H_m is obtained:

$$\overline{P_{E_m}(T)} = \sum_{T_m^e \in E_m} Q_m^e\left(T_m^e\right) P_{E_m}(T). \tag{134}$$

Further, substituting Eqs. (131) and (133) into Eq. (134), changing the order of multiplications and summations, and recognizing that the summations are performed over the entire ensembles, we obtain

$$\overline{P_{E_m}(T)} = 1 - \prod_{i=1}^{q}\left[1 - \overline{P_{E_m}^i\left(T_m^i\right)}\right], \tag{135}$$

where

$$\overline{P_{E_m}^i\left(T_m^i\right)} \triangleq \sum_{T_m^{ie}\in E_m^i} Q_m^{ie}\left(T_m^{ie}\right) P_{E_m}\left(T_m^i\right), \tag{136}$$

which is the ensemble error probability of the path H_m^i when only the trellis diagram T_m^i is considered. In other words, it is the ensemble average of the error probability of H_m^i (when only the trellis diagram T_m^i is considered) over the ensemble E_m^i with respect to the probability density function $Q_m^{ie}(\cdot)$ (Section II.H.3). Similarly, using Eqs. (132)-(134), we obtain the following ensemble bound for the error probability of the path H_m:

$$\overline{P_{E_m}(T)} \leq 1 - \prod_{i=1}^{q}\left[1 - \overline{B_m^i\left(T_m^i\right)}\right], \tag{137}$$

where

$$\overline{B_m^i\left(T_m^i\right)} \triangleq \sum_{T_m^{ie}\in E_m^i} Q_m^{ie}\left(T_m^{ie}\right) B_m^i\left(T_m^i\right), \tag{138}$$

which is an ensemble average of the bound $B_m^i\left(T_m^i\right)$ for the error probability of the path H_m^i when only the trellis diagram T_m^i is considered. Further, substituting Eq. (135) for the ensemble path error probabilities in Eq. (44) yields an ensemble average of the overall error probability for the detection of the path most likely followed by the target under consideration (i.e., the ensemble overall error probability $\overline{P_E}$), that is,

$$\overline{P_E} = \sum_{m=1}^{M} p(H_m)\left\{1 - \prod_{i=1}^{q}\left[1 - \overline{P_{E_m}^i\left(T_m^i\right)}\right]\right\}$$

$$= 1 - \sum_{m=1}^{M} p(H_m) \prod_{i=1}^{q}\left[1 - \overline{P_{E_m}^i\left(T_m^i\right)}\right]. \tag{139}$$

Similarly, substituting Eq. (137) for the ensemble path error

probabilities in Eq. (44), we obtain the following ensemble

bound for the overall error probability for the detection of

the path most likely followed by the target under consideration:

$$\bar{P}_E \leq 1 - \sum_{m=1}^{M} p(H_m) \prod_{i=1}^{q} \left[1 - \overline{B_m^i(T_m^i)} \right], \qquad (140)$$

where $p(H_m)$ is as defined in Section II.F.

III. APPLICATIONS OF THE SMOOTHING ALGORITHMS

As discussed before, the new smoothing algorithms developed

in the previous sections can be used for (linear or nonlinear)

discrete models with arbitrary independent (of each other and

from sample to sample) random interference and noise.   Sections

III.A and III.B consider applications of these smoothing algo-

rithms to some discrete models with Gaussian noise and with

or without arbitrary random interference for the time interval

[0, L].

A. AN EXAMPLE WITH GAUSSIAN DISTURBANCE AND OBSERVATION NOISES

Section III.A deals with the following discrete models:

Motion model: $x(k + 1) = f(k, x(k), u(k), w(k))$,
$$(141)$$
Observation model: $z(k) = g(k, x(k)) + v(k)$,

where $x(0)$ is an $n \times 1$ initial-state Gaussian random vector

with mean m_0 and covariance R_0; $w(k)$ is a $p \times 1$ Gaussian dis-

durbance noise vector with zero mean and covariance $R_w(k)$;

$x(k)$, $u(k)$, $z(k)$, and $f(x, x(k), u(k), w(k))$ are as described

in Section II.A; $g(k, x(k))$ is an $r \times 1$ (linear or nonlinear)

vector; and $v(k)$ is an $r \times 1$ Gaussian observation noise vector with zero mean and covariance $R_v(k)$. Moreover, the random vectors $x(0)$, $w(j)$, $w(k)$, $v(l)$, and $v(m)$ are assumed to be independent for all j, k, l, m.

1. The Metric of a Branch

The observation $z(k)$ in Eq. (141) is a linear function of the Gaussian observation noise $v(k)$. Hence given that $x(k) = x_q^i(k)$, the conditional probability density function of $z(k)$ is a multivariate Gaussian density function. Thus we have

$$p'\left(z(k) \mid x_q^i(k)\right)$$

$$\underset{=}{\triangle} p\left(z(k) \mid x(k) = x_q^i(k)\right) \tag{142}$$

$$= \frac{\exp -\left\{\left[z(k) - g\left(k, x_q^i(k)\right)\right]^T R_v^{-1}(k)\left[z(k) - g\left(k, x_q^i(k)\right)\right]/2\right\}}{(2\pi)^{r/2}[\det R_v(k)]^{1/2}}.$$

Substituting this into Eq. (29) yields the metric of the branch between the nodes $x_q^i(k - 1)$ and $x_q^i(k)$, that is

$$M\left(x_q^i(k - 1) \rightarrow x_q^i(k)\right)$$

$$= \ln \pi_k^i - \ln\left\{(2\pi)^{r/2}[\det R_v(k)]^{1/2}\right\} \tag{143}$$

$$- \left[z(k) - g\left(k, x_q^i(k)\right)\right]^T R_v^{-1}(k)\left[z(k) - g\left(k, x_q^i(k)\right)\right]/2.$$

2. The Optimum Decoding-Based Smoothing Algorithm

a. A union upper bound. As mentioned before, the bound for the overall error probability given in Section II.H.2 is very complex to evaluate; hence an ensemble bound in Section II.H.3 was considered. However, for the models in Eq. (141), a union bound (which is very easy to evaluate) for the detection of the path most probably followed by the target (from time zero to time L) by using the ODSA can be derived as follows.

Let us now consider the set Γ_i defined by Eq. (34). Substituting Eq. (142) into the metric of the path H_i [i.e., Eq. (31)] and then substituting this metric into Eq. (34), we obtain the set Γ_i as

$$\Gamma_i = \left\{ z^L : \ M'(H_j) \geq M'(H_i) \quad \text{for some} \quad j \neq i \right\}$$

$$= \bigcup_{j \neq i} \left\{ z^L : \ M'(H_j) \geq M'(H_i) \right\}, \tag{144}$$

where

$$M'(H_i) \triangleq 2 \ln \pi_0^i$$

$$+ \sum_{k=1}^{L} \left\{ 2 \ln \pi_k^i - \left[z(k) - g\left(k, \ x_q^i(k)\right) \right]^T R_v^{-1}(k) \right.$$

$$\times \left. \left[z(k) - g\left(k, \ x_q^i(k)\right) \right] \right\}. \tag{145}$$

Recall that the set Γ_i contains the complement $\overline{D}_m$ of the decision region D_m for the hypothesis H_m. From this fact and the axioms of probability, we obtain the following bound for the error probability of the path H_i

$$P_{E_i}(H_1, H_2, \ldots, H_M) \leq \text{Prob}\left\{ z^L \in \Gamma_i | H_i \right\}$$

$$\leq \sum_{j \neq i} \text{Prob}\left\{ z^L : \ M'(H_j) \geq M'(H_i) | H_i \right\}. \tag{146}$$

Substituting Eq. (145) for $M'(H_i)$, we get

$$\text{Prob}\left\{ z^L : \ M'(H_j) \geq M'(H_i) | H_i \right\}$$

$$= \text{Prob}\{ J_{ij} \geq A_{ij} | H_i \}, \tag{147}$$

where

$$J_{ij} \triangleq \sum_{k=1}^{L} 2\left[g\left(k,\ x_q^i(k)\right) - g\left(k,\ x_q^i(k)\right)\right]^T R_v^{-1}(k)\,z(k),$$

$$A_{ij} \triangleq 2\ \ell n\left[\prod_{k=0}^{L} \frac{\pi_k^i}{\pi_k^j}\right] + \sum_{k=1}^{L} \left\{g^T\left(k,\ x_q^j(k)\right) R_v^{-1}(k)\,g\left(k,\ x_q^j(k)\right)\right.$$

$$\left. - g^T\left(k,\ x_q^i(k)\right) R_v^{-1}(k)\,g\left(k,\ x_q^i(k)\right)\right\}.$$

As noted, J_{ij} is a linear function of $z(k)$, which is a multi-variate Gaussian density function when H_i is given. Hence, the conditional density function of J_{ij} given H_i is a normal density function with

Mean $= E\{J_{ij}|H_i\}$

$$= \sum_{k=1}^{L} 2\left[g\left(k,\ x_q^j(k)\right) - g\left(k,\ x_q^i(k)\right)\right]^T R_v^{-1}(k)\,g\left(k,\ x_q^i(k)\right),$$

(148)

and

$$\mathrm{Var}\{J_{ij}|H_i\} = 4\ \sum_{k=1}^{L}\left[g\left(k,\ x_q^j(k)\right) - g\left(k,\ x_q^i(k)\right)\right]^T R_v^{-1}(k)$$

$$\times \left[g\left(k,\ x_q^j(k)\right) - g\left(k,\ x_q^i(k)\right)\right].$$

(149)

Therefore, we have

$$\mathrm{Prob}\{J_{ij} \ge A_{ij}|H_i\} = \int_{A_{ij}}^{\infty} [2\pi\ \mathrm{Var}\{J_{ij}|H_i\}]^{-1/2}$$

$$\times\ \exp -\left[\frac{[J_{ij} - E\{J_{ij}|H_i\}]^2}{2\ \mathrm{Var}\{J_{ij}|H_i\}}\right]dJ_{ij}$$

$$\triangleq Q\left(\frac{A_{ij} - E\{J_{ij}|H_i\}}{[\mathrm{Var}(J_{ij}|H_i)]^{1/2}}\right),$$

(150)

where $Q(\cdot)$ is sometimes referred to as the Gaussian integral function, which is defined by

$$Q(y) \triangleq \int_y^\infty (2\pi)^{-1/2} \exp(-x^2/2)\, dx. \tag{151}$$

Combining Eqs. (150, (147), and (146), we obtain the following upper bound for the error probability of the path H_m:

$$P_{E_i}(H_1, H_2, \ldots, H_M) \le \sum_{j \ne i} Q\left(\frac{A_{ij} - E\{J_{ij}|H_i\}}{\text{Var}\{J_{ij}|H_i\}^{1/2}}\right). \tag{152}$$

This bound is sometimes referred to as the union bound for the error probability of H_i. Further, substituting this bound for the path error probabilities in Eq. (33), we obtain a union upper bound for the overall error probability.

  b. *The upper bound for the overall error probability.*
Setting $\rho = 1$ in Eq. (41), and changing the order of multiplication and integration, we get

$$P_{E_i}(H_1, H_2, \ldots, H_M) \le \sum_{j \ne i} \left(\prod_{k=0}^L \frac{\pi_k^j}{\pi_k^i}\right)^{1/2}$$

$$\times \prod_{k=1}^L \int_{z(k)} \left[p'\left(z(k)|x_q^i(k)\right)\right. \tag{153}$$

$$\left. \times\, p'\left(z(k)|x_q^j(k)\right)\right]^{1/2} dz(k).$$

From Eqs. (142) and (A.4) (Appendix A), we have

$$B\left(x_q^i(k),\, x_q^j(k)\right) \triangleq \int_{z(k)} \left[p'\left(z(k)|x_q^i(k)\right)p'\left(z(k)|x_q^j(k)\right)\right]^{1/2} dz(k)$$

$$= \exp\left\{-\frac{1}{8}\left[g\left(k,\, x_q^i(k)\right) - g\left(k,\, x_q^j(k)\right)\right]^T\right.$$

$$\left. \times\, R_v^{-1}(k)\left[g\left(k,\, x_q^i(k)\right) - g\left(k,\, x_q^j(k)\right)\right]\right\}. \tag{154}$$

Hence, substituting Eq. (154) into Eq. (153), we obtain the following bound for the error probability of the path H_i:

$$P_{E_i}(H_1, H_2, \ldots, H_M)$$

$$\leq \sum_{j \neq i} \left(\prod_{k=0}^{L} \frac{\pi_k^j}{\pi_k^i} \right)^{1/2} \prod_{k=1}^{L} B\left(x_q^i(k), x_q^j(k) \right). \tag{155}$$

Substituting this bound for the path error probabilities in Eq. (33) yields a bound for the overall error probability.

$c.$ *The ensemble upper bound for the overall error probability.* The ensemble bound in Eq. (82) is used as the performance measure of the ODSA. Using Eq. (141) and (A.4) in this bound, we obtain the following ensemble bound for the error probability for the detection of the path most likely followed by the target under consideration:

$$\overline{P}_E \leq D \prod_{k=1}^{L} \left\{ \sum_{x_1 \in X^e} \sum_{x_2 \in X^e} B(x_1, x_2) \right\}, \tag{156}$$

where D and $B(x_1, x_2)$ are given by Eqs. (83) and (154), respectively and X^e is defined in Section II.H.3. If the function $g(k,.)$ in the observation model in Eq. (141) and the statistics of the observation noise $v(k)$ [that is, $R_v(k)$] are time invariant, then the ensemble bound in Eq. (156) becomes

$$\overline{P}_E \leq D \left\{ \sum_{x_1 \in X^e} \sum_{x_2 \in X^e} B(x_1, x_2) \right\}^{L}. \tag{157}$$

3. *The Stack Sequential Decoding-Based Smoothing Algorithm*

$a.$ *The ensemble upper bound for the overall error probability.* In the bound in Eq. (122), substituting 1 for ρ and $1/N^e$ for $q(x)$ for all x, we obtain the following ensemble bound for the overall error probability:

$$\overline{P}_E \le F \left\{ \prod_{k=1}^{L} c_k + \sum_{i=1}^{L} \prod_{k=1}^{i} \left(\pi_k^{min} \right)^{-1/2} D_k \prod_{j=i+1}^{L} c_j \right\}, \tag{158}$$

where F is defined by setting $\rho = 1$ in Eq. (123), and

$$c_k \triangleq \int_{z(k)} \overline{[p'(z(k)|x_1)]} \left[\overline{p'(z(k)|x_2))^{1/2}} \right] dz(k), \tag{159}$$

$$D_k \triangleq \int_{z(k)} \left[\overline{(p'(z(k)|x))^{1/2}} \right]^2 dz(k). \tag{160}$$

Substituting Eq. (121) in c_k and then using Eq. (142) and Eq. (A.5) from Appendix A, we get

$$c_k = \left(\frac{1}{N^e} \right)^2 \frac{(2/9\pi)^{r/4}}{(\det R_v(k))^{1/4}}$$

$$\times \left[\sum_{x_1 \in X^e} \sum_{x_2 \in X^e} \exp \left\{ -\frac{1}{6} [g(k, x_1) - g(k, x_2)]^T R_v^{-1}(k) \right. \right.$$

$$\left. \left. \times [g(k, x_1) - g(k, x_2)] \right\} \right]. \tag{161}$$

Similarly, substituting Eq. (121) in D_k and using Eq. (142) and (A.4), we obtain

$$D_k = \left(\frac{1}{N^e} \right)^2 \left\{ \sum_{x_1 \in X^e} \sum_{x_2 \in X^e} B(x_1, x_2) \right\}, \tag{162}$$

where $B(x_1, x_2)$ is given by Eq. (154); N^e and X^e are defined in Section II.H.3.

If the covariance matrix $R_v(k)$ of the observation sequence $v(k)$ and the function $g(k,.)$ are time invariant, then the bound in Eq. (158) becomes

$$\overline{P}_E \le F \left\{ c_k^L + \sum_{i=1}^{L} \left[\prod_{k=1}^{i} \left(\pi_k^{min} \right)^{-1/2} \right] D_k^i c_k^{L-i} \right\}. \tag{163}$$

B. *AN EXAMPLE WITH INTERFERENCE*
    AND GAUSSIAN DISTURBANCE
    AND OBSERVATION NOISES

In Section III.B, we consider the following models:

(1) *Motion model:* $x(k + 1) = f(k, x(k), u(k), w(k)$,

(2) *Observation model:* $z(k) = g(k, x(k), I(k))$

$$+ h(k, x(k), I(k))v(k),$$

$$(164)$$

where $x(0)$, $x(k)$, $u(k)$, $w(k)$, $z(k)$ and $f(k, x(k), u(k), w(k))$
are as described in Section III.A; $g(k, x(k), I(k))$ and
$h(k, x(k), I(k))$ are $r \times 1$- and $r \times l$-dimensional (linear or
nonlinear) matrices, respectively; $v(k)$ is an $l \times 1$ Gaussian
observation noise vector with zero mean and covariance $R_v(k)$;
and $I(k)$ is an $m \times 1$ interference vector with known statistics.
Furthermore, the following assumptions are made:

(1) The random vectors $x(0)$, $w(j)$, $w(k)$, $v(l)$, $v(m)$, $I(n)$,
and $I(p)$ are independent for all j, k, l, m, n, p.

(2) $\left[h(k, x(k), I(k))R_v(k)h^T(k, x(k), I(k))\right]^{-1}$ exists for
all k.

1. *The Metric of a Branch*

Let us consider the observation model in Eq. (164). The
observation $z(k)$ is a linear function of the normal observation
noise vector $v(k)$. Therefore, the conditional probability
density function of $z(k)$, given that $x(k) = x_q^i(k)$ and $I(k)$, is
a multivariate normal density function, namely,

$$p\left(z(k)\,|x_q^i(k), I(k)\right) \triangleq p\left(z(k)\,|x(k) = x_q^i(k), I(k)\right)$$

$$= A \exp -(B/2),\qquad (165)$$

where

$$A \triangleq (2\pi)^{-r/2} \left\{ \det \left[h\left(k, \ x_q^i(k), \ I(k)\right) R_v(k) h^T\left(k, \ x_q^i(k), \ I(k)\right) \right] \right\}^{-1/2},$$

$$B \triangleq \left[z(k) - g\left(k, \ x_q^i(k), \ I(k)\right) \right]^T$$

$$\times \left[h\left(k, \ x_q^i(k), \ I(k)\right) R_v(k) h^T\left(k, \ x_q^i(k), \ I(k)\right) \right]^{-1}$$

$$\times \left[z(k) - g\left(k, \ x_q^i(k), \ I(k)\right) \right]. \tag{166}$$

From Eqs. (24a) and (24b), we have   !

$$p'\left(z(k) \,|\, x_q^i(k)\right)$$

$$= \begin{cases} \displaystyle\int_{I(k)} p\left(z(k) \,|\, x_q^i(k), \ I(k)\right) p(I(k)) dI(k), & \text{using Eq. (9),} \\[4pt] & \hspace{2cm} (167) \\[4pt] \displaystyle\sum_{l=1}^{r_k} p\left(z(k) \,|\, x_q^i(k), \ I_{dl}(k)\right) p(I_{dl}(k)), & \text{using Eq. (21),} \end{cases}$$

where $p\left(z(k) \,|\, x_q^i(k), \ I_{dl}(k)\right) \triangleq p\left(z(k) \,|\, x_q^i(k), \ I(k) = I_{dl}\right)$, which

is given by Eq. (165). Substituting Eq. (167) into Eq. (29)

yields the metric of the branch between the nodes $x_q^i(k - 1)$ and

$x_q^i(k)$; that is,

$$M\left(x_q^i(k - 1) \rightarrow x_q^i(k)\right) = \ln \pi_k^i + \ln p'\left(z(k) \,|\, x_q^i(k)\right).$$

2. *The Optimum Decoding-Based*
    Smoothing Algorithm

    a. *An upper bound for the overall error probability.* Set-

ting $\rho = 1$ in Eq. (41) yields the bound in Eq. (153). Hence

substituting Eq. (167) into Eq. (153), we obtain a bound for

the error probability of the path H_i.

    If the observation model in Eq. (164) is approximated by

Eq. (21), then an upper bound, which is very easy to evaluate,

for the error probability of H_i can be obtained as follows.

Substitute the second equality in Eq. (167) into Eq. (153) and then use the following inequality

$$\left(\sum_i a_i\right)^{\lambda} \leq \sum_i a_i^{\lambda} \qquad \text{for any } a_i \geq 0 \text{ and } \lambda \in [0, 1], \qquad (168)$$

to obtain the following bound for the error probability of the path H_i

$$P_{E_i}(H_1, \ldots, H_M)$$

$$\leq \sum_{j \neq i} \left(\prod_{k=0}^{L} \frac{\pi_k^j}{\pi_k^i}\right)^{1/2} \prod_{k=1}^{L} \left\{ \sum_{i=1}^{r_k} \sum_{j=1}^{r_k} [p(I_{di}(k))p(I_{dj}(k))]^{1/2} \right.$$

$$\times \int_{z(k)} \left[p\left(z(k) \mid x_q^i(k), I_{di}(k)\right) \right. \qquad (169)$$

$$\left. \times p\left(z(k) \mid x_q^j(k), I_{dj}(k)\right) \right]^{1/2} dz(k) \right\},$$

where r_k is the number of possible values of $I_d(k)$ (Section II.G). Using Eqs. (165) and (A.1), we can obtain the following equality for the integral in Eq. (169)

$$\int_{z(k)} \left[p\left(z(k) \mid x_q^i(k), I_{di}(k)\right) p\left(z(k) \mid x_q^j(k), I_{dj}(k)\right) \right]^{1/2} dz(k)$$

$$= A' \exp(+B'/4), \qquad (170)$$

where

$$A' \triangleq \left\{ \det\left[2\left(R_i^{-1} + R_j^{-1}\right)^{-1} \right] \right\}^{1/2} \Big/ (\det R_i)^{1/4} (\det R_j)^{1/4},$$

$$B' \triangleq \left\{ b_{ij}^T \left(R_i^{-1} + R_j^{-1}\right)^{-1} b_{ij} \right.$$

$$- g^T\left(k, x_q^i(k), I_{di}(k)\right) R_i^{-1} g\left(k, x_q^i(k), I_{di}(k)\right)$$

$$\left. - g^T\left(k, x_q^j(k), I_{dj}(k)\right) R_j^{-1} g\left(k, x_q^j(k), I_{dj}(k)\right) \right\},$$

$$R_i \triangleq h\left(k, x_q^i(k), I_{di}(k)\right) R_v(k) h^T\left(k, x_q^i(k), I_{di}(k)\right),$$

$$R_j \triangleq h\left(k, x_q^j(k), I_{dj}(k)\right) R_v(k) h^T\left(k, x_q^j(k), I_{dj}(k)\right),$$

$$b_{ij} \triangleq R_i^{-1} g\left(k, x_q^i(k), I_{di}(k)\right) + R_j^{-1} g\left(k, x_q^j(k), I_{dj}(k)\right).$$

$$(171)$$

Substituting the bound in Eq. (169) for the error probability of H_i in Eq. (33), we get an upper bound for the overall error probability.

   b. *An ensemble upper bound for the overall error probability.* Substituting Eq. (165) in Eq. (167) and then substituting Eq. (167) in Eq. (82), we obtain an ensemble bound for the detection of the path most likely followed by the target by use of the ODSA.

If the observation model in Eq. (164) is approximated by Eq. (21), then the ensemble bound mentioned above can be further upper bounded by using the inequality in Eq. (168) so that we can easily obtain the following ensemble bound, which is easy to evaluate, for the overall error probability:

$$\overline{P}_E \leq D \prod_{k=1}^{L} \left\{ \sum_{x_1 \in x^e} \sum_{x_2 \in x^e} \sum_{i=1}^{r_k} \sum_{j=1}^{r_k} [p(I_{di}(k))p(I_{dj}(k))]^{1/2} \right.$$

$$\left. \times \int_{z(k)} [p(z(k)|x_1, I_{di}(k))p(z(k)|x_2, I_{dj})]^{1/2} dz(k) \right\}, \quad (172)$$

where D is defined by Eq. (83), and the integral is given by Eq. (170).

If the functions $g(k, ., .)$ and $h(k, ., .)$ in the observation model of Eq. (164), the covariance $R_v(k)$ of the observation noise $v(k)$, and the statistics of the interference $I(k)$

are time invariant, then the bound in Eq. (172) becomes

$$\overline{P}_E \le D \Bigg\{ \sum_{x_1 \in X^e} \sum_{x_2 \in X^e} \sum_{i=1}^{r_k} \sum_{j=1}^{r_k} [p(I_{di}(k))p(I_{dj}(k))]^{1/2}$$

$$\times \int_{z(k)} [p(z(k)|x_1, I_{di}(k))p(z(k)|x_2, I_{dj}(k))]^{1/2} dz(k) \Bigg\}^L .$$

$$(173)$$

3. *The Stack Sequential Decoding-Based Smoothing Algorithm*

   a. *An ensemble upper bound for the overall error prob-ability.* In the bound in Eq. (122), substituting one for ρ and $1/N^e$ for $q(x)$ for all x, and then using Eq. (167), we obtain an ensemble bound for the overall error probability for the detection of the path most likely followed by the target being considered by the SSDSA.

   If the observation model in Eq. (164) is approximated by Eq. (21), then the ensemble bound mentioned above can be further upper bounded by using the inequality in Eq. (168) so that we can obtain an ensemble bound which is easy to evaluate as follows. Using the inequality in Eq. (168), we get

$$[p'(z(k)|x)]^{1/2} \le \sum_{i=1}^{r_k} [p(z(k)|x, I_{di}(k))]^{1/2}$$

$$\times [p(I_{di}(k))]^{1/2} . \qquad (174)$$

Then, substituting this inequality into C_k and D_k in Eq. (123), we obtain

$$
\left. C_k \right|_{\substack{\rho=1 \\ q(\cdot)=1/N^e}} \le \left(\frac{1}{N^e}\right)^2 \left\{ \sum_{i=1}^{r_k} \sum_{j=1}^{r_k} p(I_{di}(k)) \left[p(I_{dj}(k)) \right]^{1/2} \right.
$$

$$
\times \sum_{x_1 \in x^e} \sum_{x_2 \in x^e} \left[\int_{z(k)} p(z(k)|x_1, I_{di}(k)) \right.
$$

$$
\left. \left. \times (p(z(k)|x_2, I_{dj}(k)))^{1/2} dz(k) \right] \right\} \triangleq C_k^b, \qquad (175)
$$

$$
\left. D_k \right|_{\substack{\rho=1 \\ q(\cdot)=1/N^e}} \le \left(\frac{1}{N^e}\right)^2 \left\{ \sum_{i=1}^{r_k} \sum_{j=1}^{r_k} \left[p(I_{di}(k)) p(I_{dj}(k)) \right]^{1/2} \right.
$$

$$
\times \sum_{x_1 \in x^e} \sum_{x_2 \in x^e} \int_{z(k)} \left[p(z(k)|x_1, I_{di}(k)) \right.
$$

$$
\left. \left. \times p(z(k)|x_2, I_{dj}(k)) \right]^{1/2} dz(k) \right\} \triangleq D_k^b, \qquad (176)
$$

where the integral in Eq. (176) is given by Eq. (170), and the integral in C_k can be evaluated by using Eqs. (165) and (A.3) so that we can obtain

$$
\int_{z(k)} p(z(k)|x_1, I_{di}(k)) \left[p(z(k)|x_2, I_{dj}(k)) \right]^{1/2} dz(k)
$$

$$
= A' \exp(B'/4), \qquad (177)
$$

where

$$
A' \triangleq \left\{ \det\left[\left(R_1^{-1} + \tfrac{1}{2} R_2^{-1} \right)^{-1} \right] \right\}^{1/2} \Big/ (2\pi)^{r/4} (\det R_1)^{1/2} (\det R_2)^{1/4},
$$

$$
B' \triangleq b_{12}'^{T} \left[2R_1^{-1} + R_2^{-1} \right]^{-1} b_{12}' - 2g^T(k, x_1, I_{di}(k)) R_i^{-1}
$$

$$
\times g(k, x_1, I_{di}(k))
$$

$$
- g^T(k, x_2, I_{dj}(k)) R_j^{-1} g(k, x_2, I_{dj}(k)),
$$

[Eq. continues]

$$b_{12}' \triangleq 2R_1^{-1}g(k, x_1, I_{di}(k)) + R_2^{-1}g(k, x_2, I_{dj}(k)),$$

$$R_1 \triangleq h(k, x_1, I_{di}(k))R_v(k)h^T(k, x_1, I_{di}(k)),$$

$$R_2 \triangleq h(k, x_2, I_{dj}(k))R_v(k)h^T(k, x_2, I_{dj}(k)).$$

Finally, substituting these bounds for C_k and D_k in Eq. (122), we obtain the following ensemble bound for the overall error probability:

$$\bar{P}_E \leq F\left\{ \prod_{k=1}^{L} C_k^b + \sum_{i=1}^{L} \prod_{k=1}^{i} \left(\pi_k^{min} \right)^{-1/2} D_k^b \prod_{j=i+1}^{L} C_j^b \right\}, \qquad (178)$$

where F is given by setting $\rho = 1$ in Eq. (123). If the functions $g(k, ., .)$ and $h(k, ., .)$ in Eq. (164), the covariance matrix $R_v(k)$ of the observation noise $v(k)$, and the statistics of the interference $I(k)$ are time invariant, then C_k^b and D_k^b become time invariant. Hence, in this case, the bound in Eq. (178) can be rewritten as follows:

$$\bar{P}_E \leq F\left\{ \left(C_k^b \right)^L + \sum_{i=1}^{L} \left[\prod_{k=1}^{i} \left(\pi_k^{min} \right)^{-1/2} \right] \left(D_k^b \right)^i \left(C_k^b \right)^{L-i} \right\}. \qquad (179)$$

IV. NUMERICAL EXPERIMENTS

The purpose of simulation was to find out how well the smoothing algorithms developed in Section II perform both in a clear environment and in the presence of interference. In a clear environment, the aim was to compare the smoothing algorithms with the Kalman filter algorithm for linear discrete models and the extended Kalman filter algorithm for nonlinear discrete models. However, in the presence of interference, the smoothing algorithms may not be compared with the (extended) Kalman filter algorithm since it cannot handle the case of

interference. Therefore, the purpose was to discover how good
the estimates produced by the smoothing algorithms are and also
to observe the estimates obtained by the (extended) Kalman
filter algorithm (which considers only observation noise, i.e.,
with zero interference). Simulations were done for both linear
and nonlinear discrete models with or without interference.

For all simulations, the IBM Systems/370 Model 3033, For-
tran IV, and IMSL library were used. For each simulation, the
disturbance noise $w(k)$, observation noise $v(j)$, initial state
$x(0)$, and interference $I(l)$ (i.e., in the presence of interference) were taken to be white Gaussian and also independent
of each other. For a discrete random variable (with a given
number of possible values) which approximates the Gaussian
random variable with mean μ and variance σ^2, that of Eq. (B.9)
in Appendix B was used. In addition, the approximate observation model of Eq. (21) was used in all the cases of interference.

Simulation results are presented in Figs. 11a-22c. At the
top left corner of each figure, the models used, noise statistics, gate size, and numbers of possible values of the dis-
crete random variables $w_d(\cdot)$, $I_d(\cdot)$, and $x_d(0)$ [which approximate the disturbance noise $w(\cdot)$, interference $I(\cdot)$, and initial
state $x(0)$] are provided.

The following abbreviations and terms are also used in
Figs. 11a-22c:

AAEK   Average absolute error for the (extended) Kalman
filter estimates. The absolute error at time
j and the average absolute error are defined
as follows:

ABSOLUTE ERROR (at time j) $\triangleq |x(j) - \hat{x}_k(j|j)|$,

$$(180)$$

$$\text{AAEK} \triangleq (L + 1)^{-1} \sum_{j=0}^{L} |x(j) - \hat{x}_k(j|j)|, \quad (181)$$

where L is the time up to and including which the target was tracked, $\hat{x}_k(j|j)$ is the (extended) Kalman estimate of the state x(j), given the observation sequence from time 1 to time j.

AAEOP   Average absolute error for the estimates obtained by the smoothing algorithm used. The absolute error at time j and the average absolute error are defined as follows:

ABSOLUTE ERROR (at time j) $\triangleq |x(j) - \hat{x}_s(j)|$,

$$\qquad (182)$$

$$\text{AAEOP} \triangleq (L + 1)^{-1} \sum_{j=0}^{L} |x(j) - \hat{x}_s(j)|, \quad (183)$$

where L is as just defined, and $\hat{x}_s(j)$ is the estimate of the state x(j) obtained by the smoothing algorithm used.

ACTUAL   Actual values of the states

BOUND   Bound in Eq. (157) for the ODSA using an example without interference, the bound in Eq. (173) for the ODSA using an example with interference, the bound in Eq. (163) for the SSDSA using an example without interference, or the bound in Eq. (179) for the SSDSA using an example with interference.

ER.COV.   Estimation error covariance matrix for the (extended) Kalman filter algorithm. The estimation error covariance matrix at time j is defined by

$$E\{(x(j) - \hat{x}_k(j|j))(x(j) - \hat{x}_k(j|j))^T\},$$

where E{ } is the expectation. Obviously in a scalar case, this matrix reduces to the mean-square error

EX.KAL.   Extended Kalman filter algorithm used

$E(A(\cdot))$ Expectation of the random variable $A(\cdot)$

GATE SIZE   Gate size used for the quantization

KALMAN   Kalman filter algorithm used

NUM. OF DISC. FOR $A(\cdot)$

>   Number of possible values of the discrete random
>   variable (used for the simulation) which ap-
>   proximates the random variable $A(\cdot)$

OPD   ODSA used

SOD   SDSA used

SSD   SSDSA used

$VAR(A(\cdot))$ Variance of the random variable $A(\cdot)$.

A. THE OPTIMUM DECODING-BASED SMOOTHING ALGORITHM

Many examples were simulated with the ODSA and the (extended) Kalman filter algorithm. The simulation results of some of
them are presented in Figs. 11a, 11b, 11c, 12a, 12b, 12c, 13a,
13b, 13c, 14a, 14b, 14c. The simulations were stopped after
seven steps because of the exponentially growing memory requirement of the ODSA. For each example, the simulation results are presented in groups of three (Figs. 11a-11c, 12a-12c,
13a-13c, and 14a-14c).

Figures 11a, 12a, 13a, and 14a present the variations of
the actual values, the (extended) Kalman estimates, and the
ODSA estimates of the states versus time. The ODSA estimates
are the ones obtained by the ODSA. Figures 11b, 12b, 13b, and
14b present the variations of the estimation error covariance
matrix [for the (extended) Kalman filter algorithm] versus time
as well as the bound in Eq. (157) (if the example has no interference) or the bound in Eq. (173) (if the example contains
interference). This bound is used as the performance measure

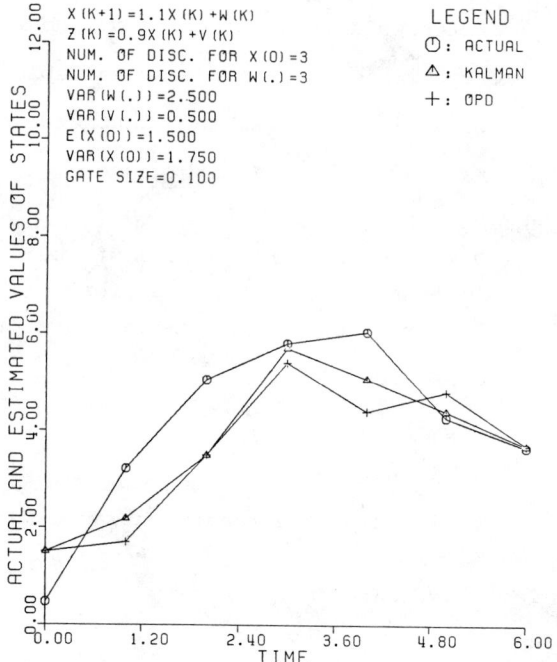

Fig. 11a.   Actual and estimated values of states.

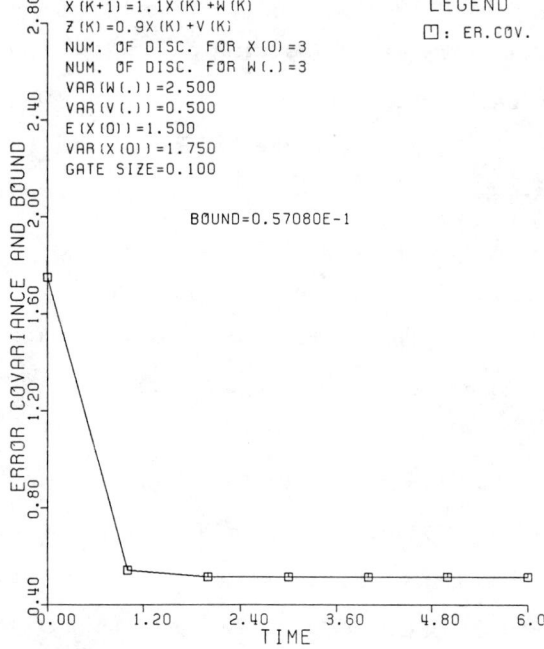

Fig. 11b.   Error covariance and bound.

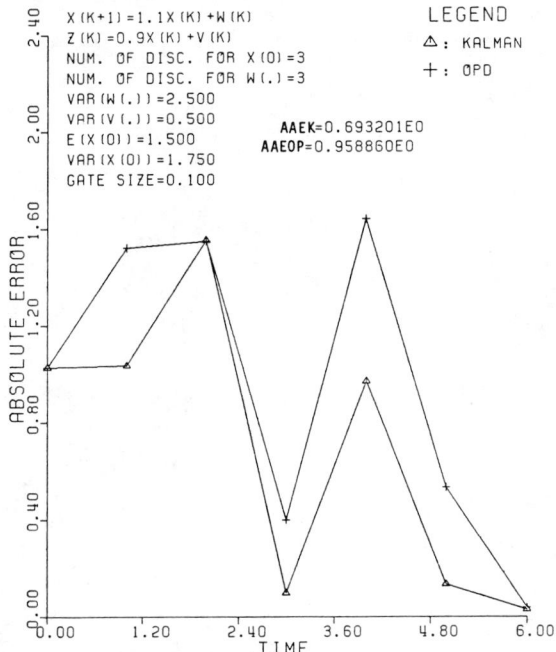

Fig. 11c. *Absolute and average absolute errors.*

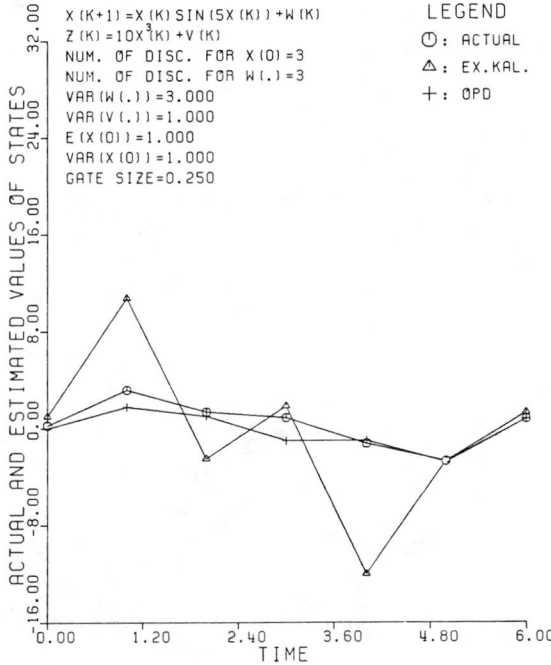

Fig. 12a. *Actual and estimated values of states.*

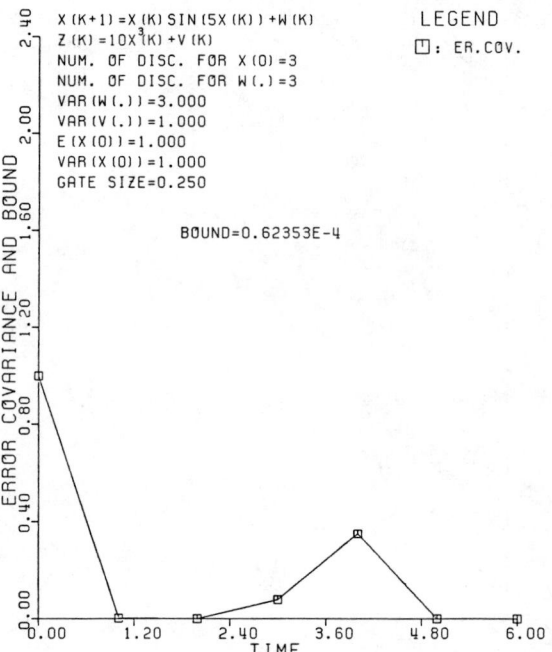

Fig. 12b.   Error covariance and bound.

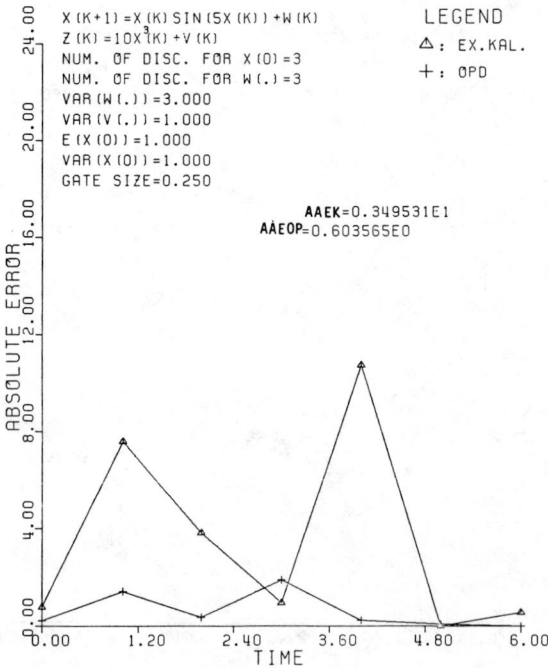

Fig. 12c.   Absolute and average absolute errors.

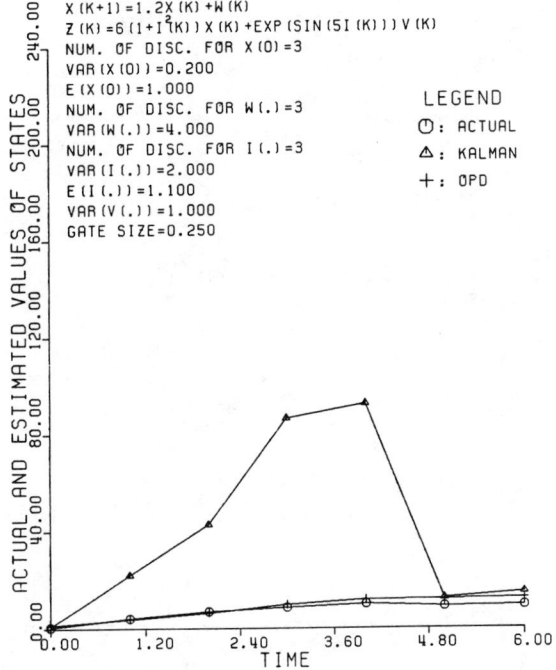

Fig. 13a.   Actual and estimated values of states.

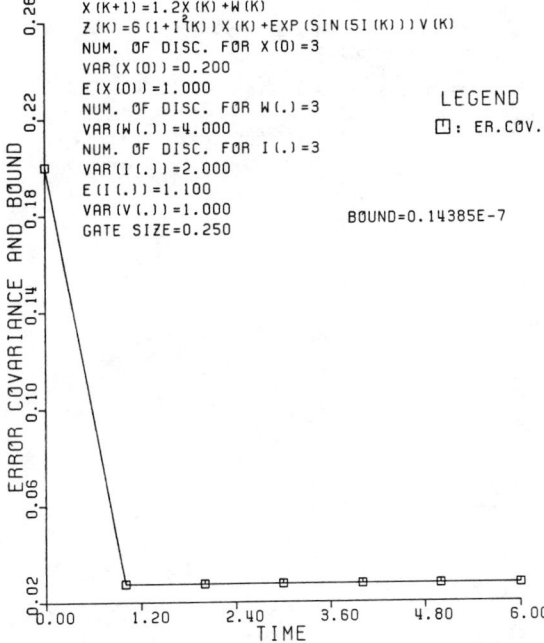

Fig. 13b.   Error covariance and bound.

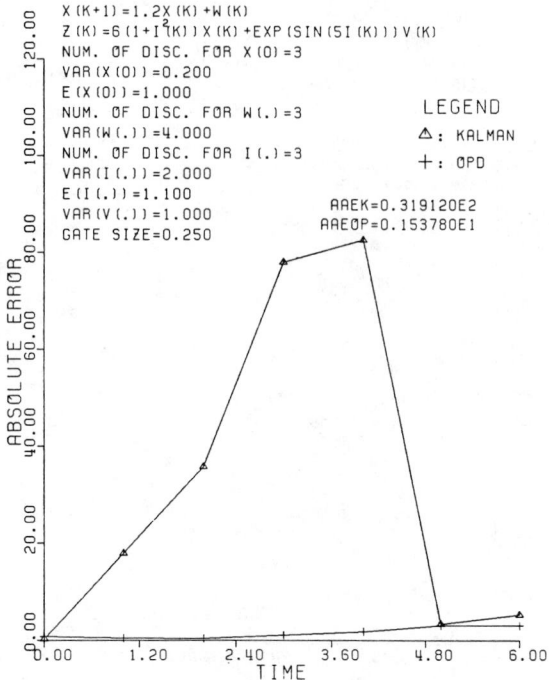

Fig. 13c. Absolute and average absolute errors.

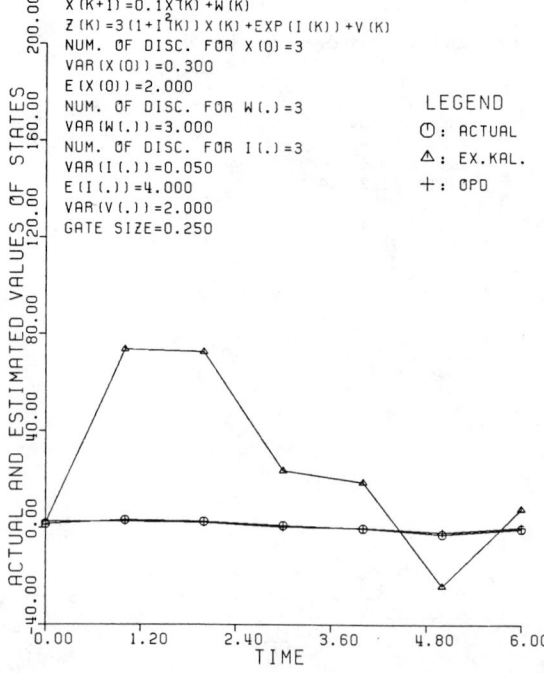

Fig. 14a. Actual and estimated values of states.

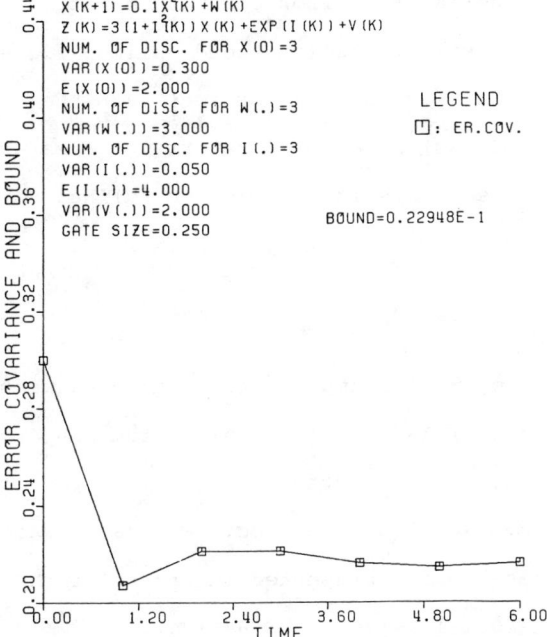

Fig. 14b. Error covariance and bound.

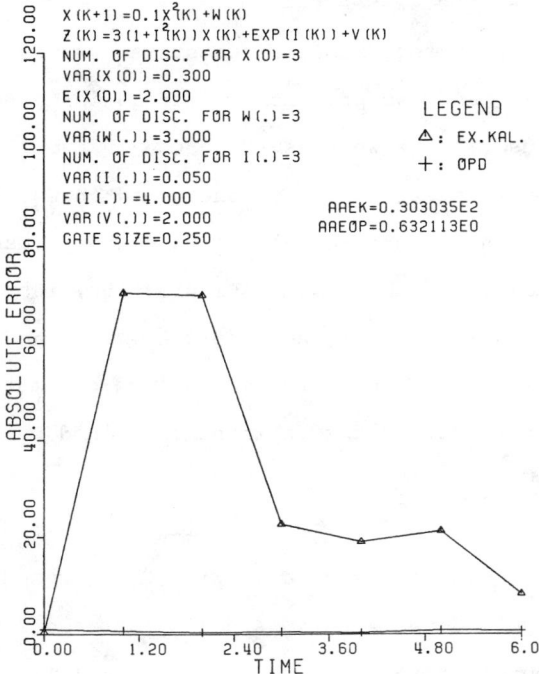

Fig. 14c. Absolute and average absolute errors.

of the ODSA, whereas the error covariance matrix is used as the performance measure of the (extended) Kalman filter algorithm. Figures 11c, 12c, 13c, and 14c present both the (extended) Kalman and the ODSA estimate absolute error curves as well as the average absolute errors for the (extended) Kalman estimates and the ODSA estimates.

B. *THE STACK SEQUENTIAL DECODING-BASED*
    SMOOTHING ALGORITHM

A large number of examples were simulated with the SSDSA and the (extended) Kalman filter algorithm, and the simulation results of some are presented in Figs. 15a, 15b, 15c, 16a, 16b, 16c, 17a, 17b, 17c, 18a, 18b, 18c. For each example, the simulation results are presented in groups of three (Figs. 15a-15c, 16a-16c, 17a-17c, and 18a-18c).

Figures 15a, 16a, 17a, and 18a present the variations of the actual values, the (extended) Kalman estimates, and the estimates obtained by the SSDSA of the states versus time. Figures 15b, 16b, 17b, and 18b present the estimation error covariance matrix versus time as well as the bound in Eq. (163) (if the example has no interference) or the bound in Eq. (179) (if the example contains interference). This bound is used as the performance measure of the SSDSA. Figures 15c, 16c, 17c, and 18c present the variations of the absolute errors and the average absolute errors for the (extended) Kalman estimates and the SSDSA estimates (i.e., the estimates obtained by the SSDSA).

C. *THE SUBOPTIMUM DECODING-BASED*
    SMOOTHING ALGORITHM

Many examples were simulated with the (extended) Kalman filter algorithm and the SDSA considering three steps at each of which six observations were used. The simulation results of

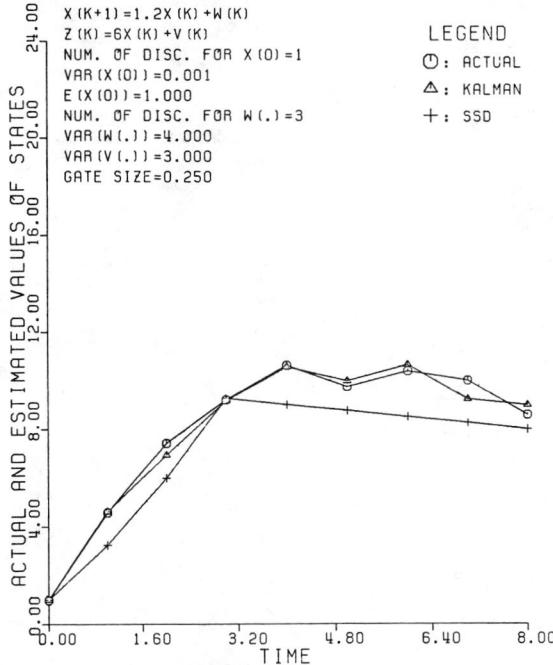

Fig. 15a. *Actual and estimated values of states.*

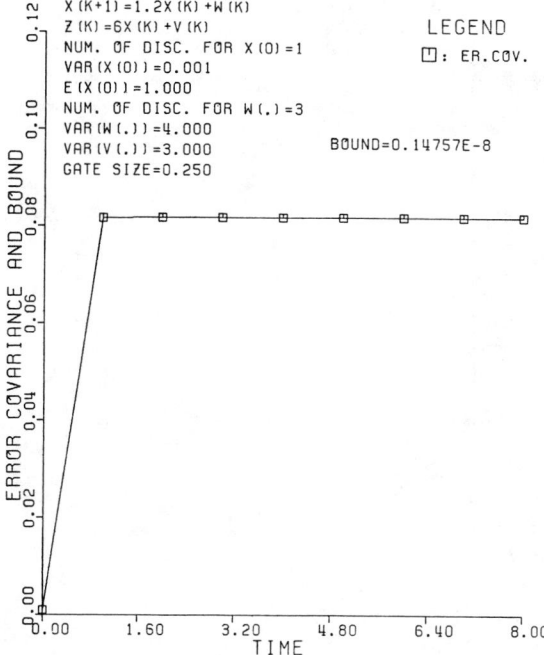

Fig. 15b. *Error covariance and bound.*

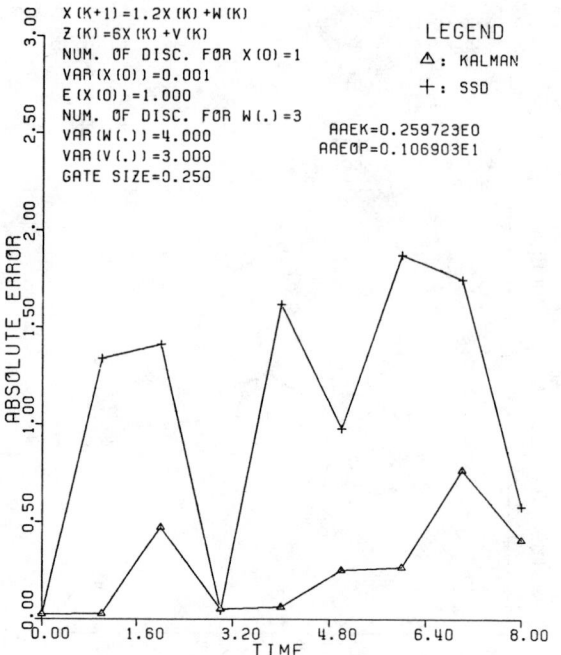

Fig. 15c.   *Absolute and average absolute errors.*

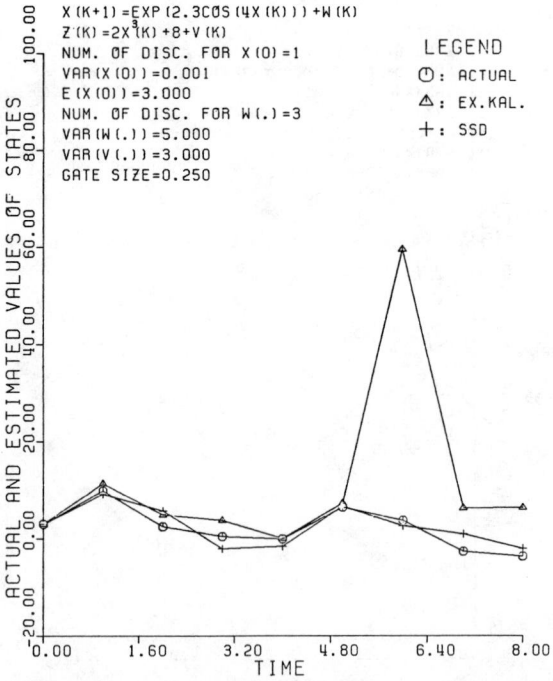

Fig. 16a.   *Actual and estimated values of states.*

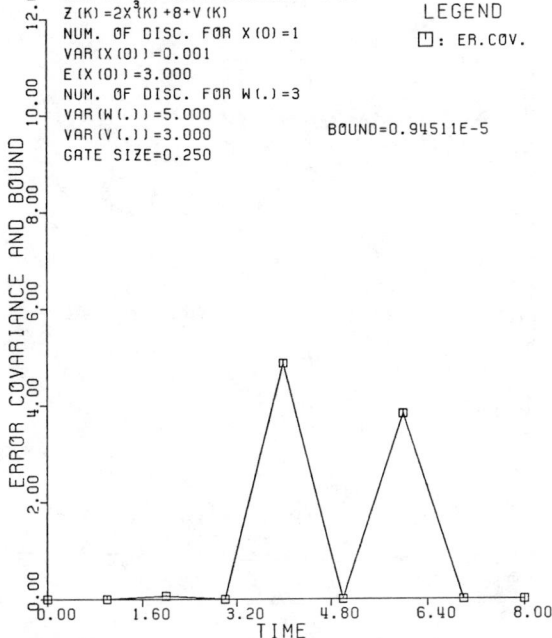

Fig. 16b. Error covariance and bound.

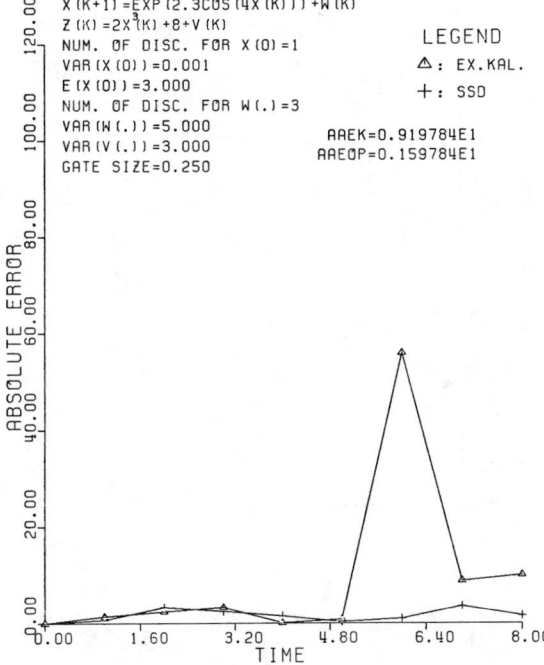

Fig. 16c. Absolute and average absolute errors.

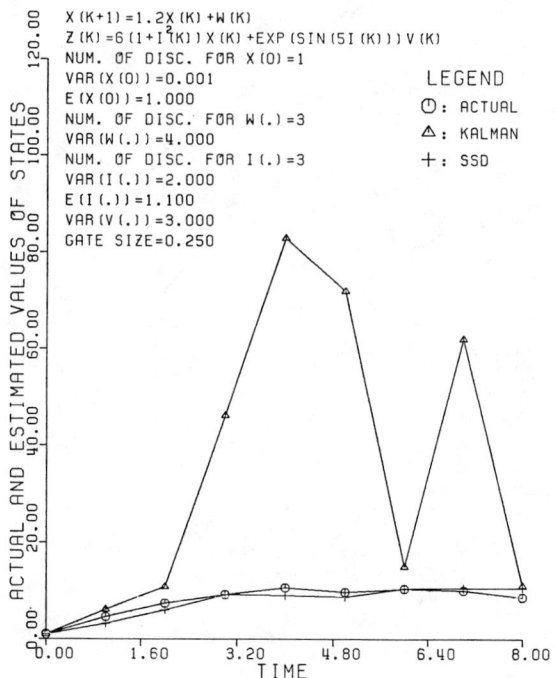

Fig. 17a.   *Actual and estimated values of states.*

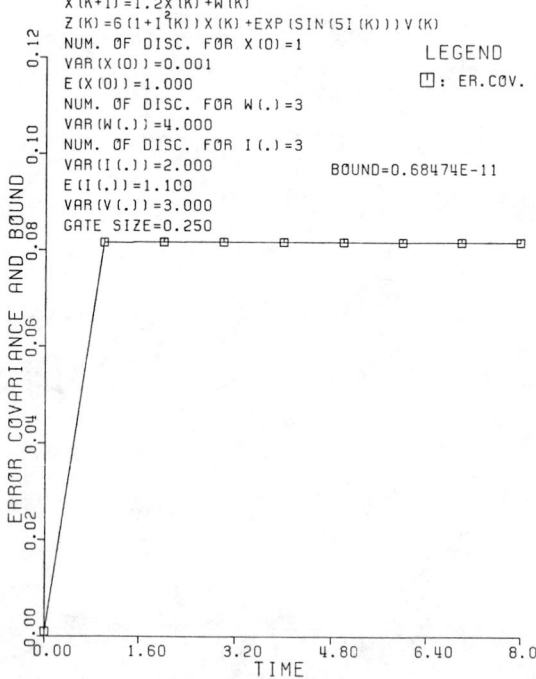

Fig. 17b.   *Error covariance and bound.*

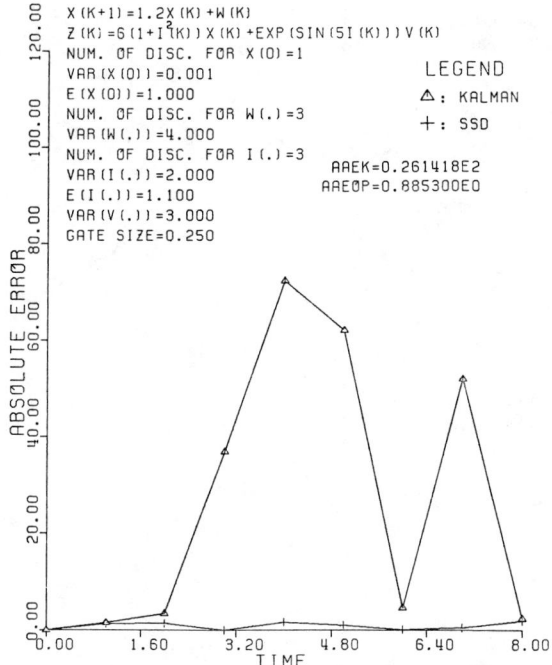

Fig. 17c.   Absolute and average absolute errors.

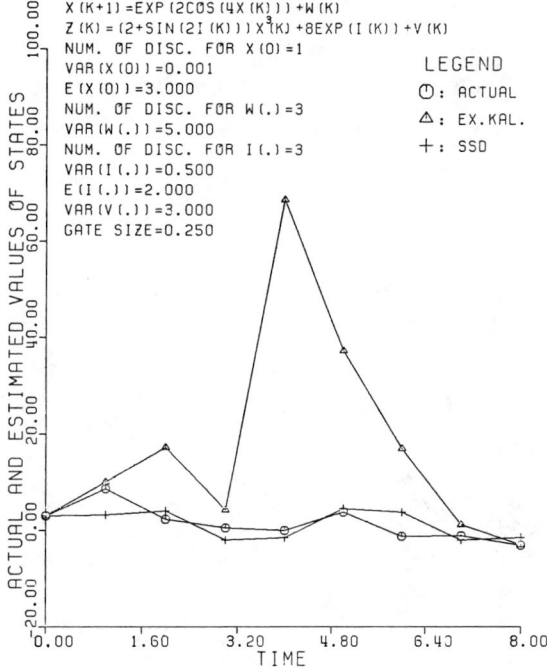

Fig. 18a.   Actual and estimated values of states.

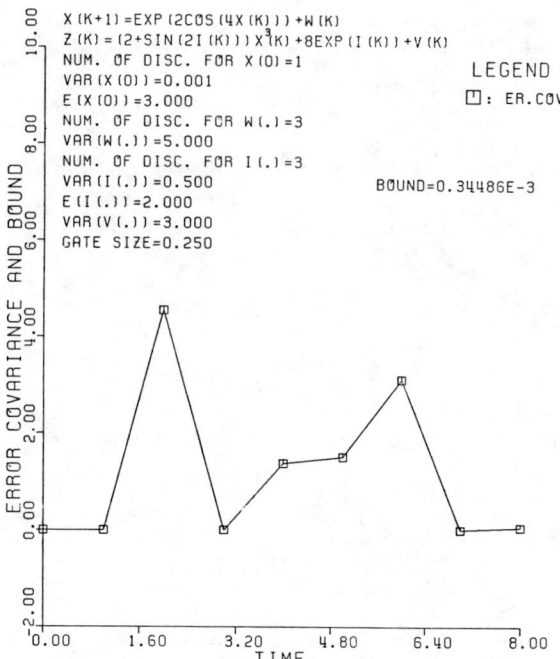

Fig. 18b. *Error covariance and bound.*

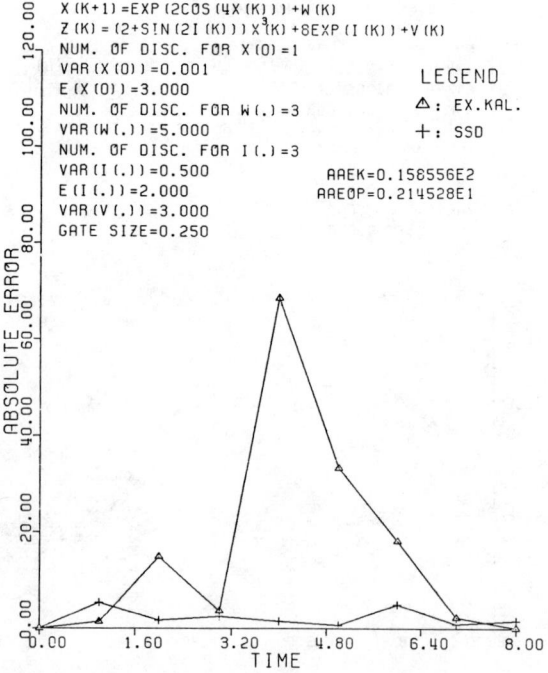

Fig. 18c. *Absolute and average absolute errors.*

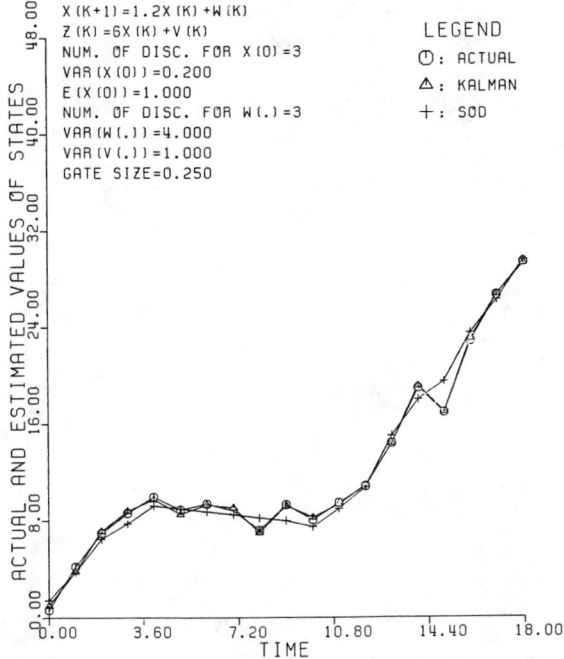

Fig. 19a. Actual and estimated values of states.

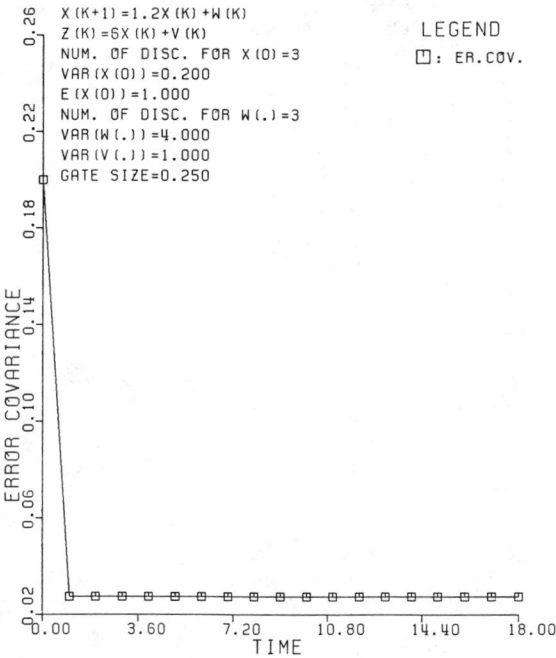

Fig. 19b. Error covariance.

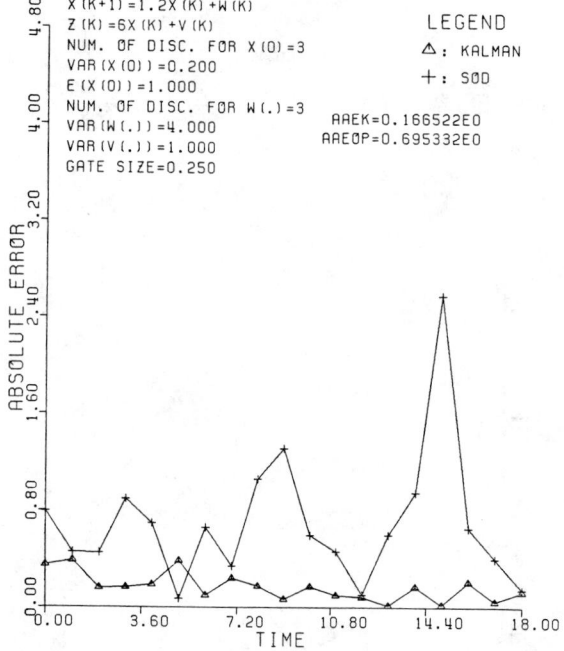

Fig. 19c.   *Absolute and average absolute errors.*

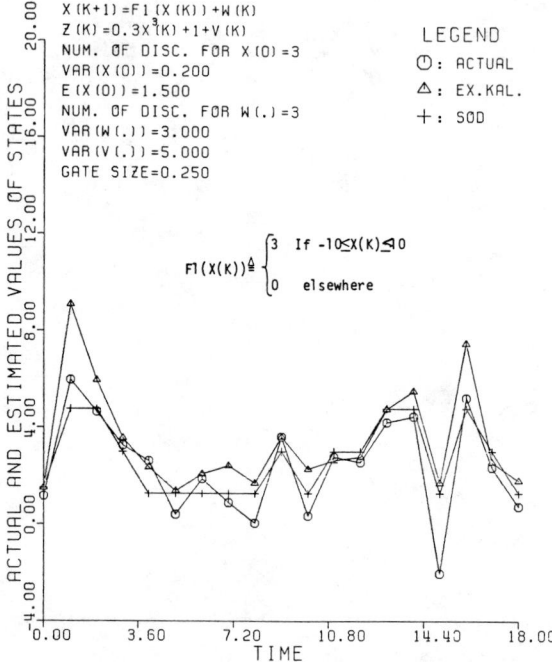

Fig. 20a.   *Actual and estimated values of states.*

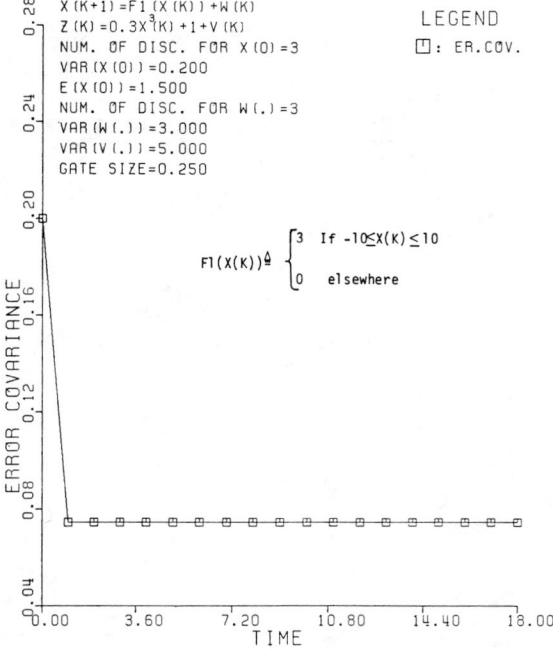

Fig. 20b.   Error covariance.

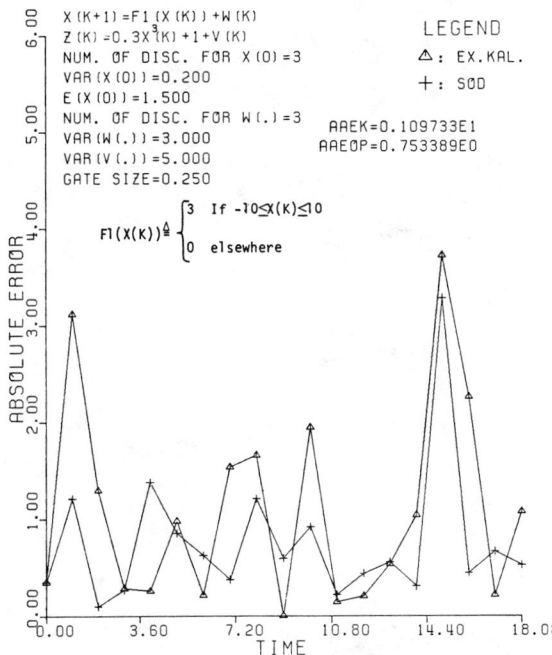

Fig. 20c.   Absolute and average absolute errors.

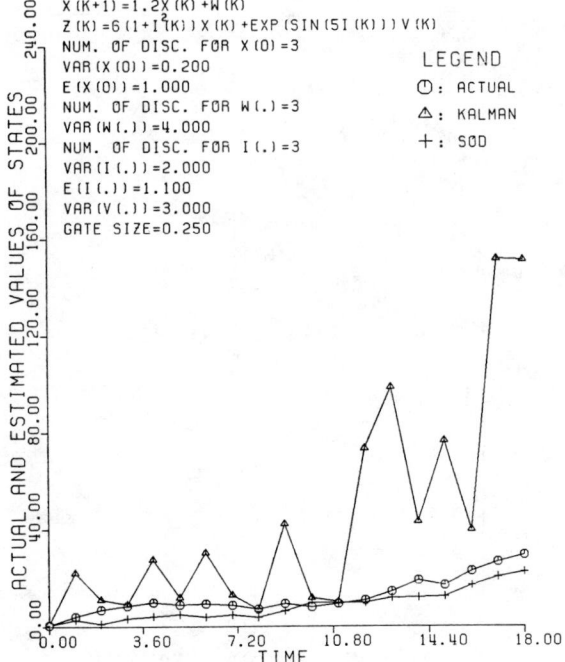

Fig. 21a.   Actual and estimated values of states.

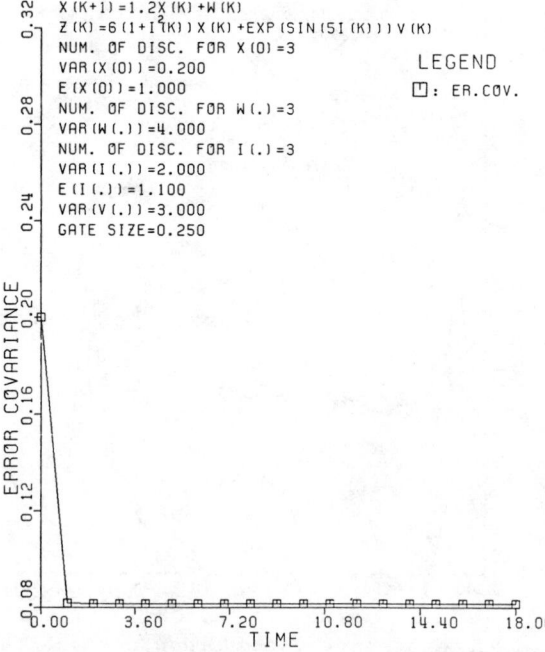

Fig. 21b.   Error covariance.

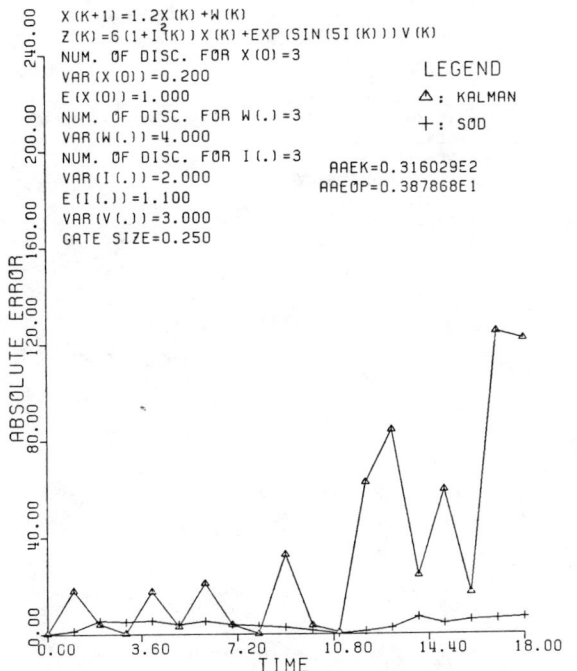

Fig. 21c. *Absolute and average absolute errors.*

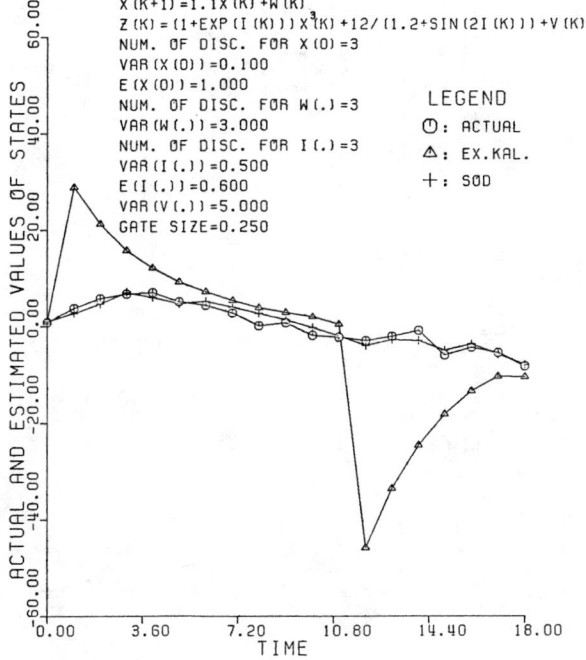

Fig. 22a. *Actual and estimated values of states.*

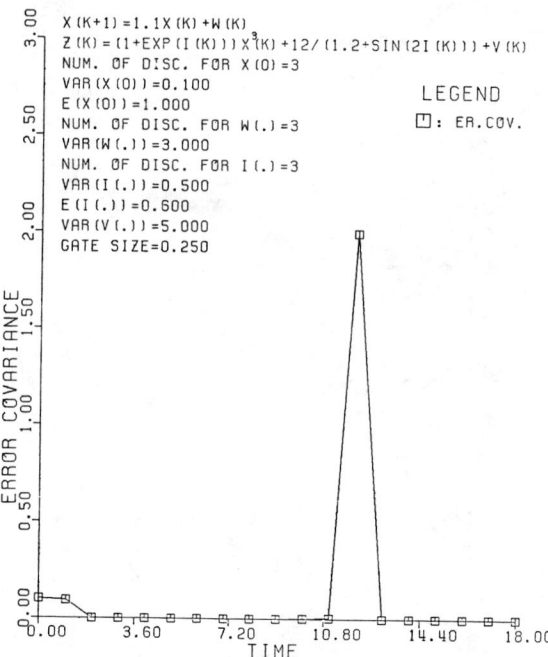

Fig. 22b.   Error covariance.

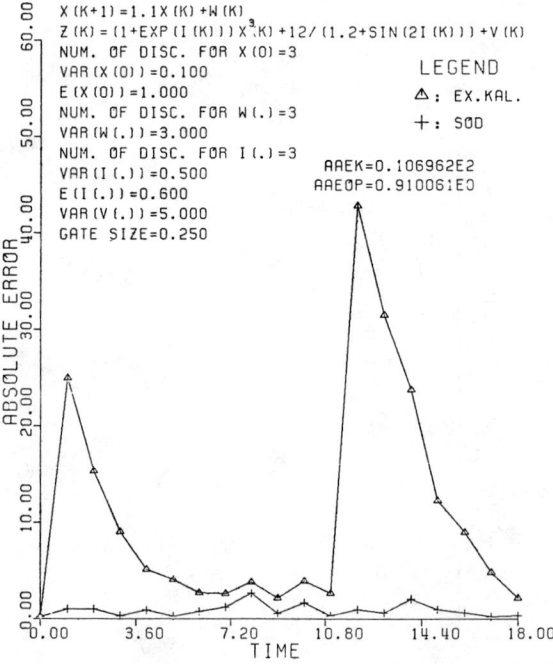

Fig. 22c.   Absolute and average absolute errors.

some of them are presented in Figs. 19a, 19b, 19c, 20a, 20b, 20c, 21a, 21b, 21c, 22a, 22b, 22c. For each example, the simulation results are presented in groups of three (Figs. 19a-19c, 20a-20c, 21a-21c, and 22a-22c).

Figures 19a, 20a, 21a, and 22a present the variations of the actual values, the (extended) Kalman estimates, and the estimates obtained by the SDSA of the states versus time. Figures 19b, 20b, 21b, and 22b present the variations of the estimation error covariance matrix for the (extended) Kalman filter algorithm. Figures 19c, 20c, 21c, and 22c present the average absolute errors and the variations of the absolute errors for the (extended) Kalman filter estimates and the SDSA estimates (i.e., the estimates obtained by the SDSA).

D. COMMENTS

Let m_k, n_0, and r_k be the numbers of possible values of the discrete random variables $w_d(k)$, $x_d(0)$, and $I_d(k)$, which approximate the disturbance noise $w(k)$, the initial state $x(0)$, and the interference $I(k)$, respectively. These numbers and gate sizes were taken to be time invariant for each example simulation. Definitely, the performance of the smoothing algorithms depends on these numbers as well as on the gate sizes used. The smoothing algorithms produce better estimates of the states for suitable gate sizes and larger m_k, n_0, and r_k. This follows from the fact that the disturbance noise, the initial state, and the interference are approximated better for larger m_k, n_0, and r_k. For large gate sizes, good estimates of the states are not expected since more quantization errors are made.

Gallager type ensemble bounds are used as the performance measure of the smoothing algorithms. The values of these bounds depend on the models as well as on the quantization (i.e., gate sizes, m_k, n_0, and r_k) used. Since some inequalities are used to derive these bounds, they can be greater than or equal to 1 for some models or quantization used. It should also be noted that even in the cases where these ensemble bounds are less than 1, they do not give complete information about the performance of the smoothing algorithms (Theorem 4).

V.   CONCLUSIONS

Three completely new smoothing algorithms have been presented for the following type of discrete models with or without random interference:

(1)   the models [in Eqs. (1) or (2)] can be linear or nonlinear;

(2)   the disturbance noise, the observation noise, and the interference can be any independent (of each other and from sample (time) to sample) not necessarily Gaussian noises;

(3)   the disturbance noise and interference can affect the motion model and the observation model in a general and not necessarily linear way;

(4) functions which define the models do not even have to be continuous, they can be any defined functions.

The new smoothing algorithms are based on the quantization of states to a finite set of states and decoding techniques of information theory. If the quantization errors are neglected, the first smoothing algorithm, which is referred to as the ODSA, is optimum with respect to the minimum error probability criterion (Bayes' decision rule). However, this requires an ever growing amount of memory with time for its implementation. Hence the second algorithm, which is referred to as the SSDSA, has been proposed. Even this algorithm requires a growing

amount of memory with time (but not as fast as the ODSA) for
its implementation. Therefore, the third smoothing algorithm,
referred to as the SDSA, has been proposed. This algorithm
requires a finite amount of memory for any time (i.e., no matter
how long the target is tracked).

In target tracking, it is difficult to determine a motion
model that is analytically tractable and at the same time
satisfactorily close to reality. Therefore, by making many
approximations, an analytically tractable motion model is obtained [2-6]. However, the smoothing algorithms presented in
this chapter use the motion model only to determine the transition probabilities of the tracked target from gate to gate. If
these transition probabilities can be determined in another
way, the proposed smoothing algorithms can be used with only
an observation model so that the motion model is no longer
necessary. Hence, the more accurate the determination of the
transition probabilities, the more accurate the estimates.

Gallager type ensemble bounds were derived and used as the
performance measure of the new smoothing algorithms. These
bounds are sometimes totally useless since they become numbers
which are greater than or equal to 1.

In order to test the smoothing algorithms, digital computer
simulations have been performed. Some of the simulation results are presented in detail. These results show that for
both linear and nonlinear discrete models with interference,
the new smoothing algorithms perform very well even though neither the Kalman nor the extended Kalman filter algorithm is ca-
pable of handling interference. Also, these smoothing algorithms
perform better than the extended Kalman filter algorithm for

some nonlinear models with Gaussian noises (and without interference), whereas they perform almost as well as the Kalman
filter algorithm for linear models with Gaussian noises (and
without interference).

APPENDIX A.   THEOREM

Let $p(z|x_i)$ and $p(z|x_j)$ be two r-dimensional (multivariate)
Gaussian density functions such that

$$p(z|x_i) = (2\pi)^{-r/2}(\det R_i)^{-1/2}$$

$$\times \exp\left\{-\frac{1}{2}\left[(z - g(x_i))^T R_i^{-1}(z - g(x_i))\right]\right\}$$

$$\triangleq N(g(x_i), R_i),$$

$$p(z|x_j) \triangleq N(g(x_j), R_j).$$

Then the following equalities hold:

(I) $\displaystyle\int_z [p(z|x_i)p(z|x_j)]^{1/2}dz = A \exp\{B/4\},$ \hfill (A.1)

where

$$A \triangleq \left\{\det\left[2\left(R_i^{-1} + R_j^{-1}\right)^{-1}\right]\right\}^{1/2} \Big/ (\det R_i)^{1/4}(\det R_j)^{1/4},$$

$$B \triangleq \left\{b_{ij}^T\left(R_i^{-1} + R_j^{-1}\right)^{-1}b_{ij} - g(x_i)^T R_i^{-1}g(x_i)\right.$$

$$\left. - g(x_j)^T R_j^{-1}g(x_j)\right\},$$ \hfill (A.2)

$$b_{ij} \triangleq R_i^{-1}g(x_i) + R_j^{-1}g(x_j);$$

(II) $\displaystyle\int_z p(z|x_i)[p(z|x_j)]^{1/2}dz = A' \exp\{B'/4\},$ \hfill (A.3)

where

$$A' \triangleq \left\{ \det\left[\left(R_i^{-1} + \frac{1}{2} R_j^{-1} \right)^{-1} \right] \right\}^{1/2} \Big/ (2\pi)^{r/4} (\det R_i)^{1/2} (\det R_j)^{1/4},$$

$$B' \triangleq b_{ij}'^{T} \left[2R_i^{-1} + R_j^{-1} \right]^{-1} b_{ij}' - 2g(x_i)^T R_i^{-1} g(x_i) - g(x_j)^T R_j^{-1} g(x_j),$$

$$b_{ij}' \triangleq 2R_i^{-1} g(x_i) + R_j^{-1} g(x_j).$$

If the covariance matrices R_i and R_j are equal, say, $R_i = R_j \triangleq R_v$, then the previous equalities become

(III) $\displaystyle \int_z [p(z|x_i) p(z|x_j)]^{1/2} dz$

$$= \exp\left\{ -\frac{1}{8} [g(x_i) - g(x_j)]^T R_v^{-1} [g(x_i) - g(x_j)] \right\}$$

(A.4)

(IV) $\displaystyle \int_z p(z|x_i) [p(z|x_j)]^{1/2} dz$

$$= G \exp\left\{ -\frac{1}{6} [g(x_i) - g(x_j)]^T R_v^{-1} [g(x_i) - g(x_j)] \right\},$$

(A.5)

where $G \triangleq (2/9\pi)^{r/4} / (\det R_v)^{1/4}$.

Proof. From the definitions of the probability density functions, we easily obtain

$$[p(z|x_i) p(z|x_j)]^{1/2} = C \exp(-D/4),$$

(A.6)

where

$$C \triangleq 1 \Big/ \left[(2\pi)^{r/2} (\det R_i)^{1/4} (\det R_j)^{1/4} \right],$$

$$D \triangleq (z - g(x_i))^T R_i^{-1} (z - g(x_i))$$

(A.7)

$$+ (z - g(x_j))^T R_j^{-1} (z - g(x_j)).$$

D can be rewritten as

$$D = (z - a)^T\left(R_i^{-1} + R_j^{-1}\right)(z - a) - B, \qquad (A.8)$$

where

$$a \triangleq \left(R_i^{-1} + R_j^{-1}\right)^{-1} b_{ij}, \qquad b_{ij} \triangleq R_i^{-1} g(x_i) + R_j^{-1} g(x_j) \qquad (A.9)$$

and B is as defined in Eq. (A.2). We have therefore

$$[p(z|x_i)p(z|x_j)]^{1/2} = \frac{\exp\left\{-\frac{1}{4}\left[(z - a)^T\left(R_i^{-1} + R_j^{-1}\right)(z - a)\right] + \frac{1}{4}B\right\}}{(2\pi)^{r/2}(\det R_i)^{1/4}(\det R_j)^{1/4}}.$$

$$(A.10)$$

Multiplying the denominator and numerator by $\left\{\det\left[2\left(R_i^{-1} + R_j^{-1}\right)^{-1}\right]\right\}^{1/2}$ gives

$$[p(z|x_i)p(z|x_j)]^{1/2}$$

$$= \left[A \exp \frac{B}{4}\right]\left\{\frac{\exp\left\{-\left[(z - a)^T \frac{1}{2}\left(R_i^{-1} + R_j^{-1}\right)\frac{1}{2}(z - a)\right]\right\}}{(2\pi)^{r/2}\left\{\det\left[2\left(R_i^{-1} + R_j^{-1}\right)^{-1}\right]\right\}^{1/2}}\right\}. \qquad (A.11)$$

Note that the term in the large braces is a multivariate
Gaussian density function with mean a and covariance
$= 2\left(R_i^{-1} + R_j^{-1}\right)^{-1}$. Hence, integrating this term over all z
yields 1, so that we obtain the equality in Eq. (A.1).

For the proof of the equality in Eq. (A.3), let us rewrite
the integral in Eq. (A.3) as

$$\int_z p(z|x_i)[p(z|x_j)]^{1/2}dz = \int_z \left[p(z|x_i)^2 p(z|x_j)\right]^{1/2} dz. \qquad (A.12)$$

On the other hand, from the equality

$$\det R_i = 2^r \det(R_i/2), \qquad (A.13)$$

we have

$$p^2(z|x_i) = Hp_1(z|x_i), \qquad (A.14)$$

where

$$p_1(z|x_i) \triangleq N\left(g(x_i), \frac{1}{2}R_i\right), \quad H \triangleq \left[(4\pi)^{r/2}(\det R_i)^{1/2}\right]^{-1}.$$

(A.15)

Substitution of Eq. (A.14) into Eq. (A.12) yields

$$\int_z p(z|x_i)[p(z|x_j)]^{1/2}dz = H^{1/2}\int_z [p_1(z|x_i)p(z|x_j)]^{1/2}dz,$$

(A.16)

and by use of the quality in Eq. (A.1) for the integral on the right-hand side of Eq. (A.16), we can easily obtain the equality in Eq. (A.3).

If $R_i = R_j \triangleq R_v$, substituting R_v for R_i and R_j in the equalities in Eqs. (A.1) and (A.3), the equalities in Eqs. (A.4) and (A.5) can readily be verified.

APPENDIX B.   APPROXIMATION OF AN ABSOLUTELY
               CONTINUOUS RANDOM VECTOR
               BY A DISCRETE RANDOM VECTOR

Let n be a given positive integer and let D^m be the set of distribution functions of all $m \times 1$ discrete random vectors with n possible values, where superscript m stands for the dimensionality of random vectors. Then the problem of approximating an absolutely continuous $m \times 1$ random vector X^m with distribution function $F_{X^m}(\cdot)$ by an $m \times 1$ discrete random vector with n possible values is to find a distribution function $F_{Y_0^m}(\cdot) \in D^m$, which minimizes the objective function $J(\cdot)$ over the set D^m:

$$J(F_{Y_0^m}(\cdot)) = \min_{F_{Y^m}(\cdot)\in D^m} J(F_{Y^m}(\cdot)).$$

(B.1)

where

$$J(_{Y^m}(\cdot)) \triangleq \int_{R^m} [F_{X^m}(a) - F_{Y^m}(a)]^2 da, \quad F_{Y^m}(\cdot) \in D^m.$$

(B.2)

note that the integration is performed over the m dimensional
Euclidean space R^m. The discrete random vector defined by $F_{y_0^m}(\cdot)$
is referred to as the optimum discrete random vector approximating the random vector X^m. Here, the approximation of an
absolutely continuous random variable X with distribution $F_X(\cdot)$
by a discrete random variable with n possible values is considered. The necessary conditions that the optimum discrete
random variable approximating X must satisfy are obtained.
Finally, discrete random variables approximating normal random
variables are obtained.

Let us now state two theorems and define some symbols which
are used. The proofs of the theorems are given in Ref. [16].

Theorem B.1 [16]

Let $f(y) \triangleq f(y_1, y_2, \ldots, y_l)$ be a real-valued function on
an open set Γ of R^l, and let $f(y)$ have finite partial derivatives $\partial f(y)/\partial y_k$, $k = 1, 2, \ldots, l$ at each point of Γ. If $f(y)$
has a local minimum at the point $y_0 \triangleq (y_{1,0}, y_{2,0}, \ldots, y_{l,0})$
in Γ, then $\partial f(y)/\partial y_k\big|_{y=y_0} = 0$ for each $k = 1, 2, \ldots, l$.

Theorem B.2 [16]

Let $f(y) \triangleq f(y_1, y_2, \ldots, y_l)$ be a real-valued function on
an open set Γ of R^l, and let $f(y)$ have continuous second-order
partial derivatives on Γ. Let $y_0 \triangleq (y_{1,0}, y_{2,0}, \ldots, y_{l,0})$ be
a point of Γ for which $\partial f(y)/\partial y_k\big|_{y=y_0} = 0$ for each $k = 1$,
$2, \ldots, l$. Assume that the determinant $G \triangleq \det\{[\nabla^2 f(y)]\big|_{y=y_0}\}$
$\neq 0$, where

$$[\nabla^2 f(y)]_{ij} \triangleq \frac{\partial^2}{\partial y_i\, \partial y_j} f(y).$$

Let G_{l-k} be the determinant obtained from G by deleting the
last k rows and columns. If the l numbers $G_1, G_2, \ldots, G_l$ are
all positive, then $f(y)$ has a local minimum at y_0.

We now define D as the set of distribution functions of all discrete random variables with n possible values, where n is a given positive integer. We next define S as the set of all stepfunctions with n steps. A stepfunction $g(\cdot)$ with n steps is, by definition, a function with n + 1 possible values in the real line R such that $g(\cdot)$ is zero at $-\infty$ and one at $+\infty$; the set of numbers which are mapped by $g(\cdot)$ to a chosen possible value is an interval in R, the intervals corresponding to the n + 1 possible values are nonoverlapping, and the union of these intervals is the real line, that is,

$$S \triangleq \{g(x): \quad g(x) = 0 \text{ for } x < y_1;$$

$$g(x) = P_i \text{ for } y_i \leq x < y_{i+1}, \ P_i \in (-\infty, \infty);$$

$$g(x) = 1 \text{ for } x \geq y_n; \ y_{i+1} > y_i, \ y_i \in (-\infty, \infty);$$

$$i = 1, 2, \ldots, n - 1\}.$$

In order to find the optimum discrete random variable with n possible values that approximates an absolutely continuous random variable X with distribution function $F_x(\cdot)$, we must find a distribution function $F_{y_0}(\cdot)$ which minimizes the objective function $J(\cdot)$ over the set D:

$$J(F_{y_0}(\cdot)) = \min_{F_y(\cdot) \in D} J(F_y(\cdot)), \tag{B.3}$$

$$= \min_{g(\cdot) \in S} J(g(\cdot)). \tag{B.4}$$

where

$$J(F_y(\cdot)) \triangleq \int_{-\infty}^{\infty} [F_x(a) - F_y(a)]^2 da, \tag{B.5}$$

The equality in Eq. (B.4) follows from the following arguments. Let a stepfunction $g_0(\cdot) \in S$ minimize $J(\cdot)$ over the set S; since the distribution function $F_x(\cdot)$ is nondecreasing, $g_0(\cdot)$ must be nondecreasing; hence it is a nondecreasing stepfunction whose range changes from zero to one; therefore $g_0(\cdot) \in D$. Thus the aim is to find a stepfunction $g_0(\cdot) \in S$ which minimizes the objective function $J(\cdot)$ over S. That is we would like to minimize the following function over $y_i \in (-\infty, \infty)$ and $P_j \in (-\infty, \infty)$ (where $i = 1, 2, \ldots, n; \ j = 1, 2, \ldots, n - 1$:

$$J(g(\cdot)) = \int_{-\infty}^{y_1} F_x^2(a)\,da + \int_{y_1}^{y_2} [F_x(a) - P_1]^2\,da$$

$$+ \int_{y_2}^{y_3} [F_x(a) - P_2]^2\,da + \cdots + \int_{y_{n-1}}^{y_n} [F_x(a) - P_{n-1}]^2\,dz$$

$$+ \int_{y_n}^{\infty} [F_x(a) - 1]^2\,da, \tag{B.6}$$

It follows from Theorem B.1 that if $g_0(x)$, which is defined by

$$g_0(x) = \begin{cases} 0, & x < y_{1,0}, \\ P_{i,0}, & \text{if } y_{i,0} \le x < y_{i+1,0}, \quad i = 1, 2, \ldots, n - 1, \\ 1, & x \ge y_{n,0}, \end{cases} \tag{B.7}$$

is a stepfunction which minimizes Eq. (B.6), this must satisfy the following set of equations:

$$P_{1,0} = 2F_x(y_{1,0});$$

$$P_{i,0} + P_{i+1,0} = 2F_x(y_{i+1,0}), \qquad i = 1, 2, 3, \ldots, n - 2;$$

$$1 + P_{n,0} = 2F_x(y_n); \tag{B.8}$$

$$P_{i,0}(y_{i+1,0} - y_{i,0}) = \int_{y_{i,0}}^{y_{i+1,0}} F_x(a)\,da, \qquad i = 1, 2, \ldots, n - 1.$$

Using Eq. (B.8) and Theorem B.2, the discrete random variables (with n possible values where $n = 1, 2, \ldots, 8$) which approximate the normal random variable with zero mean and unit variance have been numerically obtained and are tabulated in Table B.1.

TABLE B.1. *Discrete Random Variables Approximating the Gaussian Random Variable with Zero Mean and Unit Variance*

Number of possible values of y_0 [a]	i [b]	n Possible values and corresponding probabilities of y_0 [a]							
		1	2	3	4	5	6	7	8
$n = 1$	$y_{i,0}$	0.000							
	$P_{i,0}$	1.000							
$n = 2$	$y_{i,0}$	-0.675	0.675						
	$P_{i,0}$	0.500	0.500						
$n = 3$	$y_{i,0}$	-1.005	0.0	1.005					
	$P_{i,0}$	0.315	0.370	0.315					
$n = 4$	$y_{i,0}$	-1.219	-0.355	0.355	1.219				
	$P_{i,0}$	0.223	0.277	0.277	0.223				
$n = 5$	$y_{i,0}$	-1.376	-0.592	0.0	0.592	1.376			
	$P_{i,0}$	0.169	0.216	0.230	0.216	0.169			
$n = 6$	$y_{i,0}$	-1.499	-0.767	-0.242	0.242	0.767	1.499		
	$P_{i,0}$	0.134	0.175	0.191	0.191	0.175	0.134		
$n = 7$	$y_{i,0}$	-1.599	-0.905	-0.423	0.0	0.423	0.905	1.599	
	$P_{i,0}$	0.110	0.145	0.162	0.166	0.162	0.145	0.110	
$n = 8$	$y_{i,0}$	-1.683	-1.018	-0.567	-0.183	0.183	0.567	1.018	1.683
	$P_{i,0}$	0.093	0.123	0.139	0.145	0.145	0.139	0.123	0.093

[a] y_0, the discrete random variable with n possible values $y_{1,0}$, $y_{2,0}$, ..., $y_{n,0}$, which approximates the normal random variable with zero mean and unit variance.

[b] $y_{i,0}$, the ith possible value of y_0: $P_{i,0} \triangleq \text{Prob}\{y_0 = y_{i,0}\}$.

Let y_0 be the optimum discrete random variable with n possible values $y_{1,0}$, $y_{2,0}$, ..., $y_{n,0}$ which approximate the normal random variable with zero mean and unit variance, and let $P_{i,0}$ be defined by $\text{Prob}\{y_0 = y_{i,0}\}$. Let z_0 be the optimum discrete random variable with n possible values $z_{1,0}$, $z_{2,0}$, ..., $z_{n,0}$ which approximates the normal random variable with

mean μ and variance σ^2; and let $p'_{i,0}$ be defined by $\text{Prob}\{z_0 =$
$z_{i,0}\}$. By Eq. (B.8), it can easily be verified that

$$z_{i,0} = \sigma y_{i,0} + \mu, \qquad P'_{i,0} = P_{i,0}, \qquad i = 1, 2, \ldots, n. \qquad (B.9)$$

REFERENCES

1. K. DEMİRBAŞ, "Target Tracking in the Presence of Interference," Ph.D Thesis, University of California, Los
   Angeles, 1981.

2. J. S. THORP, *IEEE Trans. Aerosp. Electron. Syst. AES-9*,
   No. 4, 512-519 (1973).

3. R. A. SINGER, *IEEE Trans. Aerosp. Electron. Syst. AES-6*,
   No. 4, 473-483 (1970).

4. N. H. GHOLSON and R. L. MOOSE, *IEEE Trans. Aerosp. Electron. Syst. AES-13*, No. 3, 310-317 (1977).

5. R. L. MOOSE, H. F. VANLANDINGHAM, and D. H. McCABE, *IEEE
   Trans. Aerosp. Electron. Syst. AES-15*, No. 3, 448-456
   (1979).

6. A. FARINA and S. PARDINI, *IEEE Trans. Aerosp. Electron.
   Syst. AES-15*, No. 4, 555-563 (1979).

7. A. P. SAGE and J. L. MELSA, "Estimation Theory with
   Applications to Communications and Control," McGraw-Hill,
   New York, 1971.

8. T. KAILATH, *IEEE Trans. Autom. Control AC-13*, No. 6, 646-
   655 (1968).

9. J. MAKHOUL, *Proc. IEEE 63*, 561-580 (1975).

10. T. KAILATH, *IEEE Trans. Inf. Theory IT-20*, No. 2, 146-181
    (1974).

11. J. S. MEDICH, *Automatica 9*, 151-162 (1973).

12. H. L. VAN TREES, "Detection Estimation and Modulation,"
    Part 1, Wiley, New York, 1968.

13. G. D. FORNEY, JR., *Inf. Control 25*, 222-247 (1974).

14. A. J. VITERBI and J. K. OMURA, "Principles of Digital
    Communication and Coding," McGraw-Hill, New York, 1979.

15. R. G. GALLAGER, *IEEE Trans. Inf. Theory IT-11*, 3-18 (1965).

16. A. M. APOSTOL, "Mathematical Analysis," Addison-Wesley,
    Reading, Massachusetts, 1958.

17. R. B. ASH, "Real Analysis and Probability," Academic Press, New York, 1972.

18. S. T. FERGUSON, "Mathematical Statistics — A Decision Theoretic Approach," Academic Press, New York, 1967.

A New Approach
for Developing Practical Filters
for Nonlinear Systems

HOSAM E. EMARA-SHABAIK

School of Engineering and Applied Science
University of California
Los Angeles, California

I. INTRODUCTION

Estimation problems, including that of filtering, are basically concerned with extracting the best information from inaccurate observations of signals. Perhaps the earliest roots of this type of problem go back to the least squares estimation at the time of Galileo Galilei in 1632 and Gauss in 1795. The relatively modern and more general development of least squares estimation in stochastic processes is marked by the work of A. N. Kolmogorov and N. Wiener in the 1940s. Most recently, and due to the vast research and development of the space age,

estimation theory experienced a new outlook. This was marked
by the work of P. Swerling in 1958 and 1959 in connection with
satellite tracking, and the work of R. Kalman using state space
approach. Kalman's work [1] had the impact of greatly popu-
larizing and spreading estimation theory in various fields of
applications. Works by Stratonovich [66] and Kushner [42-44]
are also among the more recent developments of the subject.

From the control theory point of view, the problem of estimating the state of dynamical systems plays an important role.
Very often the optimal control law sought for a dynamical system is some sort of feedback of its state, for example, the
control of a chemical process, a nuclear reactor, maneuvering
of a spacecraft, guidance and navigation problems, and the
problem of control and suppression of structural vibrations.
It is sometimes of interest to know the state of a dynamic
system as well, for example, the tracking of moving objects
such as satellites in orbit and enemy missiles. These are just
a few examples of applications of this knowledge.

Fundamentally, the conditional probability density of the
state conditioned on available observations holds the key for
all kinds of state estimators. The case of the linear dynamical system, with measurements linear in the state variables,
in the presence of additive Gaussian noise and under the assumption of full knowledge of the system parameters and noise
statistics has been optimally solved. In that particular case,
the conditional probability density is Gaussian. A Gaussian
density is characterized by only two quantities, namely, its
mean and covariance. Therefore, the optimal linear filter has
a finite state, the conditional mean and the conditional covariance, and is widely known as the Kalman or the Kalman-Bucy

filter [1-4]. The Kalman filter (KF) provides the minimum
variance unbiased estimates. Also, the filter structure is
linear; its gain and covariance can be processed independently
of the estimate even before the observations are received.
These features make the KF desirable and easy to implement.

Unlike the linear case, the situation for nonlinear systems
is completely different. The conditional probability density
is no longer Gaussian even though the acting noise is itself
Gaussian. In this case the evolution of the conditional probability density is governed by a stochastic integral-partial
differential equation, Kushner's equation, or, equivalently, by
an infinite set of stochastic differential equations for the
moments of the density function [3,42,43]. Therefore, the
truly otpimal nonlinear filter is of infinite dimensionality
and is consequently of no practical interest. Practical suboptimal finite-dimensional filters are therefore very much
needed.

Inspired by Kalman's results, a great deal of research effort has been directed toward extending the linear results and
developing practical schemes for nonlinear filters. Developments have relied on two main approaches. The first approach
is based on the linearization of system nonlinearities around
a nominal trajectory by using a Taylor's series expansion.
Performing the expansion up to the first-order terms results
in the linearized filter [3, 11]. The approach can further be
improved by linearizing, again up to first order, about the
most recent estimate. Relinearization is performed as more
recent estimates become available. By so doing, the well-known
extended Kalman filter (EKF) [3] is obtained. The Taylor's
series expansion can be carried out to the second-order terms.

In this case, by making some assumptions about the conditional
probability density function, second-order filters are obtained.
Among these are the truncated second-order filter, the Gaussian
second-order filter, and the modified second-order filter (M2F).
These second-order filters are presented in [3,11].

In the second approach the conditional probability density
function is approximated by using several techniques. The
Gaussian sum approximation is used in [33,34]. In this case
the conditional probability density is approximated by a finite
w^ighted sum of Gaussian densities with different means and
covariances. Since the KF is a Gaussian density synthesizer,
the resulting Gaussian sum filter is actually a bank of KFs
working in parallel. Each one is properly tuned in terms of
system parameters, and its output is properly weighted and
summed to other filter outputs to produce the state estimate.
The approach has been used extensively by many authors to treat
the estimation problem of linear systems having unknown parameters [35-40]. The orthogonal series expansion is also used
to approximate the conditional probability density as in [41].
Also, the idea of generating a finite set of moments to replace
the infinite set for the true density has been investigated in
[44]. A more detailed account and discussion of the previously
mentioned techniques is given by the author in [61].

In all of these approaches for developing suboptimal finite-
dimensional filters, the goal of theoretical assessment of
such filters in the sense of providing a measure of how far a
suboptimal filter is from being truly optimal has remained very
hard to achieve. It inherits the very same practical difficulty of the optimal filter, infinite dimensionality, that one
is trying to avoid. Therefore, the support of any such scheme

has to rely heavily on computer simulation, and for that same
reason not a single scheme can be claimed as always superior.
For example, in [11], the truncated second-order filter, the
Gaussian second-order filter, the M2F, the EKF, and the linearized filter were considered in numerical simulation. The
linearized filter had the poorest performance, but no conclusion was evident about which one of the other filters was superior. The EKF was favored for its relatively simple structure compared to the other filters. The final judgement is
therefore left to experience and the special case at hand.
Consequently, the development of new practical filters will enhance the contributions to this field.

The main theme of this chapter is to consider the nonlinear
filtering problem from a different approach. The approach
taken here is to consider the problem as the combination of
approximating the system description and to solve the filtering
problem for the approximate model. As a result some new schemes
are developed. The problem formulation and the proposed solutions are given next followed by some numerical results.

II. PROBLEM FORMULATION

Consider the general nonlinear dynamical system whose state
$x(t)$ evolves in time according to the following differential
equation:

$$dx(t) = [A(t)x(t) + f(x(t), t)]dt + Q^{1/2}(t)dW(t),$$
$$x(t_0) = x_0, \qquad t \geq t_0,$$

(1)

where $x(t) \in R^n$ is an n-dimensional state vector; $A(t)$ is an
$n \times n$ real matrix; $f(x(t), t)$ is an n-dimensional vector-valued

real function; and $x_0 \in R^n$ is an n-dimensional Gaussian random vector (GRV) with

$$E\{x_0\} = \bar{x}_0{}^{1},\tag{2}$$

$$Cov(x_0, x_0) \triangleq E\left\{(x_0 - \bar{x}_0)(x_0 - \bar{x}_0)'\right\} = P_0{}^{2},\tag{3}$$

and where $W(t) \in R^n$ is an n-dimensional Wierner process with $dW(t) = W(t + dt) - W(t)$. Therefore,

$$E\{dW(t)\} = 0 \qquad \text{for all } t \geq t_0,\tag{4}$$

$$Cov(dW(t), dW(t)) \triangleq E\{dW(t)dW'(t)\} = Idt,\tag{5}$$

where I is the $n \times n$ unit matrix; $Q^{1/2}(t)$ is a real matrix; and $Q(t) \triangleq Q^{1/2}(t)Q^{1/2'}(t)$ is a positive semidefinite $n \times n$ matrix. Consider also the observation process $dy(t)$ to be given by

$$dy(t) = [C(t)x(t) + h(x(t), t)]dt + R^{1/2}(t)dv(t),\tag{6}$$

where $\zeta_{\gamma}(t) \in R^m$ is an m-dimensional observation vector; $C(t)$ is an $m \times n$ real matrix; $h(x(t), t)$ is an m-dimensional vector-valued real function; $v(t) \in R^m$ is an m-dimensional Wiener process; and $dv(t) = V(t + dt) - V(t)$. Therefore,

$$E\{dv(t)\} = 0 \qquad \text{for all } t \geq t_0,\tag{7}$$

$$Cov(dv(t), dv(t)) \triangleq E\{dv(t)dv'(t)\} = Idt,\tag{8}$$

where $R^{1/2}(t)$ is a real matrix, and $R(t) \triangleq R^{1/2}(t)R^{1/2'}(t)$ is a positive definite $(n \times n)$ matrix.

We assume that x_0, $w(t)$, and $v(t)$ are all independent of each other for all values of $t \geq t_0$. We also assume that Eq. (1) satisfies the conditions for existence and uniqueness of solution given in [3, 23, 57]. This means that our dynamical

---

[1] *Note that $E\{\cdot\}$ denotes the expected value of $(\cdot)$.*
[2] *Note that $Cov(\cdot, \cdot)$ denotes the covariance of $(\cdot)$.*

system (1) permits only one solution $x(t)$, $t \geq t_0$ to be its state trajectory in the mean-square sense. Furthermore, it is assumed that both $f(x(t), t)$ and $h(x(t), t)$ are continuous in $x(t)$.

As noted from Eqs. (1) and (6), the system structure is considred to be composed of two parts: a linear part plus a nonlinear part. We further assume that the system behavior is dominated by its linear part; that is to say,

$$\| f(x(t), t) \|_i < \| A(t)x(t) \|_i, \qquad \forall \, i \qquad (9)$$

$$\| h(x(t), t) \|_j < \| C(t\_x(t) \|_j, \qquad \forall \, j \qquad (10)$$

where $\| z \|_i$ is the absolute value of the ith component of the vector z. Equations (1) and (6), along with conditions (9) and (10), can be the description of the original system. It can also be a representation obtained by linearization of a non-linear system, where $f(x(t), t)$ and $h(x(t), t)$ represent second-order and higher terms. In this case conditions (9) and (10) are valid as long as the system state $x(t)$ remains within a small neighborhood of the nominal (linearizing) trajectory.

Accordingly, conditions (9) and (10) suggest that for a good guess of the system state $x^*(t)$, the following approximate equations for the dynamics and observations can be written as

$$dx_1(t) = \left[A(t)x_1(t) + f(x^*(t), t) \right] dt + Q^{1/2}(t)dw(t), \qquad (11)$$

$$dy(t) = \left[C(t)x_1(t) + h(x^*(t), t) \right] dt + R^{1/2}(t)dv(t). \qquad (12)$$

By virtue of the continuity of the nonlinearities in $x(t)$, we should note the following. As $x^*(t)$ approaches $x_1(t)$, the approximate description given in Eqs. (11) and (12) approaches

the true description given in Eqs. (1) and (6). In fact,
Eq. (13)

$$dx_1(t) = [A(t)x_1(t) + f(x_1(t), t)]dt + Q^{1/2}(t)dw(t),$$

$$x(t_0) = X_0, \qquad t \geq t_0,$$

(13)

and Eq. (1) have the same solution both in the mean-square
sense and with probability one.

It thus follows that the filtering problem of the system of
Eqs. (1) and (6) can be considered as a unification of model
approximation and a state estimation of the approximate model.
In other words, we first approximate the system description by
finding a suitable $x^*(t)$. We then solve the optimal filtering
problem of the approximate model. The purpose of optimal filtering is basically to seek the minimum mean-square error estimate of the state $x(t)$ based on the available observations
$Y_t = [y(s), t_0 \leq s \leq t]$.

Generally, according to Theorem 6.6 of [3, 184] and its
specialization to linear systems [3, Theorem 7.3, 219], the
otpimal filter imitates the dynamics of the system and is
linearly driven by the net observations. Therefore, guided by
these results, we shall seek the optimal filter for the system
in Eqs. (11) and (12) as a linear dynamic system driven linearly by the net observations. The optimality of the filter is
in the sense of achieving minimum mean-square error.

If we therefore define the estimation error $e_1(t)$ as

$$e_1(t) \triangleq x_1(t) - \hat{x}_1(t),$$

(14)

and the covariance matrix $P(t)$ as

$$P(t) \triangleq E\left\{e_1(t) - \bar{e}(t))(e_1(t) - \bar{e}_1(t))'\right\},$$

(15)

where $\hat{x}_1(t)$ is an estimate of $x_1(t)$ based on Y_t and

$$\overline{e}_1(t) \triangleq E\{e_1(t)\}, \tag{16}$$

then

$$J(e_1(t)) = tr\left(E\left\{e_1(t)e_1'(t)\right\}\right) = tr(P(t)) + tr\left(\overline{e}_1(t)\overline{e}_1'(t)\right)$$

$$\tag{17}$$

is to be minimized.

III. PROPOSED SOLUTIONS

A. *DERIVATION OF THE E1 FILTER*

According to the approximate model in Eqs. (11) and (12), the minimum variance unbiased estimate $\hat{x}_1(t)$ is given by a KF having the following expression:

$$d\hat{x}_1(t) = \left[A(t)\hat{x}_1(t) + f(x^*(t), t)\right]dt$$

$$+ K(t)\left[dy(t) - C(t)\hat{x}_1(t)dt - h(x^*(t), t)dt\right],$$

$$\hat{x}_1(t_0) = \overline{X}_0, \qquad K(t) = P(t)C'(t)R^{-1}(t), \tag{18}$$

$$dP(t) = [A(t)P(t) + P(t)A'(t)$$

$$- P(t)C'(t)R^{-1}(t)C(t)P(t) + Q(t)]dt,$$

$$P(t_0) = P_0.$$

A well-known property of the KF is that $\hat{x}_1(t)$ is the conditional expectation of $x_1(t)$ given the measurements Y_t; i.e., $\hat{x}_1(t) = E_{Y_t}\{x_1(t)\}$.

According to the argument following Eqs. (11) and (12), $x^*(t)$ is required to provide the optimal solution of the following minimization problem:

$$\min_{x^*(t)} J(x^*(t)) = E_{Y_t}\left\{\left(x_1(t) - x^*(t)\right)'\left(x_1(t) - x^*(t)\right)\right\}, \tag{19}$$

then for every $t \geq t_0$, setting $\partial J(x^*(t))/\partial x^*(t) = 0$ we get

$$x^*(t) = E_{Y_t}\{x_1(t)\} = \hat{x}_1(t). \tag{20}$$

Combining the results of Eqs (18) and (20), we therefore obtain the following filter, hence forth denoted as ElF:

$$d\hat{x}(t) = [A(t)\hat{x}(t) + f(\hat{x}(t), t)]dt$$
$$+ K(t)[dy(t) - C(t)\hat{x}(t)dt - h(\hat{x}(t), t)dt], \tag{21}$$
$$\hat{x}(t_0) = \overline{x}_0,$$

$$K(t) = P(t)C'(t)R^{-1}(t), \tag{22}$$

$$dP(t) = [A(t)P(t) + P(t)A'(t) - P(t)C'(t)R^{-1}(t)C(t)P(t)$$
$$+ Q(t)]dt, \qquad P(t_0) = P_0. \tag{23}$$

It is a straightforward matter to recognize that in the case of a linear system [i.e., where $f(x(t), t)$ and $h(x(t), t)$ are identically zero or only functions of time], Eqs. (21)-(23) reduce to the well-known KF.

The EKF [3] of Eqs. (1) and (6) is given by Eqs. (24)-(26):

$$d\hat{x}(t) = [A(t)\hat{x}(t) + f(\hat{x}(t), t)]dt$$
$$+ K(t)[dy(t) - C(t)\hat{x}(t)dt$$
$$- h(\hat{x}(t), t)dt], \qquad \hat{x}(t_0) = x_0, \tag{24}$$

$$K(t) = P(t)[C(t) + h_x(\hat{x}(t), t)]'R^{-1}(t), \tag{25}$$

$$dP(t) = \left\{[A(t) + f_x(\hat{x}(t), t)]P(t) + P(t)[A(t) + f_x(\hat{x}(t), t)]'\right.$$
$$- P(t)(C(t) + h_x(\hat{x}(t), t))'R^{-1}(t)(C(t)$$
$$\left. + h_x(\hat{x}(t), t))P(t) + Q(t)\right\}dt, \qquad P(t_0) = P_0, \tag{26}$$

where

$$f_x(\hat{x}(t), t) = [\partial f(x(t), t)/\partial x(t)]\Big|_{x(t)=\hat{x}(t)},$$

$$h_x(\hat{x}(t), t) = [\partial h(x(t), t)/\partial x(t)]\Big|_{x(t)=\hat{x}(t)}.$$

The ElF bears a close relationship to the EKF. The equations for the state estimate of both the ElF and the EKF [Eqs. (21) and (24)] have the same structure, whereas the equations for the gain and covariance of the ElF [Eqs. (22) and (23)] are different from those for the EKF [Eqs. (25) and (26)]. Equations (22) and (23) are no longer state-estimate dependent. Thus, unlike·the EKF, the gain and covariance for the ElF can be processed off-line and prior to receiving the observations, like the KF. Therefore, the ElF will have more advantages than the EKF when on-line computations of gain and covariance are not practical due to capacity limitations of on-line computers. This is usually the case for airborne and spaceborne computers.

Furthermore, although the EKF has to be strictly interpreted in the Itô sense [62], it is not the case with the ElF. This is so because the gain $K(t)$ as given by Eq. (22) is not estimate dependent.

B. NUMERICAL EXPERIMENT FOR THE ElF

The Van der Pol oscillator is chosen to compare the ElF, the KF, and the EKF. The Van der Pol oscillator is characterized by the following differential equation [24]:

$$\ddot{x}(t) - \epsilon\dot{x}(t)(1 - x^2(t)) + x(t) = 0, \tag{27}$$

which describes a dynamical system with a state-dependent damping coefficient equal to $- (1 - x^2(t))$ where ϵ is a positive parameter. The damping in the system goes from negative to zero to positive values as the value of $x^2(t)$ changes from less than to greater than unity. The response of the oscillator is characterized by a limit cycle in the $x(t)$, $\dot{x}(t)$ plane (the phase plane). The limit cycle approaches a circular shape as ϵ becomes very small: it has a maximum value for $x(t)$ equal to

2.0 regardless of the value of ϵ. This type of oscillation occurs in electronic tubes, which also exhibit what is known as thermal noise. Denoting $x(t)$ as $x_1(t)$ and $\dot{x}(t)$ as $x_2(t)$, Eq. (27) can be rewritten in a state-space formulation. Also, considering the existence of some noise forcing on the system, we get the following representation for the Van der Pol oscillator:

$$\begin{bmatrix} dx_1(t) \\ dx_2(t) \end{bmatrix} = \begin{bmatrix} 0 & 1 \\ -1 & \epsilon \end{bmatrix} \begin{bmatrix} x_1(t) \\ x_2(t) \end{bmatrix} dt + \begin{bmatrix} 0 \\ -\epsilon x_1^2(t) x_2(t) \end{bmatrix} dt + Q^{1/2} \begin{bmatrix} dW_1(t) \\ dW_2(t) \end{bmatrix},$$

(28)

Also, suppose that the following measurement is taken:

$$dy(t) = \left[x_1(t) + x_1^3(t) \right] dt + R^{1/2} dv(t). \tag{29}$$

In Eqs. (28) and (29), $[W_1(t) W_2(t)]^T$ is considered to be a two-dimensional Wiener process. Also, $v(t)$ is a one-dimensional Wiener process, R is a positive nonzero real value, and Q is a 2×2 matrix. Table I lists the values for noise statistics considered here. In the numerical experiments the system non-linearities $-x_1^2(t) x_2(t)$, and $x_1^3(t)$ are first expanded in Taylor's series using the trajectory described by a circular limit cycle of radius 2 in the phase plane. All the terms of order higher than one are grouped to correspond with the $f(x(t), t)$ and $h(x(t), t)$ terms in Eqs. (1) and (6).

TABLE I.   Values for Noise Statistics[a]

Van der Pol case No.	Q_{11}	Q_{12}	Q_{22}	R	Fig. No.
1	0.5	0.0	0.5	4.0	1 and 2
2	5.0	2.0	5.0	10.0	3 and 4

[a] ϵ is taken to be 0.2.

As indicated by Figs. 1-4, in cases 1 and 2 both the E1F and EKF provide very accurate tracking of the system states, whereas the KF provides only crude estimates.

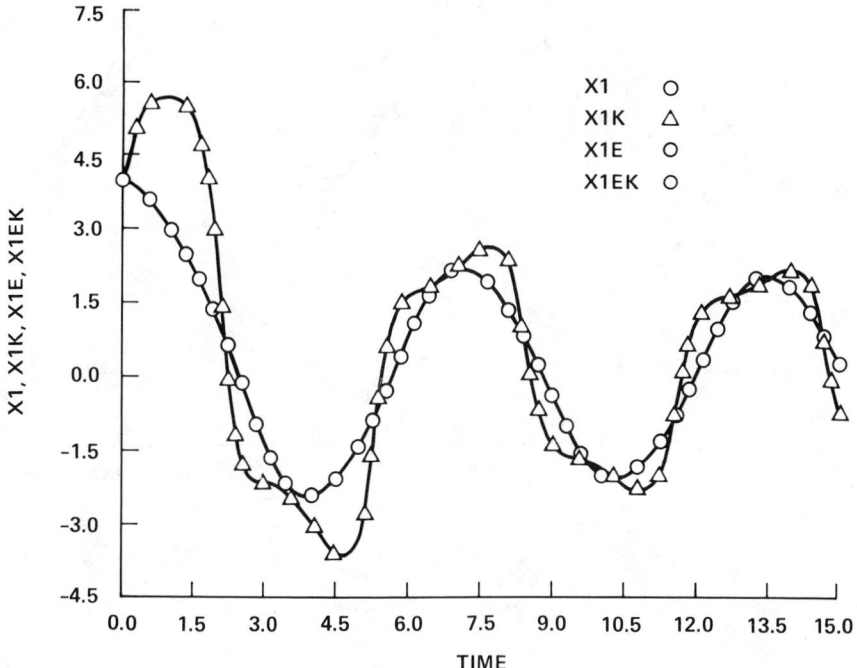

Fig. 1. First state and estimates by the KF, the E1F, and the EKF: X1 is the 1st state; X1K, X1E, and X1EK are the estimates of the state provided by the KF, the E1F, and the EKF, respectively.

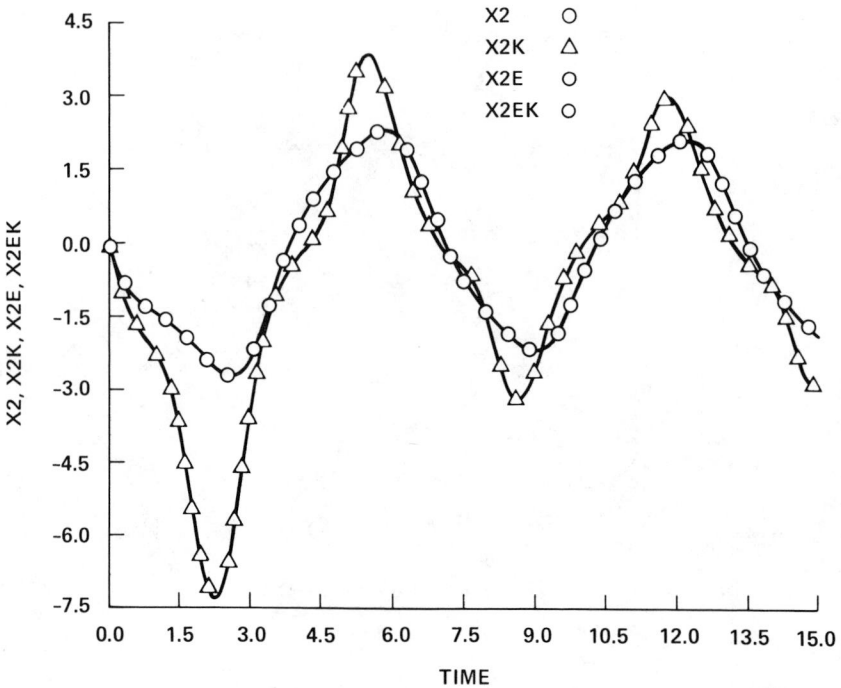

Fig. 2. Second state and estimates by the KF, the E1F, and the EKF: X2 is the 2nd state; X2K, X2E, and X2EK are the estimates of the 2nd state provided by the KF, the E2F, and the EKF, respectively.

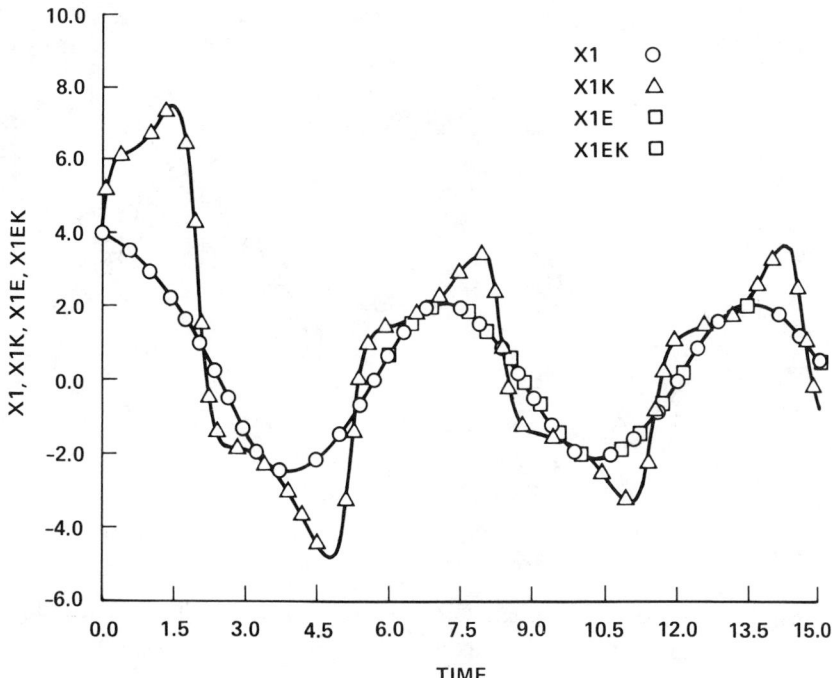

Fig. 3. First state and estimates by the KF, the E1F, and the EKF. (X1, X1K, X1E, and X1EK are as defined in Fig. 1 legend.)

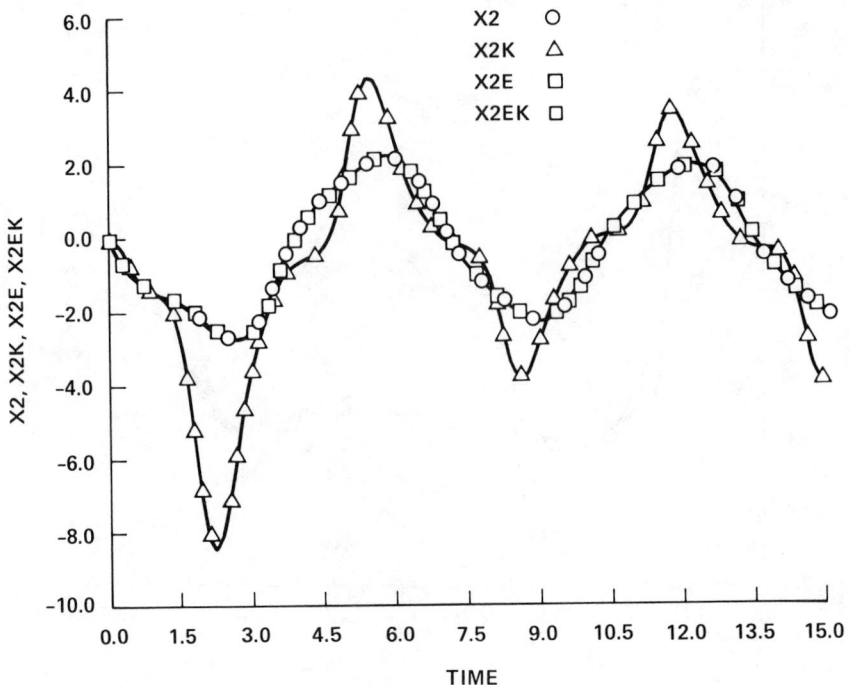

Fig. 4. Second state and estimates by the KF, the E1F, and the EKF. (X2, X2K, X2E, and X2EK are as defined in Fig. 2 legend.)

C. DERIVATION OF THE E2 AND E2N FILTERS

We have

$$dx_1(t) = \left[A(t)x_1(t) + f(x^*(t), t)\right]dt + Q^{1/2}(t)dw(t), \quad (30)$$

$$dy(t) = \left[C(t)x_1(t) + h(x^*(t), t)\right]dt + R^{1/2}(t)dv(t), \quad (31)$$

as our approximate model for some given good guess of the system state $x^*(t)$. We then seek a filter which is a linear dynamic system, linearly driven by the available observations as follows:

$$d\hat{x}_1(t) = [B(t)\hat{x}_1(t)]dt + K(t)dy(t), \quad (32)$$

where $B(t)$ is an $n \times n$ matrix and $K(t)$ is an $n \times m$, the gain
matrix of the filter. In order to evaluate the accuracy of
this filter in estimating the state $x_1(t)$, we define the estimation error $e_1(t)$ as

$$e_1(t) \triangleq x_1(t) - \hat{x}_1(t). \tag{33}$$

From Eqs. (30)-(32) we therefore obtain

$$de_1(t) = [(A(t) - K(t)C(t) - B(t))x_1(t) + B(t)e_1(t)$$

$$+ f(x^*(t), t) - K(t)h(x^*(t), t)]dt \tag{34}$$

$$+ Q^{1/2}(t)dw(t) - K(t)R^{1/2}(t)dv(t),$$

$$e_1(t_0) = x_0 - \hat{x}_1(t_0).$$

It is desirable to have the estimation error independent of
the state. In this case large state variables can be estimated
as accurately as small state variables. We may therefore
choose

$$B(t) = A(t) - K(t)C(t). \tag{35}$$

Hence, the dependence of the estimation error on the state is
eliminated. Also, the initial minimum variance estimate is
the mean of the initial state x_0. Therefore,

$$\hat{x}_1(t_0) = \overline{x}_0. \tag{36}$$

Equation (34) thus reduces to

$$de_1(t) = \left[(A(t) - K(t)C(t))e_1(t) + f(x^*(t), t)\right.$$

$$\left. - K(t)h(x^*(t), t)\right]dt + Q^{1/2}(t)dw(t)$$

$$- K(t)R^{1/2}(t)dv(t), \quad e_1(t_0) = x_0 - \overline{x}_0. \tag{37}$$

Accordingly, the equation for the mean value of the error $\overline{e}_1(t)$
is following:

$$d\overline{e}_1(t) = \left[(A(t) - K(t)C(t))\overline{e}_1(t) + f(x^*(t), t)\right.$$

$$\left. - K(t)h(x^*(t), t)\right]dt, \quad \overline{e}_1(t_0) = 0. \tag{38}$$

Due to the term $[f(x^*(t), t) - K(t)h(x^*(t), t)]$, it is clear

that Eq. (38) will have a nonzero solution, i.e.,

$$\bar{e}_1(t) \equiv E\{e_1(t)\} \neq 0. \tag{39}$$

Hence our estimate is biased unless the term $[f(x^*(t), t) -$

$K(t)h(x^*(t), t)]$ is identically equal to zero for all values

of $t \geq t_0$.

From Eqs. (37) and (38), we have

$$de_1(t) - d\bar{e}_1(t) = [A(t) - K(t)C(t)](e_1(t) - \bar{e}_1(t))dt$$

$$+ Q^{1/2}(t)dw(t) - K(t)R^{1/2}(t)dv(t), \tag{40}$$

$$e_1(t_0) - \bar{e}_1(t_0) = x_0 - \bar{x}_0.$$

By definition the covariance matrix $P(t)$ is

$$P(t) = E\left\{(e_1(t) - \bar{e}_1(t))(e_1(t) - \bar{e}_1(t))'\right\}. \tag{41}$$

Straightforward mathematical manipulations therefore show that

$P(t)$ is given by the following differential equation:

$$dP(t) = [(A(t) - K(t)C(t))P(t) + P(t)(A(t) - K(t)C(t))'$$

$$+ Q(t) + K(t)R(t)K'(t)]dt, \qquad P(t_0) = P_0. \tag{42}$$

We next seek the gain $K(s)$, $t_0 \leq s \leq t$, that will provide the

minimum mean-square error. We therefore formulate the following

optimization problem:

$$\begin{array}{c} \min \\ K(s) \\ t_0 \leq s \leq t \end{array} \quad \mathrm{tr}(P(t))$$

$$+ \int_{t_0}^{t} [f(x^*(s), s) - K(s)h(x^*(s), s)]'$$

$$\times [f(x^*(s), s) - K(s)h(x^*(s), s)]ds, \tag{43}$$

subject to the constraint given by Eq. (42). This can be

rewritten as the following minimization problem:

$$\min_{\substack{K(s) \\ t_0 \leq s \leq t}} \int_{t_0}^{t} tr([A(s) - K(s)C(s)]P(s) + P(s)[A(s) - K(s)C(s)]'$$

$$+ K(s)R(s)K'(s) + [f(x^*(s), s) - K(s)h(x^*(s), s)]$$

$$\times [f(x^*(s), s) - K(s)h(x^*(s), s)]')ds. \qquad (44)$$

The integrand in Eq. (44) is a convex quadratic in $K(t)$. According to the theory of calculus of variations [19], the minimizing $K(s)$, $t_0 \leq s \leq t$, is given as the solution of the Euler's equation, which reduces to a simple algebraic equation in the present case, namely:

$$\partial/\partial K(t) \ tr([A(t) - K(t)C(t)]P(t) + P(t)[A(t) - K(t)C(t)]'$$

$$+ K(t)R(t)K'(t) + [f(x^*(t), t) - K(t)h(x^*(t), t)]$$

$$\times [f(x^*(t), t) - K(t)h(x^*(t), t)]') = 0. \qquad (45)$$

Using the concept of gradient matrices and the formulas developed in [52], we get

$$\partial/\partial K(t) \ tr(K(t)C(t)P(t)) = P(t)C'(t), \qquad (46)$$

$$\partial/\partial K(t) \ tr(P(t)C'(t)K'(t)) = P(t)C'(t), \qquad (47)$$

$$\partial/\partial K(t) \ tr(K(t)R(t)K'(t)) = 2K(t)R(t), \qquad (48)$$

$$\partial/\partial K(t) \ tr([f(x^*(t), t) - K(t)h(x^*(t), t)]$$

$$\times [f(x^*(t), t) - K(t)h(x^*(t), t)]')$$

$$= -2f(x^*(t), t)h'(x^*(t), t) + 2K(t)h(x^*(t), t)h'(x^*(t), t). \qquad (49)$$

Substituting Eqs. (46)-(49) in Eq. (45), the optimal gain is found to satisfy the following equation:

$$K(t)[R(t) + h(x^*(t), t)h'(x^*(t), t)]$$

$$= P(t)C'(t) + f(x^*(t), t)h'(x^*(t), t). \qquad (50)$$

The solution to the filtering problem of the approximate model is therefore given by

$$d\hat{x}_1(t) = A(t)\hat{x}_1(t)dt + K(t)[dy(t) - C(t)\hat{x}_1(t)dt], \tag{51}$$

$$\hat{x}_1(t_0) = \bar{x}_0,$$

$$K(t) = [P(t)C'(t) + f(x^*(t), t)h'(x^*(t), t)]$$
$$\times [R(t) + h(x^*(t), t)h'(x^*(t), t)]^{-1}, \tag{52}$$

$$dP(t) = [(A(t) - K(t)C(t))P(t) + P(t)(A(t) - K(t)C(t))'$$
$$+ Q(t) + K(t)R(t)K'(t)]dt, \qquad P(t_0) = P_0. \tag{53}$$

It is clear that the inverse in the gain equation [Eq. (52)] exists because $R(t)$ is a positive definite matrix and $h(x^*(t), t)h'(x^*(t), t)$ is always a positive semidefinite matrix.

Although the bias term $[f(x^*(t), t) - K(t)h(x^*(t), t)]$ has been minimized by choosing the gain $K(t)$ according to Eq. (50), it is not identically zero. The bias can be eliminated by modifying the state estimate equation such that the filter will be as follows:

$$d\hat{x}_1(t) = \left[A(t)\hat{x}_1(t) + f(x^*(t), t)\right]dt$$
$$+ K(t)\left[dy(t) - \left(C(t)\hat{x}_1(t) + h(x^*(t), t)\right)dt\right], \tag{54}$$

$$\hat{x}_1(t_0) = \bar{x}_0,$$

$$K(t) = [P(t)C'(t) + f(x^*(t), t)h'(x^*(t), t)]$$
$$\times [R(t) + h(x^*(t), t)h'(x^*(t), t)]^{-1}, \tag{55}$$

$$dP(t) = [(A(t) - K(t)C(t))P(t) + P(t)(A(t) - K(t)C(t))$$
$$+ Q(t) + K(t)R(t)K'(t)]dt, \qquad P(t_0) = P_0. \tag{56}$$

Next, the nominal trajectory $x^*(t)$ has to be updated optimally so as to drive it as close as possible to $x_1(t)$. Hence, the following minimization problem is formulated:

$$\min_{x^*(t)} \quad J(x^*(t)) = E_{Y_t}\left\{\left(x_1(t) - x^*(t)\right)'\left(x_1(t) - x^*(t)\right)\right\}. \tag{57}$$

Then for every $t \geq t_0$, setting $\partial J(x^*(t))/\partial x^*(t) = 0$, we get

$$x^*(t) = E_{Y_t}\{x_1(t)\} = \hat{x}_1(t). \tag{58}$$

Now, by combining the results in Eqs. (51)-(53) and (58) we obtain the E2 filter (E2F) as follows:

$$d\hat{x}_1(t) = A(t)\hat{x}_1(t)dt + K(t)[dy(t) - C(t)\hat{x}_1(t)]dt, \tag{59}$$

$$\hat{x}_1(t_0) = \overline{x}_0,$$

$$K(t) = \left[P(t)C'(t) + f(\hat{x}_1(t), t)h'(\hat{x}_1(t), t)\right]$$
$$\times \left[R(t) + h(\hat{x}_1(t), t)h'(\hat{x}_1(t), t)\right]^{-1}, \tag{60}$$

$$dP(t) = [(A(t) - K(t)C(t))P(t) + P(t)(A(t) - K(t)C(t))'$$
$$+ K(t)R(t)K'(t) + Q(t)]dt, \qquad P(t_0) = P_0. \tag{61}$$

And, by combining the results in Eqs. (54)-(56) and (58) we obtain the E2N filter (E2NF) as follows:

$$d\hat{x}_1(t) = [A(t)\hat{x}_1(t) + f(\hat{x}_1(t), t)]dt + K(t)$$
$$\times [dy(t) - (C(t)\hat{x}_1(t) + h(\hat{x}_1(t), t))dt], \tag{62}$$

$$\hat{x}_1(t_0) = \overline{x}_0,$$

$$K(t) = \left[P(t)C'(t) + f(\hat{x}_1(t), t)h'(\hat{x}_1(t), t)\right]$$
$$\times \left[R(t) + h(\hat{x}_1(t), t)h'(\hat{x}_1(t), t)\right]^{-1}, \tag{63}$$

$$dP(t) = [A(t) - K(t)C(t))P(t) + P(t)(A(t) - K(t)C(t))'$$
$$+ K(t)R(t)K'(t) + Q(t)]dt, \qquad P(t_0) = P_0. \tag{64}$$

A few points should be mentioned in commenting on the results given by Eqs. (59)-(61) and (62)-(64). It is easy to recognize that both the E2F and the E2NF will reduce to the standard KF when there are no nonlinearities in the system structure. The E2F has a linear structure for the state estimate equation. However, the gain matrix $K(t)$ and the covariance matrix $P(t)$ for both filters are state-estimate dependent, a common feature in many of the suboptimal nonlinear filters. The results indicate that the measurement nonlinearities have an effect on the filter gain similar to that of adding to the measurement noise by increasing its covariance. On the other hand, both the dynamics and the measurement nonlinearities have a combined effect similar to $P(t)C'(t)$. If there are no measurement nonlinearities [$h(x(t), t) \equiv 0$], then the E2F will reduce to the standard KF without compensating for the dynamics nonlinearities, whereas the E2NF will reduce to the E1F given by Eqs. (21)-(23).

D. *NUMERICAL EXPERIMENTS*
   FOR THE E2F AND THE E2NF

As was done earlier, the Van der Pol oscillator is chosen to compare the following filters: the E2F and the KF in one experiment; and the E2NF, the EKF, and the M2F in a second experiment. Table II gives the values for noise statistics considered.

TABLE II. *Values for Noise Statistics*[a]

Van der Pol case No.	Q_{11}	Q_{12}	Q_{22}	R	Fig. No.
1	0.5	0.0	0.5	4.0	5-8
2	5.0	0.0	5.0	10.0	9-12
3	10.0	0.0	10.0	20.0	13-16

[a]ϵ is taken to be 0.2.

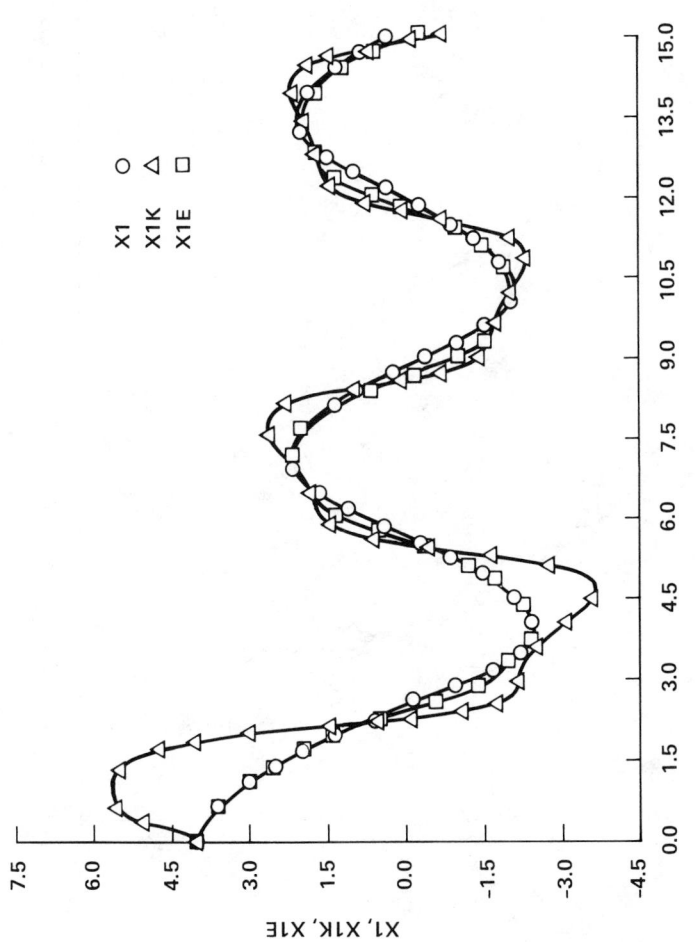

Fig. 5. First state and estimates by the KF and the E2F: X1 is the 1st state; X1K and X1E are the estimates of the 1st state provided by the KF and E2F, respectively.

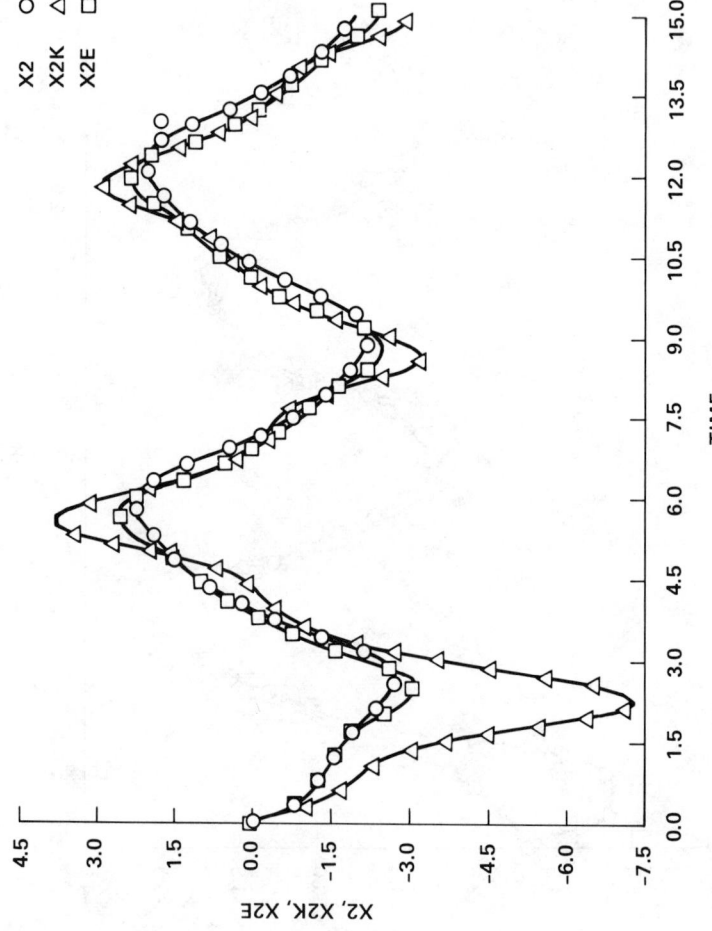

Fig. 6. Second state and estimates by the KF and the E2F: X2 is the 2nd state, X2K and X2E are the estimates of the 2nd state provided by the KF and the E2F, respectively.

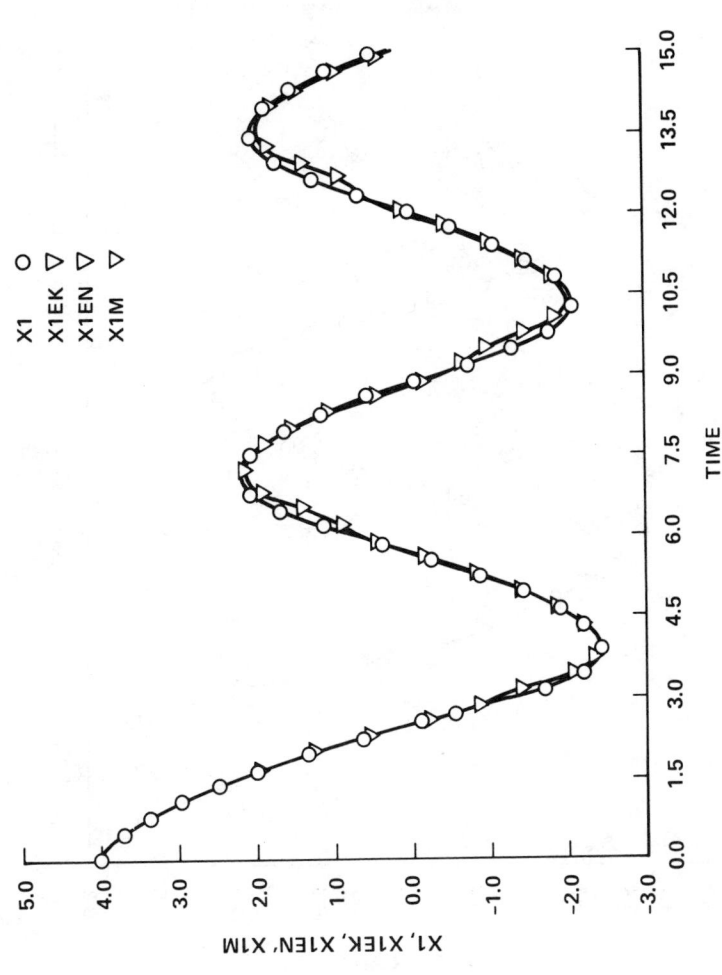

Fig. 7. First state and estimates by the EKF, the E2N, and the M2F: X1 is the 1st state; X1EK, X1EN, and X1M are the estimates of the 1st state provided by the EKF, the E2NF, and the M2F, respectively.

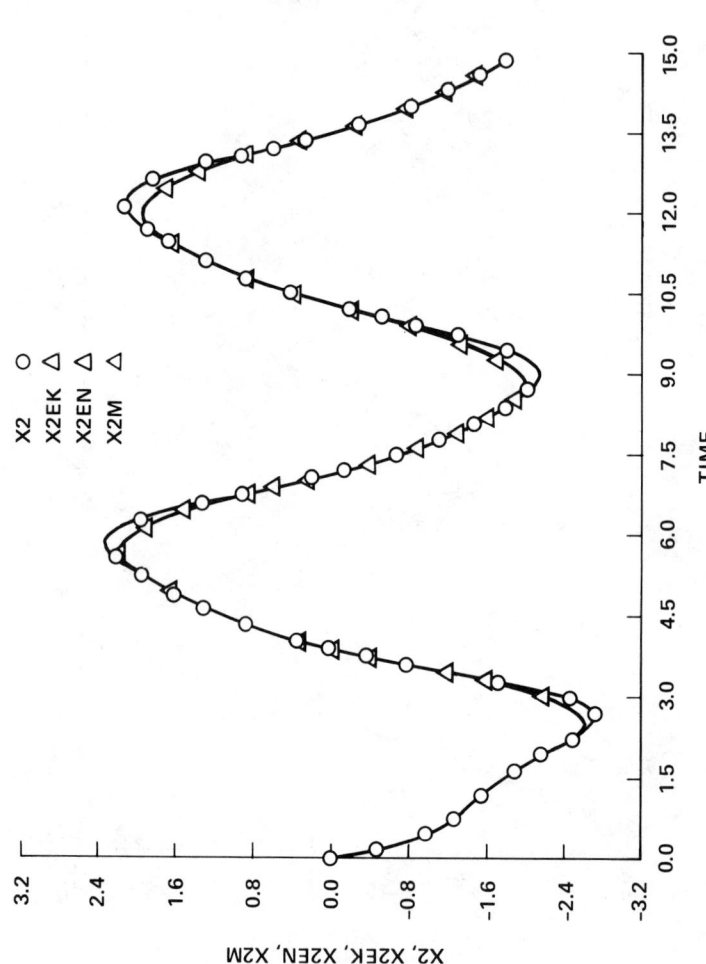

Fig. 8. Second state and estimates by the EKF, the E2N, and the M2F: X2 is the 2nd state; X2EK, X2EN, and X2M are the estimates provided by the EKF, the E2NF, and the M2F, respectively.

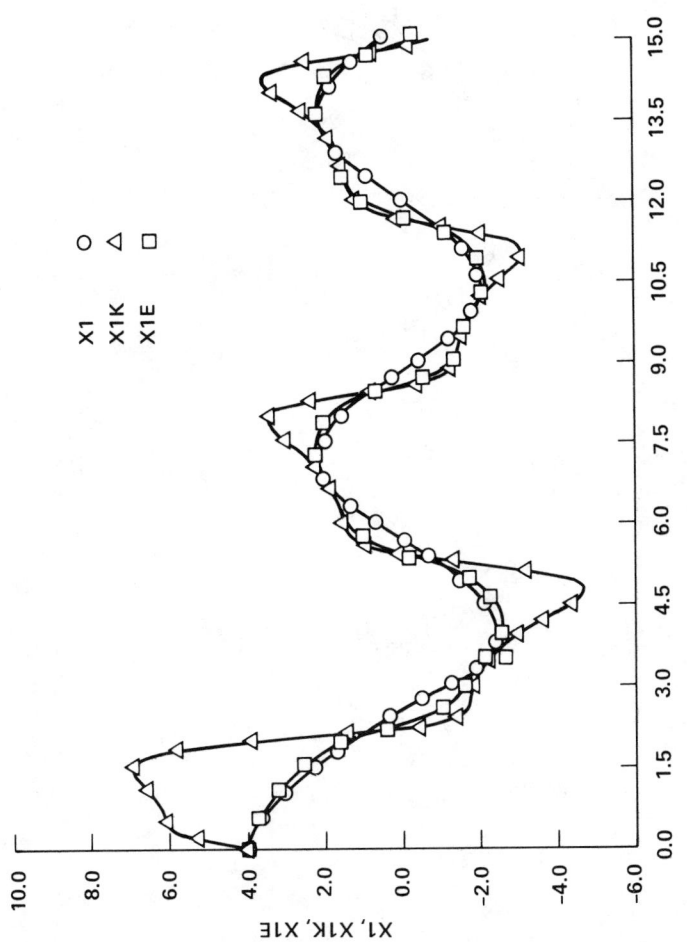

Fig. 9. First state and estimates by the KF and the E2F. (X1, X1K, and X1E are as defined in Fig. 5 legend.)

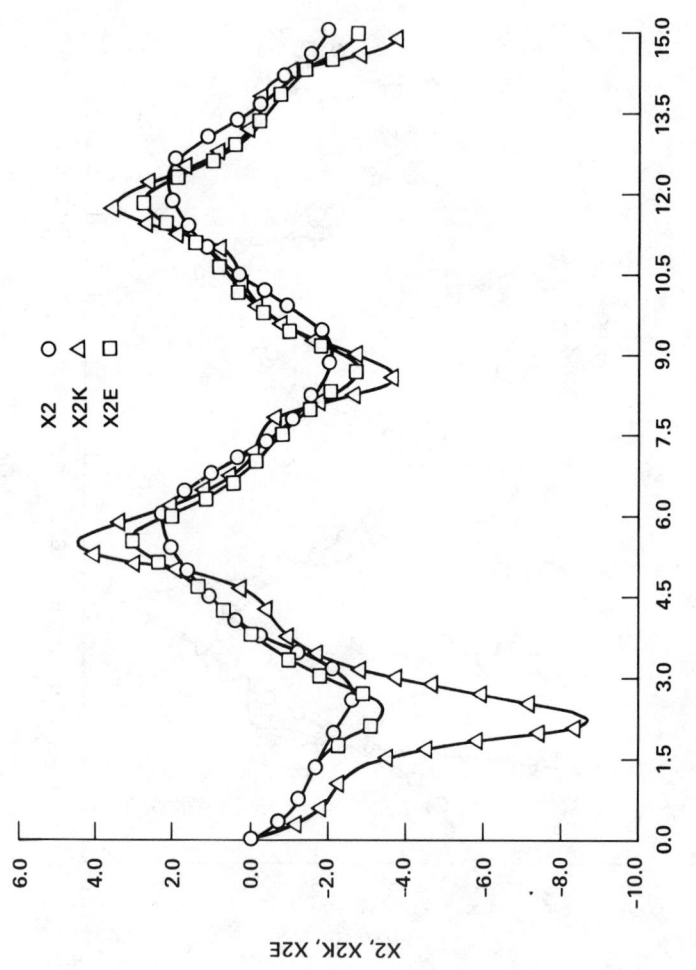

Fig. 10. Second state and estimates by the KF and E2F. (X2, X2K, and X2E are as defined in Fig. 6 legend.)

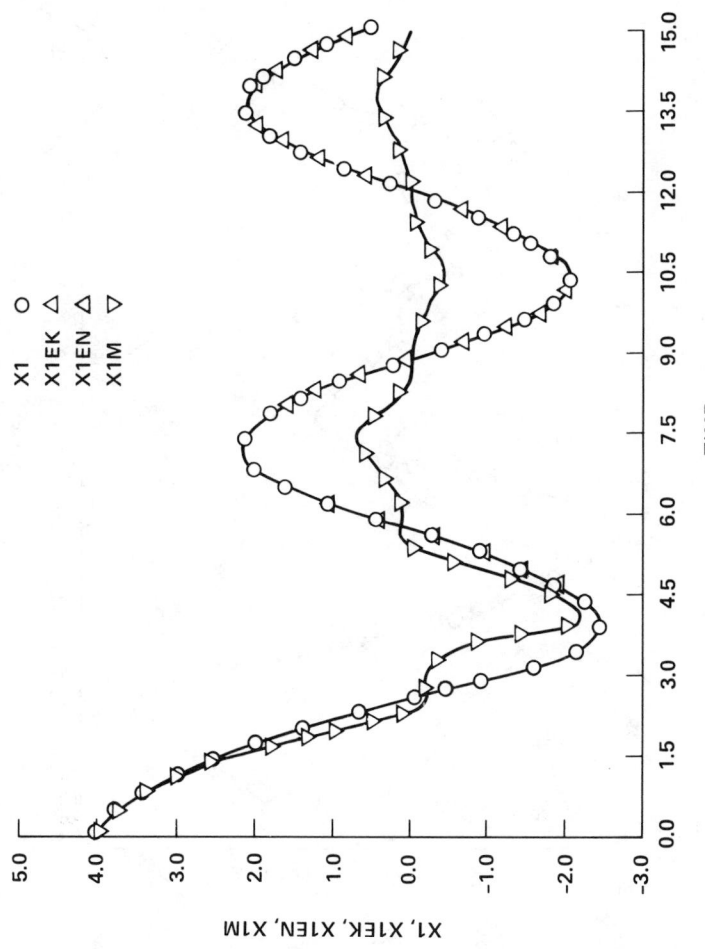

Fig. 11. First state and estimates by the EKF, the E2N, and the M2F. (X1, X1EK, X1EN, and X1M are as defined in Fig. 7 legend.)

325

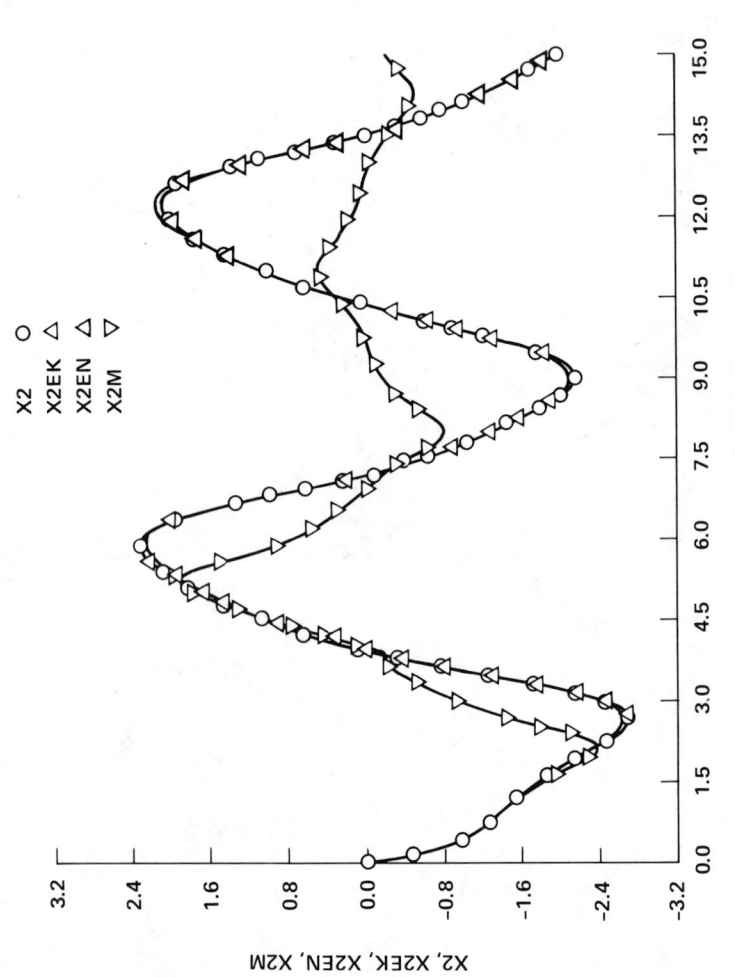

Fig. 12. Second state and estimates by the EKF, the E2N, and the M2F. (X2, X2EK, X2EN, and X2M are as defined in Fig. 8 legend.)

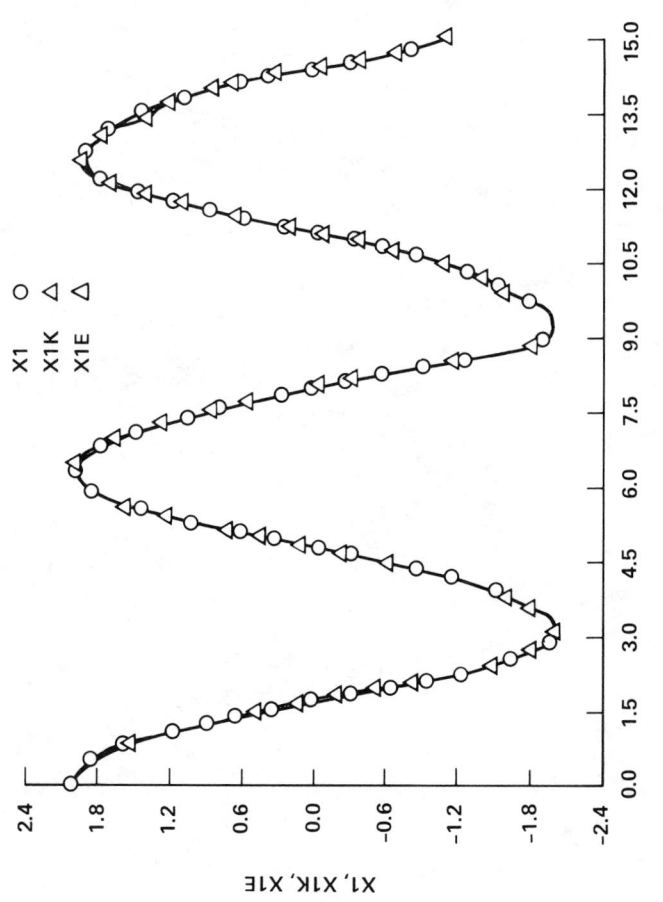

Fig. 13. First state and estimates by the KF and the E2F. (X1, X1K, and X1E are as defined in Fig. 5 legend.)

327

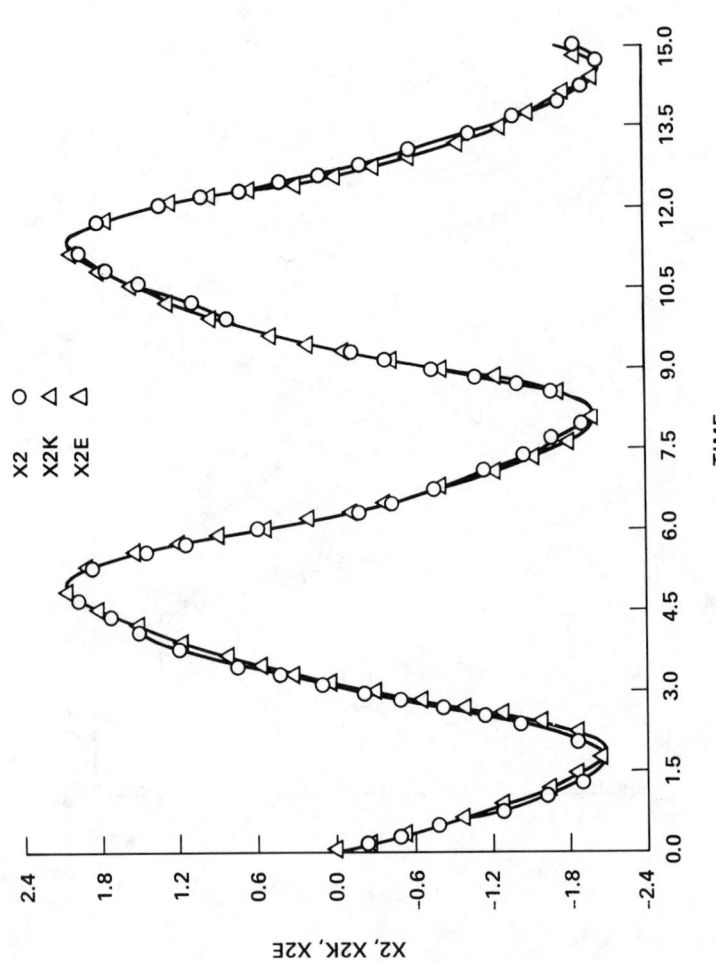

Fig. 14. Second state and estimates by the KF and the E2F.   (X2, X2K, and X2E are defined in Fig. 6 legend.)

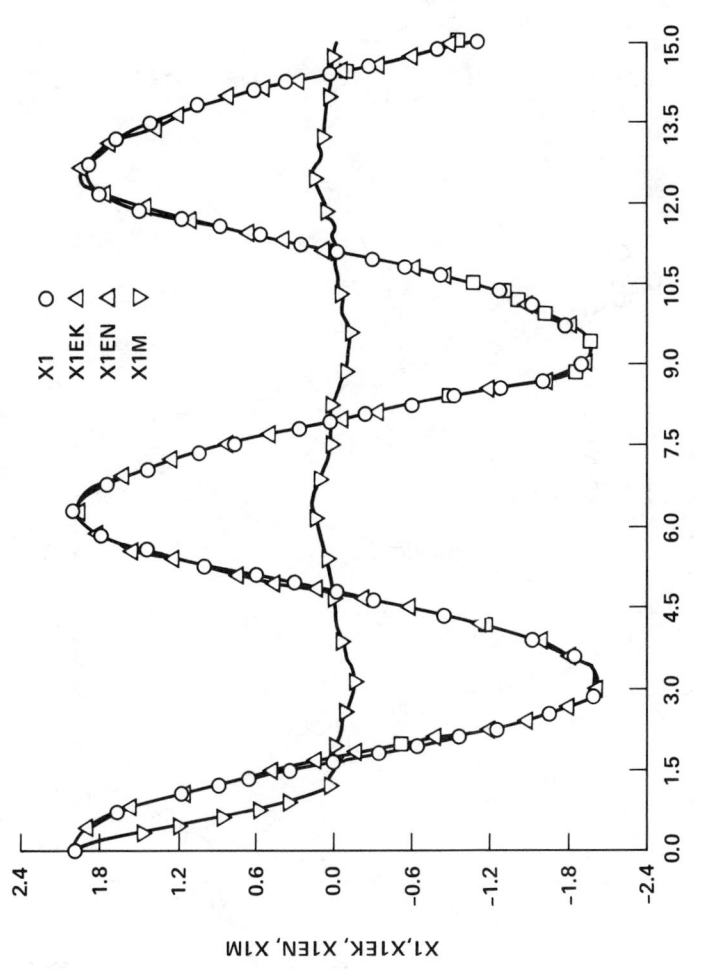

Fig. 15. First state and estimates by the EKF, the E2N, and the M2F. (X1, X1EK, X1EN, and X1M are as defined in Fig. 7 legend.)

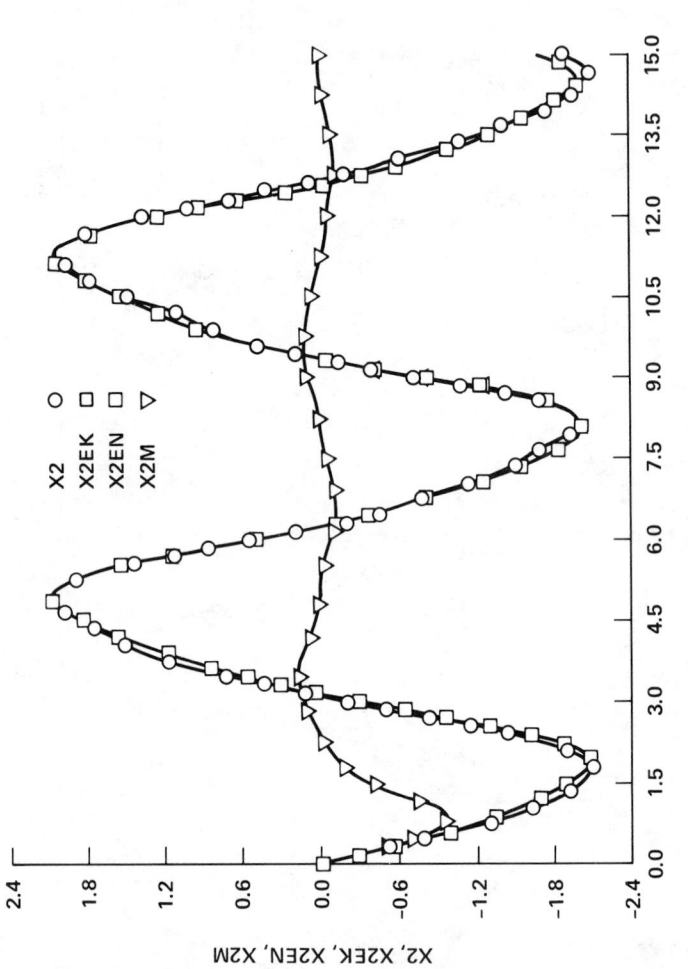

Fig. 16. Second state and estimates by the EKF, the E2N, and M2F. (X2, X2EK, X2EN, and X2M are as defined in Fig. 8 legend.)

From the results the following can be pointed out. In the
first two sets of results, Van der Pol 1 and 2, the E2F provides better tracking of the system states than the KF. This
is evident from Figs. 5, 6, 9, and 10. It is clear that the
E2NF has an accuracy similar to that of the EKF, which is better
than the M2F as indicated by Figs. 7, 8, 11, and 12. In the
third set of results, Van der Pol 3, Figs. 13-16, the noise
level is high enough to cover the effect of the system nonlinearities. Therefore, all filters except the M2F are similar
in performance. The M2F is badly degraded and provides a crude
estimate of the system state.

IV. CONCLUSIONS

The nonlinear filtering problem is treated by using a new
approach. The approach consists of unifying a system model
approximation technique with the filtering solution based on
the approximate model. As a result, several filters are
developed.

The first filter (the E1F) structurally fits into the gap
between the KF and the EKF. On one hand it enjoys the same
computational facility enjoyed by the KF, namely, the off-
line computations of its gain matrix. On the other hand it
provides state estimates on the same level of accuracy as those
provided by the EKF. In this sense the E1F therefore provides
a missing link between the KF and the EKF.

The other two filters are referred to as the E2F and the
E2NF. The state estimate provided by the E2F has a structure
similar to KF, whereas that of the E2NF has a structure similar
to the EKF. Both filters have new formulas for the gain which
provide further insight into the effects of the system

nonlinearities. Specifically, measurement nonlinearities have
the effect of increasing the measurements noise level. Moreover, the dynamics nonlinearities and also the measurement nonlinearities have a combined effect similar to the $P(t)C'(t)$
term in the KF.

In conclusion, the contribution of this chapter is to provide three new practically implementable filters for stochastic
dynamic systems that include nonlinearities in their structure.

REFERENCES

1. R. E. KALMAN, *J. Basic Eng. 82*, 35-45 (1960).

2. F. C. SCHWEPPE, "Uncertain Dynamic Systems," Prentice-Hall,
    Englewood Cliffs, New Jersey, 1973.

3. A. H. JAZWINSKI, "Stochastic Processes and Filtering
    Theory," Academic Press, New York, 1970.

4. J. S. MEDITCH, "Stochastic Optimal Linear Estimation and
    Control," McGraw-Hill, New York, 1969.

5. R. E. BELLMAN, H. H. KAGIWADA, R. E. KALABA, and R. SRIDHAR,
    J. Astronaut. Sci. 13, 110-115 (1966).

6. D. M. DETCHMENDY and R. SRIDHAR, *Proc. Joint Automatic Con-
    trol Conf.*, pp. 56-63 (1965).

7. H. COX, *IEEE Trans. Autom. Control AC-9*, 5-12 (1964).

8. V. O. MOWERY, *IEEE Trans. Autom. Control AC-10*, 399-407
    (1965).

9. M. ATHANS, and E. TSE, *IEEE Trans. Autom. Control AC-12*,
    690-698 (1967).

10. M. ATHANS, R. P. WISHNER, and A. BERTOLINI, *IEEE Trans.
    Autom. Control AC-13*, 504-514 (1968).

11. LAWRENCE SCHWARTZ and EDWIN B. STEAR, *IEEE Trans. Autom.
    Control AC-13*, 83-86 (1968).

12. D. F. LIANG and G. S. CHRISTENSEN, *Automatica 11*, 603-612
    (1975).

13. W. FELLER, "An Introduction to Probability Theory and Its
    Applications," erd ed., Vol. 1, Wiley, New York, 1970.

14. J. E. POTTER and J. C. DECKERT, *SIAM J. Control 8*, 513-525
    (1970).

15. E. J. BAUMAN, C. T. LEONDES, and D. A. WISMER, *Int. J. Control 8*, 473-481 (1968).

16. M. ZAKAI and J. ZIV, *IEEE Trans. Autom. Control AC-18*, 325-331 (1972).

17. C. M. FRY and A. P. SAGE, *Comput. Elec. Eng. 1*, 361-389 (1973).

18. H. COX, "Estimation of State Variables Via Dynamic Programming," *Preprints Joint Autom. Control Conf. 5th Stanford*, pp. 376-381 (1964).

19. D. E. KIRK, "Optimal Control Theory," Prentice-Hall, Englewood Cliffs, New Jersey, 1970.

20. R. COURANT, "Calculus of Variations," Courant Inst. of Mathematical Sci., New York University, New York, 1962.

21. I. M. GELFAND and S. V. FOMIN, "Calculus of Variations," Prentice-Hall, Englewood Cliffs, New Jersey, 1963.

22. NAUM I. AKHIEZER, "The Calculus of Variations," Random House (Blasidell), New York, 1962.

23. LUDWIG ARNOLD, "Stochastic Differential Equations: Theory and Applications," Wiley (Interscience), New York, 1974.

24. W. J. CUNNINGHAM, "Introduction to Nonlinear Analysis," McGraw-Hill, New York, 1958.

25. W. FELLER, "An Introduction to Probability Theory and Its Applications," 2nd ed., Vol. 2, Wiley, New York, 1970.

26. J. L. SNYDER and I. B. RHODES, *Automatica 8*, 747-753 (1972).

27. J. S. LEE and F. F. TUNG, *IEEE Trans. Autom. Control AC-15*, 74-81 (1970).

28. R. J. FITZGERAL, *IEEE Trans. Autom. Control AC-16*, 736-747 (1971).

29. A. L. C. QUIGLEY, *Int. J. Control 17*, 741-746 (1973).

30. W. F. DENHAM and S. PINES, *AIAA J. 4*, 1071-1076 (1966).

31. A. S. GILMAN and I. B. RHODES, *IEEE Trans. Autom. Control AC-18*, 260-265 (1973).

32. I. B. RHODES and A. S. GILMAN, *IEEE Trans. Autom. Control AC-20*, 632-642 (1975).

33. H. W. SORENSON and D. L. ALSPACH, *Automatica 7*, 465-479 (1971).

34. D. L. ALSPACH and H. W. SORENSON, *IEEE Trans. Autom. Control AC-17*, 439-448 (1972).

35. D. T. MAGILL, *IEEE Trans. Autom. Control AC-10*, 434-439 (1965).

36. F. L. SIMS and D. G. LAINIOTIS, *IEEE Trans. Autom. Control AC-14*, 215-218 (1969).

37. D. G. LAINIOTIS and F. L. SIMS, *IEEE Trans. Autom. Control AC-15*, 249-250 (1970).

38. C. G. HILBORN, JR. and D. G. LAINIOTIS, *IEEE Trans. Syst. Sci. Cybernetics SSC-5*, 38-43 (1969).

39. D. G. LAINIOTIS, *IEEE Trans. Autom. Control AC-16*, 160-170 (1971).

40. M. ATHANS and C. B. CHANG, "Adaptive Estimation and Parameter Identification Using Multiple Model Estimation Algorithm," Tech. Note 1976-28, Massachusetts Institute of Technology, Lexington, Massachusetts, June 1976.

41. H. W. SORENSON and A. R. STUBBERUD, *Int. J. Control 8*, 33-51 (1968).

42. H. J. KUSHNER, *J. Math. Anal. Appl. 8*, 332-344 (1964).

43. H. J. KUSHNER, *J. SIAM Control Ser. A 2*, 106-119 (1962).

44. H. J. KUSHNER, *IEEE Trans. Automatic Control AC-12*, 546-556 (1967).

45. A. K. BEJCZY and R. SRIDHAR, "Approximate Nonlinear Filters and Deterministic Filter Gains." ASME, 1970, New York, N. Y., Nov. 29-Dec. 3, 1970.

46. A. K. BEJCZY and R. SRIDHAR, *J. Math. Anal. Appl. 36*, 477-505 (1971).

47. W. M. WONHAM, *J. SIAM Control Ser. A 2*, 347-369 (1965).

48. T. KAILATH, *IEEE Trans. Autom. Control AC-13*, (1968).

49. T. KAILATH and P. FROST, *IEEE Trans. Autom. Control AC-13*, 655-660 (1968).

50. P. A. FROST and T. KAILATH, *IEEE Trans. Autom. Control AC-16*, 217-226 (1971).

51. W. C. MARTIN and A. R. STUBBERUD, in "Control and Dynamic Systems," "Advances in Theory and Applications," ed. C. T. Leondes, Vol. 12, pp. 173-258, Academic Press, New York, 1976.

52. M. ATHANS and F. C. SCHWEPPE, "Gradient Matrices and Matrix Calculations," Tech. Note 1965-53, Massachusetts Institute of Technology, Lincoln Laboratory, Lexington, Massachusetts, November 1965.

53. M. ATHANS, *Inf. Control 11*, 592-606 (1968).

54. A. GELB (ed.), "Applied Optimal Estimation," M.I.T. Press, Cambridge, Massachusetts, 1974.

55. D. G. LAINIOTIS (ed.), "Estimation Theory," American Elsevier, New York, 1974.

56. A. PAPOULIS, "Probability, Random Variables, and Stochastic Processes," McGraw-Hill, New York, 1965.

57. E. WONG, "Stochastic Processes in Information and Dynamical Systems," McGraw-Hill, New York 1971.

58. H. P. McKEAN, JR., "Stochastic Integrals," Academic Press, New York, 1969.

59. A. T. BHARUCHA-REID, "Elements of the Theory of Markov Processes and Their Applications," McGraw-Hill, New York, 1960.

60. T. KAILATH, *Inf. Theory IT-20*, 145-181 (1974).

61. H. E. EMERA-SHABAIK, "Filtering of Systems with Non-linearities," Ph.D. Dissertation, University of Californa, Los Angeles, 1979.

62. H. E. EMARA-SHABAIK and C. T. LEONDES, *Automatica 17*, 411-412 (1981).

63. P. P. GUSAK, *Telemekhanika* No. 4, pp. 70-76 (1981).

64. M. H. A. DAVIS, *Proc. IEE 128*, 166-172 (1981).

65. C. B. CHANG, *IEEE Trans. Autom. Control AC-26*, 1294-1297 (1981).

66. R. L. STRATONOVICH, "Application of the Theory of Maikov Processes for Optimum Filtration of Signals," Radio Eng. Elecron. Phys. (USSR), Vol. 1, pp. 1-19, Nov. 1960.

Applications of Filtering Techniques to Problems in Communication–Navigation Systems

TRIVENI N. UPADHYAY

JERRY L. WEISS

Radio Communication and Navigation Division
The Charles Stark Draper Laboratory, Inc.
Cambridge, Massachusetts

I. INTRODUCTION

There exists enough motivation in the avionics community for serious attention to be paid to integrated-systems concepts. One of the widely advanced reasons for developing an integrated system underscores human limitations in responding, managing, and utilizing multiple-sensor output, in real time, to achieve a mission's objectives. Additional limitations of space on board the vehicle (e.g., a tactical aircraft) for accommodating new sensors (to meet new mission requirements) have also motivated development of integrated-system concepts. Finally, system reliability with regard to a failure in one or more sensor, or equivalently, system availability necessitates inclusion of integration concepts into the development of avionics systems.

Considerable success has been achieved toward developing integrated-system design concepts that involve some of these considerations. Reduction of size (volume) and weight through functional integration of all communication, navigation, and identification (CNI) sensors on an aircraft is the theme of the work described by Botha and Smead [18], Ward and Reilly [19], and Camana, *et al.* [20]. Their work describes new approaches for achieving reduction in both size and weight while maintaining existing levels of individual system performance and reliability. Ward and Reilly [19] advocate the merits of a fully digital programmable receiver to support the integration requirements of CNI sensors and have proposed a means of achieving it through employing very high speed charge-coupled device (CCD) transversal filters as radio frequency (RF) selective filters. This approach is attractive because it alleviates the need for developing a very high speed, high accuracy A/D converter (e.g., 16 bits A/D at 3 GHz for this application) and very high speed digital filters for the integrated CNI receiver. The technology to realize very high speed CCD transversal filters, for this application, however, is not yet here and will require extensive device development effort. Camana *et al.* [20] also recommend the trend toward a digital integrated receiver, however, their approach is based on a common RF front end followed by a centralized digital processing subsystem which takes advantage of multitasking attributes of emerging very large scale integration (VLSI) signal processors. The highlights of their design in terms of supporting CNI system integration requirements are the extensive use of common modules, RF large-scale integration (LSI) technology and VLSI digital signal processor technology (up to 100 million instructions per second [MIPS]).

In addition to CNI sensor integration, a significant amount of work has gone into improving system reliability through integration [21-23]. System integration provides a vehicle for incorporation of redundancy (through both hardware replication and analytic relationships) and fault detection, identification, and reconfiguration methodology while keeping the system complexity and cost to a minimum. The successful implementation of fault detection, identification, and reconfiguration methodologies in an avionics system requires a highly reliable processor. Recent progress in developing a highly reliable, fault-tolerant processor is presented in [22], and is application to an integrated advanced aircraft avionics system is detailed in [23].

Although the integration approaches just discussed (namely, the improvement of system reliability through incorporation of fault-tolerant concepts and the reduction of size and weight through the use of common modules or a generalized RF front end) are important in their own right, they have failed to take advantage of the additional performance benefits that can be derived from the integration of the signal-processing algorithms which exploit the synergism between the various navigation and communication sensors. This article attempts to fill this gap.

Until recently [1-5] very little effort was devoted toward the goal of developing an unified theoretical framework for an optimal integration of the various sensor signal-processing functions, primarily because it was felt that the individual radio navigation and communication sensor performance was adequate to meet the mission goals, and therefore the need for integration over subsystems was not clearly established. In addition, the available mathematical theory and tools were considered too inadequate to address the difficult problem of performance optimization over several, preferably nonlinear, subsystems. Furthermore, it was implicitly understood that the optimization results might lead the way to redesigning (hardware) some of the sensors, which might add additional complexity (cost) to the problem. These considerations may have discouraged development of unified integration approaches in the past. Recent advances, however, in approximation theory and nonlinear filtering, coupled with advances in microprocessor technology have made performance optimization tractable and attractive in terms of implementation. The whole field of applications of filtering theory to improve performance, through optimal integration of information, is still relatively

new. This article addresses a limited subset of the problem
in which the integration concepts and results are applied to
navigation sensors comprising only the global positioning system (GPS) and the inertial navigation system (INS). However,
the approach is considered general and has wider applications
(e.g., failure detection-estimation, maneuver target detection,
and tracking). The optimism in this statement is based on the
work of several researchers [24-27,49-50] in which it has been
shown that the detection, estimation, and system identification
problems arising in communication and navigation systems can be
viewed as a general problem in filtering theory.

Section II reviews the theoretical background on the detection, estimation, and identification problems in communication-
navigation systems. It is shown here that problems of detection,
estimation, and system identification can be formulated and
solved as problems in optimal nonlinear estimation. The rest
of the article is organized as follows: Section III reviews
optimal nonlinear filtering theory and presents some important
theoretical results in this field, namely, Fokker-Planck equations and Kushner's equations, [33,34]. The approach taken in
Section III utilizes the Markov property of the outputs of con-
tinuous-time, finite-dimensional dynamic systems when excited
by independent increment processes. Implementation problems
associated with optimal nonlinear filters are discussed. Several approximation techniques used to obtain suboptimal but
computable, finite-dimensional filters are discussed in Section
IV. Section V deals with applications. It outlines some of
the performance limitations of individual communication-navigation systems and presents an optimal integration approach based
on nonlinear filtering theory. Specific nonlinear estimation

results for the GPS signal-tracking problems are described in
Section V. An alternative to the minimum mean-square error
(MMSE) criterion, namely, the least mean square (LMS), has also
been applied to optimize the GPS signal-tracking performance.
The LMS result is contrasted with the MMSE solution. Extensions and applications of these concepts to the integration of
the combined signal acquisition, tracking, and navigation problem and to a general target-tracking problem are also discussed
in Section V. The final topic dealt with in this article focuses on the realization (implementation) issues of information integration of these sensors in light of recent advances
in VLSI/VHSIC processing technology (Section VI). Section VII
summarizes the status of the current research trend in nonlinear
estimation and presents an outlook for the future.

II. A UNIFIED APPROACH TO DETECTION, ESTIMATION, AND SYSTEM IDENTIFICATION

A fundamental, and common, problem in the design of radio
communication-navigation systems involves the coupling between
the major functions of these systems: (1) detection of signals
in noise (signal classification), (2) estimation or extraction
of the signal in noise, and (3) system identification or
parameter estimation. The signal-detection problem for the
most part concerns itself with a statistical decision based on
observations from a random process with regard to the presence
or absence of a signal. In the estimation or filtering problem, the observations are related to the value of some random
process of interest (e.g., signal amplitude), and one is concerned with estimating the value of the random process. The

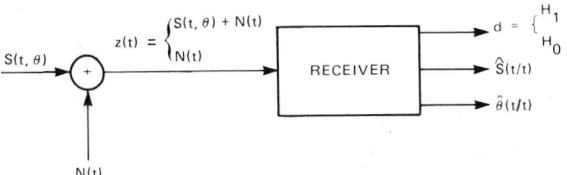

Fig. 1.   Joint detection-estimation-system identification problem.

parameter estimation problem, on the other hand, attempts to estimate the value of the system parameter (system realization) that may have produced the random process of interest.

The general problem of joint detection-estimation and system identification is depicted in Fig. 1. In many applications, the well-known binary decision problem with regard to the presence or absence of the signal (hypothesis H_1 or H_0) based on noisy observations $Z(t_1) = \{z(\tau);\ t_0 \leq \tau \leq t_1\}$ is of interest. A communication/navigation receiver, however, may be required to provide an estimate of the signal waveform $S(t)$ and(or) an estimate of the parameter θ that characterize the unknown system generating the signal as well.

Traditionally, these problems are addressed independently in a system design. Consequently, the signal-detection problem implicitly assumes that the value of the parameter or random process that constitutes the signal is known. Alternatively, in a signal parameter (or waveform) estimation design, it is assumed that the signal in question is always surely present (with probability 1) in the noisy observations.

In many practical applications (e.g., radar, communications, fault detection), the situation frequently arises where the properties of the signal source as well as the signal presence or absence are jointly desired. Attempting to solve the detection and estimation problems independently compromises

system performance due to the fact that different cost functions and couplings are associated with these functions.

In spite of the possible performance benefits that can be
realized by a joint consideration (optimization) of the detection, estimation, and identification problems, very little
seems to have been accomplished until the fundamental work of
Middleton and Esposito [27]. Previous attempts in this direction by Price [28] and Kailath [29] were only partially successful. Price and Kailath have suggested the idea of an optimum receiver (a detector) as a combination of an estimator
followed by a detector that treats the signal estimate as its
true value.

Middleton and Esposito [27] formulated the problem of joint
detection and estimation of signals in noise and have obtained
optimum structures under the Bayes criterion of minimum average
risk for detection and a general cost function for estimation.
It was correctly pointed out [27] that from a purely mathematical point of view, the "detection" problem can be viewed as a
special case of "estimation" of a scale factor of the signal,
when 0 and 1 are the only admissible values. Alternatively,
estimation can be viewed as a generalized detection process
with continuum of values. In spite of this simplicity of appearance, the solution of the coupled problem is complex and
was first derived by Middleton and Esposito [27]. We shall
first review this important result and later discuss its gen-
eralizations and extensions.

A. JOINT DETECTION-SIGNAL ESTIMATION

Consider the single-shot[1] compound detection-estimation problem where the observation vector is given by

$$z(t) = s(t) + v(t) = \beta y(t) + v(t). \tag{1}$$

In Eq. (1), $v(t)$ is a white Gaussian zero-mean observation noise random process with covariance matrix $R(t)$, and $\beta(t)$ is the indicator variable which takes values 1 or 0 depending on whether H_1 (signal present) or H_0 (signal absent) is true. Furthermore, assume that a priori probabilities of H_1 and H_0 are p_1 and p_0, respectively.

The problem is to seek an optimum (Bayes) solution for the detection ($\hat{\beta}$) and estimation ($\hat{s}$) problems with an appropriate choice of cost functions and coupling between the two. It is well known [59] that if the detection and estimation are done independently and in parallel, then the Bayes solution for the detection problem is the generalized likelihood ratio (GLR) test. The solution for the estimation problem, however, is different from the classical estimation problem (where it is assumed that the signal is surely present, i.e., $p_1 = 1$) and involves optimization to include the uncertainty regarding the signal presence. If, on the other hand, the detection and estimation processes are coupled, then the solution of the detection problem is not the well-known GLR test.

Middleton and Esposito [27] have shown that for discrete data and a quadratic cost function (MMSE) estimation of $s(t)$

---

[1]Single shot is meant to imply a single set of terminal decisions. In this case β is not a function of time, although its estimate will be a function of time.

$(p_1 \neq 1)$, the following fundamental relationship holds:

$$\hat{s}(t/Z_n) = [\Lambda_n/(1 + \Lambda_n)]\hat{s}(t/Z_n, \beta = 1),$$

$$\hat{s}(t/z_n) = E\{s(t)/Z_n\};$$

$$Z_n = \{z(t_1), z(t_2), \ldots, z(t_n)\}, \tag{2}$$

$$\hat{s}(t/Z_n, \beta = 1) = E\{s(t)/Z_n, \beta = 1\},$$

and Λ_n is the GLR defined as

$$\Lambda_n = \frac{\rho p(Z_n/\beta = 1)}{p(Z_n/\beta = 0)}; \qquad \rho = \frac{p_1}{p_0}. \tag{3}$$

Equation (2) shows that to obtain an optimal estimate of the signal, when the signal is not known surely (with probability 1) to be present, one calculates the optimal estimate assuming that the signal is present $(p_1 = 1)$ and weights this estimate by a scalar factor that is a simple function of the GLR Λ_n operating on the same data. Although the above result holds for all types of coupling between detection and estimation, Middleton and Esposito [27] first obtained it for the no-coupling case (i.e., $R_{total} = R_{detection} + R_{estimation})$.

Lainiotis [25] later established the fundamental fact that detection is indeed a mean-square estimation of the indicator variable β for the model of Eq. (1). He also showed that the optimal Bayes decision is to compare $\hat{\beta}(t_n/Z_n)$ with a threshold where the threshold depends on the a priori cost (risk) for each decision. The optimal MMSE estimate of β is given as

$$\hat{\beta}(t_n/Z_n) = \Lambda_n/(1 + \Lambda_n), \tag{4}$$

where Λ_n is defined by Eq. (3), and $\hat{\beta}(t_n/Z_n) = E\{\beta(t_n)/Z_n\}$. Note that $\hat{\beta}(t_n/Z_n) = p(\beta = 1/Z_n)$ (which is the desired quantity in the detection problem), and $\hat{\beta}(t_n/Z_n, \beta = 1) = 1$.

Combining Eqs. (2) and (4), an interesting interpretation for the joint detection-estimation problem can be obtained as follows:

$$\hat{s}(t_n/Z_n) = \hat{\beta}(t_n/Z_n)\hat{s}(t_n/Z_n, \beta = 1). \tag{5}$$

The optimal estimate of the signal, in the joint detection-estimation problem can be calculated by first obtaining the signal estimate assuming that the signal is present, i.e., solving a classical estimation problem and then multiplying this signal estimate by a scalar factor $\hat{\beta}$, which is the sufficient statistic needed for solving the optimal Bayes decision problem. Equation (5) shows the connection between the estimation and detection problems.

The results just obtained for the discrete data case can be easily extended to the case of continuous observations. We next consider the problem of simultaneous detection-estimation and system identification.

B. JOINT DETECTION-ESTIMATION AND SYSTEM IDENTIFICATION

In this section we shall be dealing with the problem of detection and estimation of random signals in white Gaussian noise and identification of the system that generates the random signal. Furthermore, the results of Section II.A are extended for the case of continuous observations. The results tection-estimation-identification, the optimal solution to the signal detection problem is intimately related to mean-square estimation of a discrete valued scale factor of the signal and that the system identification problem is equivalent to multihypothesis testing with a continuum or finite sequence of hypotheses depending on whether the parameter space is discrete or continuous. Let us proceed with the problem statement. The problem statement and results follow the work of Lainiotis [26] and Park and Lainiotis [24].

Let the signal random process y(t) [Eq. (1)] be described by the following state vector model:

$$z(t) = \beta y(t) + v(t), \qquad y(t) = h(x(t), t, \theta), \qquad (6)$$

$$dx(t)/dt = f(x(t), t, \theta) + g(x(t), t, \theta)w(t), \qquad (7)$$

where x(t) is the n-dimensional state random process, and w(t) is a zero-mean white Gaussian noise process. The functionals h(·), f(·), and g(·) are time-varying nonlinear functionals of the state vector x(t), and the parameter vector θ. The parameter vector θ is assumed to be time invariant and, if known, specifies the model completely.

Furthermore, assume that w(t) and v(t) are statistically independent. Given the continuous observation record Z(t) = $\{z(\tau): t_0 \le \tau \le t\}$, determine the optimal mean-square estimate $\hat{x}_a(t/t)$ of the augmented state vector $x_a(t)$, where $x_a(t)$ is defined as follows. Let

$$x_a^T(t) = [x^T(t) \, \theta^T \, \beta],$$

$$f_a(x_a(t), t) = [f^T(x(t), t, \theta) \quad 0 \quad 0]^T,$$

$$g_a(x_a(t), t) = [g^T(x(t), t, \theta) \quad 0 \quad 0]^T,$$

$$h_a(x_a(t), t) = \beta h(x(t), t, \theta).$$

Then the model described by Eqs. (6) and (7) can be written in terms of the augmented state vector $x_a(t)$ as

$$z(t) = h_a(x_a(t), t) + v(t), \qquad (8)$$

$$dx_a(t)/dt = f_a(x_a(t), t) + g_a(x_a(t), t)w(t). \qquad (9)$$

From these definitions, it is evident that the optimal mean-square estimate of $x_a(t)$ given Z(t) contains all the quantities for the joint minimum Bayes-risk detection, minimum mean square

estimation and system identification. The solution is given

by Lainiotis [26] and Park and Lainiotis [24] and is presented

here without proof. The reader is referred to [24,26] for the

proof of this important result. The joint detection-estimation-

identification problem is solved as a minimum mean-square non-

linear estimation problem. The solution is now given:

$$\hat{\beta}(t) = \frac{\rho \int \Lambda(t/\theta) p(\theta) d\theta}{1 + \rho \int \Lambda(t/\theta) p(\theta) d\theta}, \tag{10}$$

$$\hat{s}(t/t) = \hat{\beta}(t) \hat{s}_1(t/t), \tag{11}$$

$$\hat{\theta}(t) = \hat{\theta}(t_0) p(\beta = 0/t) + \frac{\rho \int \theta \Lambda(t/\theta) p(\theta) d\theta}{1 + \rho \int \Lambda(t/\theta) p(\theta) d\theta}, \tag{12a}$$

$$\hat{x}(t/t) = \int \hat{x}(t/t, \theta) p(\theta/t) d\theta \tag{12b}$$

where

$$\hat{\beta}(t) = E\{\beta/Z_t\} = p(\beta = 1/Z_t) = 1 - p(\beta = 0/Z_t),$$

$$\hat{s}(t/t) = E\{s(t)/Z_t\}, \qquad \hat{s}_1(t/t) = E\{s(t)/Z_t, \beta = 1\},$$

$$\hat{\theta}(t) = E\{\theta(t)/Z_t\}, \qquad \hat{\theta}(t_0) = E\{\theta/t_0\} = \text{the a priori mean}$$

and where

$$p(\theta/t) = \frac{1 + \rho \Lambda(t/\theta)}{1 + \rho \int \Lambda(t/\theta) p(\theta) d\theta} \cdot p(\theta) \tag{13}$$

$$\Lambda(t/\theta) = \exp\left\{ \int_{t_0}^{t} \hat{h}_1^T(\sigma/\sigma, \theta) R^{-1}(\sigma) z(\sigma) d\sigma \right.$$

$$\left. - \frac{1}{2} \int_{t_0}^{t} \|\hat{h}_1(\sigma/\sigma, \theta)\|^2_{R^{-1}(\sigma)} d\sigma \right\}, \tag{14}$$

$$\hat{h}_1(\sigma/\sigma, \theta) = E\{h_a(x_a(\sigma))/\sigma, \theta, \beta = 1\}.$$

The proof of these results is given in [26].

From Eqs. (1) and (6), it readily follows that

$$\hat{s}_1(t/t) = \hat{y}_1(t/t) = \hat{h}_1(t/t).$$ (15)

From Eqs. (11) and (15), it also follows that

$$\hat{s}(t/t) = \hat{\beta}(t)\hat{h}_1(t/t).$$ (16)

The following additional comments are in order.

The results just given suggest that the optimal solution of the simultaneous detection-estimation-identification problem involves the GLR $\Lambda(t) = \rho \int \Lambda(t/\theta)p(\theta)d\theta$, where $\Lambda(t/\theta)$ is the likelihood ratio for the detection problem (H_1 or H_0) when the model is specified by the parameter θ. This suggests that the identification problem is essentially a multihypothesis testing problem. If the parameter space is discrete and finite, then the system identification problem involves testing with a finite set of hypotheses. Note that to solve for $\Lambda(t/\theta)$, one needs to calculate $\hat{h}_1(t/t)$, which is given by Eq. (14).

Even though the solution just presented is in closed form, numerical results are impossible to obtain for the general case of a nonlinear system of equations and continuous parameter space, as formulated in this section. The difficulty primarily arises in evaluating the integrals. There are, of course, some special cases where explicit solutions can be obtained. These cases will be described in the following. Approximate (sub-optimal) solutions to the general problem are described later.

It is evident at the outset that if the parameter θ is known (i.e., $\theta = \theta^*$), then the state vector model described in Eqs. (6) and (7) is completely defined, and therefore the simultaneous estimation-detection-identification problem described here reduces to the problem described in Section II.A for which the solution is given by Eqs. (2)-(4). On the other hand, if $\beta = 1$ (i.e., the signal is known to be present in the data), then this is a problem in adaptive estimation. Results for this case are presented next.

C. JOINT STATE ESTIMATION AND PARAMETER IDENTIFICATION

In this case, we assume that the observations contain the signal (hypothesis H_1 is true or, equivalently, $\beta = 1$), and the problem is to estimate the signal and some parameters of the dynamical model that generates the signal process. This is a problem in adaptive estimation. Following the approach similar to one undertaken in Section II.B, namely, augmenting the state vector $x(t)$ with the parameter vector $\theta(t)$, it can be seen that the adaptive estimation problem constitutes a class of non-linear estimation problems.

If the dynamic and measurement models [Eqs. (8) and (9)] are assumed to be linear, then an explicit closed-form solution to the optimal adaptive estimation problem can be obtained. The final results [31] are now presented.

Once again, assume that the measurement process $z(t)$ is described by

$$z(t) = y(t) + v(t), \tag{17}$$

$$y(t) = H(t, \theta)x(t),$$

and the state vector $x(t)$ is described by the following linear stochastic differential equation:

$$dx(t)/dt = F(t, \theta)x(t) + G(t, \theta)w(t), \tag{18}$$

where $F(\cdot)$, $G(\cdot)$, and $H(\cdot)$ are matrices of appropriate dimension. The observation noise $v(t)$ and the process noise $w(t)$ are once again assumed to be independent zero mean white Gaussian random processes with covariance matrices $R(t)$ and I, respectively. Note that the system model described in Eqs. (17)

and (18) is completely defined except for the parameter vector θ. It is assumed, for simplicity, that θ is time invariant and has a priori probability density of $p(\theta)$.

The adaptive estimation problem, therefore, is to obtain an optimal estimate of the state vector $x(t)$ and the parameter vector θ, given the observation record $Z(t) = \{z(\tau), t_0 \leq \tau \leq t\}$. The optimal estimate $\hat{x}(t/t)$ of the state vector $x(t)$ is given by its conditional expectation

$$\hat{x}(t/t) = \int \hat{x}(t/t, \theta) p(\theta/t) d\theta, \tag{19}$$

where $\hat{x}(t/t, \theta) = E\{x(t)/Z(t), \theta\}$, and the a posteriori probability density of the parameter vector θ is

$$p(\theta/t) = \frac{\Lambda(t/\theta)}{\Lambda(t)} p(\theta). \tag{20}$$

The optimal parameter estimate

$$\hat{\theta}(t) = \int \theta(t) p(\theta/t) d\theta. \tag{21}$$

Note that the optimal solution $\hat{x}(t/t)$ involves the θ conditional optimal estimate $\hat{x}(t/t, \theta)$ and the a posteriori probability density $p(\theta/t)$ of the parameter vector θ. The model conditional (θ conditional) estimate $\hat{x}(t/t, \theta)$ is given by the well-known Kalman-Bucy filter [32] matched to the model specified by θ. The a posteriori probability density $p(\theta/t)$ is given by the ratio of two likelihood ratios, $\Lambda(t/\theta)$ and $\Lambda(t)$.

The model conditional likelihood ratio $\Lambda(t/\theta)$ is the likelihood ratio of the detection problem

$$H_1: z(t) = H(t, \theta)x(t) + v(t), \qquad H_0: z(t) = v(t), \tag{22}$$

and is given by

$$
\Lambda(t/\theta) = \exp\left\{\int_{t_0}^t \hat{x}^T(\tau/\tau,\ \theta) H^T(\tau,\ \theta) R^{-1}(\tau) z(\tau) d\tau \right.
$$

$$
\left. - \frac{1}{2}\int_{t_0}^t \| H(\tau,\ \theta)\hat{x}(\tau/\tau,\ \theta) \|^2_{R^{-1}(\tau)} d\tau \right\}. \tag{23}
$$

The unconditional likelihood ratio $\Lambda(t)$ is the likelihood ratio of another detection problem

$$
H_2: \quad z(t) = y(t) + v(t), \qquad H_0: \quad z(t) = v(t), \tag{24}
$$

and is given by

$$
\Lambda(t) = \exp\left\{\int_{t_0}^t \hat{y}^T(\tau/\tau) R^{-1}(\tau) z(\tau) d\tau \right.
$$

$$
\left. - \frac{1}{2}\int_{t_0}^t \| \hat{y}(\tau/\tau) \|^2_{R^{-1}(\tau)} d\tau \right\}. \tag{25}
$$

At this point several interesting observations can be made. It is evident that the solution to the previous joint parameter and state estimation problem is easy to implement if the parameter space is discrete and finite. One case where it is definitely so is the case of the joint detection and estimation problem where the scalar β can be treated as the unknown parameter θ admitting two distinct values 0 and 1. Therefore, the joint detection-estimation problem of Section II.A can be treated as a special case of the adaptive estimation problem. Furthermore, it is fairly obvious that the adaptive estimation problem of this section is a special case of the joint detection-estimation-identification problem discussed in Section II.B. To see this, note that $\beta = 1 = >\rho = \infty$, Eq. (13) reduces to Eq. (20), and Eq. (12) reduces to Eq. (21).

This completes the general argument advanced earlier in Section II that the joint detection-estimation-identification problem commonly arising in communication-navigation systems can be formulated and solved as a problem in nonlinear filtering theory. The solution for some special cases, (e.g., linear dynamical models with unknown parameters) were explicitly presented. These results can be readily implemented. The solution for the general nonlinear problem was presented in Section II.B. However, it was stated the results cannot be easily implemented. In Section III, we briefy review the current status of optimal nonlinear estimation theory. In Section IV we present some results from approximation theory which can be applied to arrive at suboptimal nonlinear filters. Section V focuses on a specific application of these suboptimal results to a satellite signal-tracking and navigation problem.

III.   OPTIMAL NONLINEAR ESTIMATION

Section II has shown that the problems in communication-navigation systems, namely, signal detection, estimation, and identification, can be formulated and solved as problems in nonlinear estimation. In this section, we review some recent results in the optimal nonlinear estimation. The optimal estimation approach discussed here follows closely the work of Fisher [9].

Let us rewrite the augmented state vector model of Eqs. (8) and (9) (we have dropped the subscript "a" for convenience):

$$z(t) = h(x, t) + \dot{\eta}(t), \tag{26}$$

$$dx(t)/dt = f(x, t) + \dot{\zeta}(t), \tag{27}$$

where $x(t)$ is the augmented n-dimensional state vector; $z(t)$ is m-dimensional measurement vector; $f(x, t)$ and $h(x, t)$ are piecewise continuous functions; $\zeta(t)$ and $\eta(t)$ are, respectively,

n- and m-dimensional vector stochastic independent-increment

processes. In other words, $\dot{\zeta}(t)$ and $\dot{\eta}(t)$ are white noises, but

they are not necessarily Gaussian. (Note that the formulation

of Section II.B is included in the dynamical model Eqs. (26)

and (27), since the class of independent-increment processes[2]

includes the Gaussian independent increment process). It is

further assumed that the probability distribution of the process

noise $\zeta(t)$ and of the measurement noise $\eta(t)$ is known.

Let $Z(t)$ denote the set of random variables $z(\tau)$ for

$t_0 \leq \tau \leq t$: $Z(t) = \{z(\tau): t_0 < \tau < t\}$. Let $Z(t)$ be a separable

process [45] and assume that the conditional probability density

of $x(t)$ given $Z(t)$, denoted by $p(x(t)/Z(t))$ exists. The solu-

tion of the optimal nonlinear estimation problem requires

$p(x(t)/Z(t))$. Fisher [9] has derived a dynamical equation for

this density function under the assumptions previously stated.

The final results and some observations are presented here.

We need to define some new quantities. Following Doob [45],

the class of independent-increment processes can be represented

as the sum of a Gaussian independent-increment process and a

generalized Poisson process such that the Gaussian process is

independent of the Poisson process. Let

$$\zeta(t) = \zeta_g(t) + \zeta_p(t), \tag{28}$$

$$\eta(t) = \eta_g(t) + \eta_p(t), \tag{29}$$

For the Gaussian part, we further assume that ζ_g and η_g are

---

[2] *An important property of the independent-increment processes is the statistical independence of the past values,
i.e., if $\zeta(t)$ is of independent increments, then $d\zeta(t)$ is statistically independent of the past values of $\zeta(\tau)$ for $\tau \leq t$.*

statistically independent, and that

$$E\left\{d\zeta_g(t)\,d\zeta_g^T(t)\right\} = Q(t)\,dt, \tag{30}$$

$$E\left\{d\eta_g(t)\,d\eta_g^T(t)\right\} = R(t)\,dt. \tag{31}$$

Let $\lambda(t)$ be the probability per unit time of a jump occurring in at least one component of $d\zeta_p(t)$, and let $\omega(t)$ be the probability of a jump in $d\eta_p(t)$. Let $\nu(\mu,\ t)$ be the probability density of the amplitudes of the increments in the vector components of $\zeta_p(t)$ given that a change has occurred in some component. Let $\omega(\nu,\ t)$ be the corresponding probability density for $d\eta_p(t)$. In order to further simplify the results, it is assumed that $\eta(t)$ is a Gaussian independent-increment process, i.e., $\eta_p(t) \equiv 0$.

A. DIFFERENTIAL EQUATION FOR THE A POSTERIORI DENSITY FUNCTION

Let $p(x(t)/Z(t))$ be denoted for simplicity by $p(\mu,\ t)$ where μ is the random variable associated with the process $x(t)$. The solution for $p(\mu,\ t)$ is given by the following integrodifferential equation [9]:

$$\frac{d}{dt}p(\mu,\ t) = \lambda(t)\int_{-\infty}^{\infty} p(\tilde{\mu},\ t)\nu(\mu - \tilde{\mu},\ t)\,d\tilde{\mu}$$

$$+ \left\{[z(t) - \hat{h}(t)]^T R^{-1}(t)[h(\mu,\ t) - \hat{h}(t)]\right.$$

$$\left. - \lambda(t)\right\}p(\mu,\ t) - \left(\frac{\partial}{\partial\mu}\right)^T [f(\mu,\ t)p(\mu,\ t)]$$

$$+ \frac{1}{2}\,\mathrm{Tr}\left\{\left(\frac{\partial}{\partial\mu}\right)\left(\frac{\partial}{\partial\mu}\right)^T [Q(t)p(\mu,\ t)]\right\} + \epsilon(\mu,\ t), \tag{32}$$

where $\hat{h}(t)$ indicates the expected value of $h(x,\ t)$ with respect to $x(t)$ conditioned on $z(t)$, i.e.,

$$\hat{h}(t) = \int h(\mu,\ t)p(\mu,\ t)\,d\mu, \tag{33}$$

and Tr is the trace operator. The last term in Eq. (32), namely, $\epsilon(\mu, t)$ is given as

$$\epsilon(\mu, t) = -\frac{1}{2} \text{Tr} \{ R^{-1}(t) [R(t) - zz^T] R^{-1}(t) \{ [h(\mu, t) - \hat{h}(t)]$$
$$\times [h(\mu, t) - \hat{h}(t)]^T - [\hat{hh}^T - \hat{h}(t)\hat{h}^T(t)] \} p(\mu, t) \},$$

$$(34)$$

with the initial condition

$$p(\mu, t_0) = p(x_0, t_0). \tag{35}$$

Equations (32)-(35) are somewhat simpler than Fisher's [9] Eqs. (128-130). In the previous results, $\zeta(t)$ and $\eta(t)$ are assumed to be uncorrelated, i.e., $E[d\zeta(t)d\eta^T(t)] = 0$, implying that $s(t)$ in [9] is identical to zero. Note that the results of Section III.A are very general since we have only required that the noise process $\zeta(t)$ be of independent increments and that $\eta(t)$ be a Gaussian independent increment process. Fisher [9] has shown that the well-known results of Kushner [33,34], the famous Fokker-Planck equation [8], and the Kalman-Bucy results [32] can all be derived as the special case of Eqs. (32)-(35). To see this, let us consider these important special sults [32] can all be derived as special cases of Eqs. (32)-(35). Let us now consider these important special cases:

(1) If the process noise $\zeta(t)$ is strictly a Gaussian independent increment, then $\lambda = 0$. In this case Eq. (32) reduces to

$$\frac{dp(\mu, t)}{dt} = [z(t) - \hat{h}(t)]^T R^{-1}(t) [h(\mu, t) - \hat{h}(t)] p(\mu, t)$$

$$- \left(\frac{\partial}{\partial\mu}\right)^T [f(\mu, t) p(\mu, t)]$$

$$+ \frac{1}{2} \text{Tr} \left\{ \left(\frac{\partial}{\partial\mu}\right) \left(\frac{\partial}{\partial\mu}\right)^T [Q(t) p(\mu, t)] \right\}. \tag{36}$$

Rearranging Eq. (36), we have[3]

$$\frac{dp(\mu, t)}{dt} = -\frac{\partial[p(\mu, t)f(\mu, t)]}{\partial\mu} + \frac{\partial^2[p(\mu, t)Q(t)]}{\partial\mu^2}$$

$$+ [h(\mu, t) - \hat{h}(t)]R^{-1}(t)[z(t) - \hat{h}(t)]p(\mu, t).$$

(37)

Note that Eq. (37) is the well-known Kushner's [33,34] equation that describes the evolution in time t of the conditional probability density function $p(\mu, t)$. In contrast to Fisher's [9] stochastic integrodifferential Eq. (32) for the general case, Kushner's equation is a stochastic differential equation.

(2) If $\zeta(t)$ is a Gaussian independent-increment process and $R^{-1}(t) = 0$, then the measurements contribute no information about the state vector $x(t)$. This is equivalent to the case of no measurements. In this case Eq. (32) further reduces to the famous Fokker-Planck equation:

$$\frac{dp(\mu, t)}{dt} = -\frac{\partial[p(\mu, t)f(\mu, t)]}{\partial\mu} + \frac{1}{2}\frac{\partial^2[p(\mu, t)Q(t)]}{\partial\mu^2} \quad (38)$$

for propagating the density function $p(\mu, t)$ in the absence of measurements.

B. DIFFERENTIAL EQUATIONS
    FOR CONDITIONAL MEAN
    AND COVARIANCE

Utilizing Kushner's equation [Eq. (37)] for the conditional probability density function, Jazwinski [8] developed the equations for the conditional mean and covariance matrix for the state vector $x(t)$ of Eqs. (26) and (27). These equations [Eqs. (39) and (40)] are very useful for implementation and

---

[3]It has been shown by Fisher [9] that $\epsilon(\mu, t)$ in Eq. (32) is of O(1) for the Gaussian independent increment process $\zeta(t)$, so that it contributes only O(1) to dp/dt. For this reason, this term has been dropped in Eq. (37).

are used extensively in Section V. We shall also refer to these
equations in Section IV when we discuss the suboptimal nonlinear
filters. Equations (39) and (40) are now presented:

$$d\hat{x}(t)/dt = \hat{f}(x,\ t) + [\widehat{x(t)h}^T(t) - \hat{x}(t)\hat{h}^T(t)]R^{-1}(t)$$
$$\times\ (z(t) - \hat{h}(t)), \tag{39}$$

$$dP_{ij}(t)/dt = \left\{\left[\widehat{x_i(t)f_j}(t) - \hat{x}_i(t)\hat{f}_j(t)\right]\right.$$

$$+ \left[\widehat{f_i(t)x_j}(t) - \hat{f}_i(t)\hat{x}_j(t)\right]$$

$$+ \hat{Q}_{ij}(t) - \left[\widehat{x_i(t)h}(t) - \hat{x}_i(t)\hat{h}(t)\right]^T$$

$$\times\ R^{-1}(t)\left[\widehat{h(t)x_j}(t) - \hat{h}(t)\hat{x}_j(t)\right]\right\}$$

$$+ \left[\widehat{x_i(t)x_j(t)h}(t) - \widehat{x_i(t)x_j}(t)\hat{h}(t)\right.$$

$$- \hat{x}_i(t)\widehat{x_j(t)h(t)} - \hat{x}_j(t)\widehat{x_i(t)h(t)}$$

$$+ 2\hat{x}_i(t)\hat{x}_j(t)\hat{h}(t)\right]^TR^{-1}(t)[z(t) - \hat{h}(t)], \tag{40}$$

where $P_{ij}(t)$ is the (i, j)th element of the covariance matrix
$P(t)$.

In closing Section III, it should be noted that there are
other approaches that have been employed to solve the optimal
nonlinear estimation problem. Famous among these are the Bucy
representation theorem [35] for the conditional density function and the work of Kailath [36] and Frost and Kailath [37]
utilizing innovation processes. In [36,37], it has been shown
that the innovation process $\nu(t)$ defined by $\nu(t) = z(t) -$
$\hat{y}(t/t)$ [see Eq. (17)] is white, and furthermore, if the measurement noise is Gaussian, then the innovations are also Gaussian for the general nonlinear problem discussed here in Section III. Frost and Kailath [37] have also proved the causal

equivalence of $z(t)$ and $\nu(t)$ for the general case. Even though the innovations approach is intuitively appealing, it should be stressed that it has not yielded any simpler or more practical nonlinear estimators. Some recent attempts [38,39] to develop finite-dimensionally computable[4] filters by applying estimation algebra (an infinite-dimensional Lie algebra) to the Zakai equation for the normalized probability density function have also not been successful. Therefore, attention must be focused toward suboptimal (approximate) nonlinear estimators which are implementable. This is the topic of Section IV.

IV. SUBOPTIMAL NONLINEAR FILTERS

In our discussion in Section III on the optimal nonlinear estimators, we mentioned the difficulty of computing the conditional mean and covariance matrix. We noted that the difficulty arises because the equations for the mean and covariance matrix require all the moments of the conditional density function. In general, the conditional density function cannot be characterized by a finite set of parameters. Some approximation of the density function is therefore needed in order to develop computable nonlinear estimators.

In Section IV we present some approximate nonlinear filters that are computable and have been shown to yield good performance. We also mention some new research approaches [40,41] directed toward this goal. The approximation techniques to nonlinear filtering problem that are discussed here are (1) the quasi-moment approximate filter, (2) the truncated

---

[4]*A nonlinear estimator is finite-dimensionally computable if it can be expressed as the output of a finite-dimensional system of stochastic differential equations driven by the observations $z(t)$.*

second-order density filter, (3) the Gaussian second-order nonlinear filter, and (4) the asymptotic nonlinear filter.

A. *QUASI-MOMENT FUNCTIONS*

A parametrization of the density function by a finite (and
small) set of parameters is attractive for engineering applications. One such parametrization is an orthogonal series expansion of the density function by certain moment functions,
which have been called in the literature quasi-moment functions.
Fisher [9] and Jazwinski [8] have derived the expressions for
the quasi-moment functions and have indicated that in many
practical applications, the fourth-order (i.e., the first four)
quasi-moment function will suffice.

Let the conditional probability density function[5] $p(x, t)$
admit an orthogonal series expansion

$$p(x, t) \simeq \sum_{i=1}^{N} C_i \phi_i (x),$$ (41)

where

$$\int \phi_i (x) \phi_j (x) \, dx = \delta_{ij}.$$ (42)

From Eqs. (41) and (42) it follows that the coefficient C_i is

$$C_i = \int \phi_i (x) p(x, t) \, dx.$$ (43)

---

[5]*Fisher [9] in his work on quasi-moments expands the ratio
of $p(x, t)$ and a Gaussian density $p_g(x, t)$ by the orthogonal
series; i.e.,*

$$\rho(x, t) = p(x, t)/p_g(x, t) \simeq \sum_{i=1}^{N} C_i \phi_i'(x).$$

*Therefore, in his formulation the first two quasi-moments are
always zero.*

In order for the expansion in Eq. (41) to be valid for small N, the orthogonal functions $\phi_i(x)$ have to be selected properly. One such set of functions are the Hermite polynominals. An important property of the Hermite polynominals [9] is that they form a complete set of eigenfunctions with a zero mean Gaussian density weighting function over the n-dimensional Euclidean space. Therefore, if the density function to be approximated is nearly Gaussian, then the Hermite polynominals are attractive as orthogonal function $\phi_i(x)$ in Eq. (41). In this case, the coefficients C_i in the series expansion for $p(x, t)$ are called the quasi-moment functions. The quasi-moment functions are closely related to moments, and there exists a one-to-one relationship between central moments and quasi-moments. In fact, the first two quasi-moments are nothing more than the mean and variance, and the third quasi-moment is proportional to the third central moment [8]. Even though there is direct association between the quasi-moments and the central moments, the quasi-moments have the advantage over the central moments in that both the characteristic function and the probability density functions are easily expressed in terms of quasi-moments. Furthermore, dropping higher order quasi-moments in the expansion for $p(x, t)$ does not lead to any unusual assumption on the density function; namely, that it contains derivatives of all orders of the Dirac delta function as is the case when higher order central moments are dropped.

Fisher [9] has derived a series expansion for the conditional density function $p(x, t)$ in terms of quasi-moment functions and the Hermite polynomials. Inspite of their analytical appeal, the quasi-moment functions have not been applied to practical problems because it is not feasible to carry

third-order and higher functions due to the computational complexity of the moment evolution equation. Therefore, attention
is focused toward obtaining only the first two moment functions
of any arbitrary conditional density function.

B. *TRUNCATED SECOND-ORDER*
    DENSITY FILTER

   If the conditional density is symmetric and concentrated
near its mean, then a reasonable approximation is to work with
the first two moments only. Equations for the evolution of
mean and covariance of a state vector have been given previously
in Eqs. (39) and (40). It is important to point out that assuming that third-order and higher central moments are zero,
and therefore neglecting them, may present a problem in that
the approximate density function which results is not really a
density function. It can, for example, assume negative values
[$p(x, t) < 0$ on a set with a nonzero measure] or not integrate
to unity [$\int p(x, t)dx \neq 1$].

   Jazwinski [42] has developed the equations for the truncated
second-order density filter. This involves evaluating the conditional expectations in Eqs. (39) and (40). For the case of
first-order dynamic system and scalar measurements, the equations for conditional mean and variance are given in Eqs. (44)
and (45). For the general case of vector processes, the results
are available in [8, Appendix 9A].

   The equations for the conditional mean and variance are

$$d\hat{x}(t)/dt = f(\hat{x}, t) + \frac{1}{2}P(t)f_{xx}(\hat{x}, t) + P(t)h_x(\hat{x}, t)R^{-1}(t)$$

$$\times \left\{ z(t) - \left[h(\hat{x}, t) + \frac{1}{2}P(t)h_{xx}(\hat{x}, t) \right] \right\}, \qquad (44)$$

$$dP(t)/dt = \left[2P(t)f_x(\hat{x},\ t) + Q - P(t)h_x(\hat{x},\ t)R^{-1}(t)h_x(\hat{x},\ t)P(t) \right]$$

$$- \frac{1}{2}P^2(t)h_{xx}(\hat{x},\ t)R^{-1}(t)$$

$$\times \left\{ z(t) - \left[h(\hat{x},\ t) + \frac{1}{2}P(t)h_{xx}(\hat{x},\ t) \right] \right\}, \tag{45}$$

where $h_x = \partial h/\partial x$ and $h_{xx} = \partial^2 h/\partial x^2$. Note that the covariance equation [Eq. (45)] is no longer deterministic because it includes a random forcing term $z(t)$. Potentially it may cause problems by making the covariance negative. This is more likely to happen when $R(t)$ is small. It is comforting to know here, that if we neglect the second partial derivatives of the nonlinearity f and h, namely, set $f_{xx} = h_{xx} = 0$, then Eqs. (44) and (45) reduce to the well-known extended Kalman filter [7,10] equations.

C. *GAUSSIAN TRUNCATED SECOND-ORDER NONLINEAR FILTER*

It is sometimes reasonable to assume that the conditional density is nearly Gaussian or that it can be approximated [46] by a finite number of Gaussian density functions, i.e.,

$$p(x,\ t) = \sum_{i=1}^{N} \alpha_i p_{gi}(x,\ t), \tag{46}$$

where $p_{gi}(\cdot)$ is the ith-member Gaussian density function in the mixture and α_i is the weighting function, i.e., $\sum_{i=1}^{N} \alpha_i = 1$. In either case, invoking the Gaussian assumption simplifies the equations for the conditional mean and covariance matrix. This approach has recently been applied [2-4] successfully to a pseudo-random-noise (PRN) code-tracking problem discussed in Section V.

Depending on the type of the nonlinearity $f(x, t)$ and $h(x, t)$, it may still (in spite of the Gaussian assumption) be difficult to evaluate the conditional expectations involving $f(\cdot)$ and $h(\cdot)$. Therefore, to simplify further the implementation,[6] only the quadratic nonlinearity [up to the second partial derivative of $f(\cdot)$ and $h(\cdot)$] is maintained, and all others are neglected. This approximation is called the Gaussian truncated second-order nonlinear filter or, equivalently, the Gaussian second-degree filter. Jazwinski [8] and Fisher [9] were the first to derive the nonlinear filter equations using the Gaussian assumption on the conditional density function. Assuming the nonlinearity to be of the second degree, namely, only keeping the second-order terms in the Taylor series around the conditional mean, we have

$$f(x, t) = f(\hat{x}, t) + f_x(\hat{x}, t)(x - \hat{x}) + \frac{1}{2}f_{xx}(\hat{x}, t)(x - \hat{x})^2.$$

The equations for conditional mean and covariance equations for a first-order system with scalar measurements can be written as [8] follows:

$$d\hat{x}(t)/dt = \left[f(\hat{x}, t) + \frac{1}{2}P(t)f_{xx}(\hat{x}, t)\right]$$

$$+ P(t)h_x(\hat{x}, t)R^{-1}\left\{z(t) - \left[h(\hat{x}, t) + \frac{1}{2}P(t)h_{xx}(\hat{x}, t)\right]\right\},$$

$$\tag{47}$$

$$dP(t)/dt = \left[2P(t)f_x(\hat{x}, t) + Q - P(t)h_x(\hat{x}, t)R^{-1}(t)h_x(\hat{x}, t)P(t)\right]$$

$$+ P^2(t)h_{xx}(\hat{x}, t)R^{-1}(t)$$

$$\times \left\{z(t) - \left[h(\hat{x}, t) + \frac{1}{2}P(t)h_{xx}(\hat{x}, t)\right]\right\}. \tag{48}$$

---

[6]*It is somewhat awkward to justify Gaussian assumption for an arbitrary nonlinearity. Therefore, any simplification of the nonlinear function also helps to justify the Gaussian assumption.*

Note that Eq. (47) for the conditional mean is identical to Eq. (44) derived under the truncated second-order density approximation. The covariance Eq. (48), however, is different. In the Gaussian second-degree filter above, the term multiplying the innovation process, namely, $[z(t) - (h(\hat{x}, t) + \frac{1}{2} P(t) \times h_{xx}(\hat{x}, t))]$ is $P^2(t) h_{xx}(\hat{x}, t) R^{-1}(t)$, whereas the corresponding term in Eq. (45) is $-\frac{1}{2} P^2(t) h_{xx}(x, t) R^{-1}(t)$. Note also the change in sign. It is assuring to note, once again, that Eq. (47) and (48) reduce to the standard extended Kalman filter if we set the second partial derivatives of the nonlinearity to zero, namely, $f_{xx}(\cdot) = h_{xx}(\cdot) = 0$.

Equations (47)-(48) were derived by assuming that the conditional density is Gaussian. If, on the other hand, it is represented by a mixture density consisting of a finite weighted sum of Gaussian density functions, as shown in Eq. (46), the computational requirements will increase. In order to keep the computations to a minimum, only a small number of terms in Eq. (46) need to be maintained. This is achieved by either combining two or more member densities of a Gaussian sum density or by disregarding some of the member densities. A criterion that has been applied to make decisions to drop a member density or to combine two or more densities of a Gaussian sum density is based on a distance measure derived from the so-called Bhattacharrya coefficient ρ_{ij} [24]. The B-coefficient between two probability densities $p_i(x)$ and $p_j(x)$ is defined as

$$\rho_{ij} = \int_{-\infty}^{\infty} [p_i(x) p_j(x) dx]^{1/2}. \tag{49}$$

Note that if $p_i(x) = p_j(x)$, then $\rho_{ij} = 1$. For Gaussian densities, the Bhattacharrya coefficient can be easily computed.

Let $p_i(x) = N(m_i, P_i)$ and $p_j(x) = N(m_j, P_j)$. Then $\rho_{ij} = \exp\{-B_{ij}\}$ where the Bhattacharya distance B_{ij} is given by

$$B_{ij} = \frac{1}{8}(m_i - m_j)^T P_{ij}^{-1}(m_i - m_j)$$

$$+ \frac{1}{2} \log n\left\{|P_{ij}|/(|P_i||P_j|)^{1/2}\right\}, \qquad (50)$$

$$P_{ij} = \frac{1}{2}(P_i + P_j).$$

Note that this method of approximation ignores the impact that the α's (Eq. (46)) have on the quality of the approximation.

D. ASYMPTOTIC NONLINEAR FILTER

The last approximate filter to be discussed is the asymptotic nonlinear filter. The basic idea is to use perturbation theory to express the nonlinearity and/or the conditional density in terms of ϵ, where ϵ is a small parameter, and then to collect similar terms in ϵ, ϵ^2, and so on. Applying perturbation theory to nonlinear filtering problem implies developing a set of stochastic differential equations such that it yields an optimal estimate of the state x for $\epsilon = 0$ and, in addition, that it is asymptotically optimal; i.e., the estimation error goes to zero faster than ϵ.

Let us consider the stochastic differential equation of the form

$$dx(t)/dt = Ax(t) + \epsilon f(x(t)) + w(t), \qquad (51)$$

where f has a certain amount of smoothness (f is differentiable); ϵ is small; A is an n × n matrix with eigenvalues in the left half-plane; and w(t) is a white Gaussian noise process. If w(t) = 0 (i.e., deterministic system), then the solution of

Eq. (51) up to first order in ϵ, can be written as[7]

$$x(t) = e^{At}x(0)$$

$$+ \epsilon \int_0^t A(t - \sigma)f(e^{A\sigma}x(0))d\sigma + n(t), \tag{52}$$

with $n(t)/\epsilon$ going to zero as ϵ goes to zero.

Brockett [41] has completed a similar analysis for the stochastic Eq. (51) and has derived the results for the auto-correlation function of $x(t)$. His analysis concentrated on the steady state and showed the steps involved to develop an equivalent linear filter equation for a corresponding nonlinear problem. In general, the dimensionality of the linear filter is larger than the dimensionality of the nonlinear problem.

Brockett [41] suggested a three-step procedure for steady-state filter design for small ϵ. The steps are

1. Compute the power spectrum of $x(t)$ in terms of ϵ.
2. Find a linear system that realizes the given power spectrum driven by white noise.
3. The optimal linear filter [32] for (2) is the solution for the original nonlinear problem.

Questions regarding the performance of the asymptotic filter for some value of ϵ, compared to a linearized filter have not been yet been addressed, and this promises to be an interesting area of investigation.

An alternative approach to [41] is to assume that the conditional density function can be expressed as

$$p(x, t) = p_0(x, t) + \epsilon p_1(x, t), \tag{53}$$

---

[7]*Similar expression involving ϵ^2 terms can be easily developed.*

and to substitute $p(x, t)$ from Eq. (53) into the Kushner's equation [Eq. (37)]. Rearranging terms will, in theory, solve for p_0 and p_1 for the steady state. This approach has been carried out in [5] for a specific two-dimensional phase-tracking problem, but the results are too complicated to be useful for implementation.

V. APPLICATIONS

In order to fix the ideas discussed in earlier sections, practical applications related to signal tracking, navigation, and communication systems are now described. The purpose here is to show the progress already made and to indicate the direction where further investigation will pay off. The first example to be discussed is the GPS signal-tracking problem. In this case, an explicit solution to the nonlinear estimation problem has been obtained [1-5]. The first set of results [2-4] to be presented were developed using the minimum mean-square error (MMSE) criterion, whereas the second result [1] is derived using the least mean-square (LMS) technique. Extension of these results to a combined GPS signal-tracking and navigation problem is discussed next from the viewpoint of optimal integration of GPS and inertial navigation system (INS). An application involving a general maneuver target-tracking problem is also discussed.

A. GPS SIGNAL-TRACKING PROBLEM: MMSE ESTIMATION

The Global Positioning System (GPS) is a satellite-based radio navigation system. Each GPS satellite transmits navigation signals which consist of pseudo-random-number (PRN) codes

modulating a carrier wave.[8]   The GPS user determines his range
to the satellite by slewing an identical PRN code, generated
locally in the receiver, until it coincides (maximum correlation voltage) with that received from the satellite. The amount
of slew (the number of code chip offset) between the received
and local code phase is a measure of the transit time (slant
range) of the signal. The range-rate measurements in the receiver are made by tracking the carrier signal phase and count-
ing the carrier Doppler over a specified time period.

    As long as user dynamics and/or noise-to-signal ratios are
not excessive, the receiver can provide accurate range and
range-rate measurements that can be used to estimate accurately
the user position and velocity. In many applications, the user
dynamics may be severe for certain parts of the mission (e.g.,
during severe aircraft manuevers or in a missile boost phase),
and the signal-to-noise power may be degraded either by inten-
tional or unintentional jamming noise. Under either of these
conditions, the accuracy of a GPS receiver range and range-rate
measurements tend to degrade. At some levels of dynamics and/or
noise, the receiver will lose the signal,[9] making GPS measurements unavailable until the environment conditions improve
sufficiently for a successful reacquisition of the signal.
The reason for loss-of-signal-tracking function in the receiver
is primarily due to its small tracking range provided by the
standard delay-lock-loop design and the fast loop response-time
requirements under high dynamics. We shall have more to say
about these problems later in this section.

[8]*Many of the modern ranging and communication systems use
a PRN code to modulate a carrier.*

[9]*This event is commonly known as tracking-loop loss-of-lock
condition.*

In order to alleviate loss-of-lock conditions and thereby to improve the tracking performance of a GPS receiver, one can envision doing the following[10]: (1) adaptively extend (i.e., adding more correlators) the detector tracking range (instead of the standard ±1 chip delay-lock detector) to contain the signal during severe dynamic conditions; and (2) reduce the tracking-loop noise bandwidth to tolerate more noise. The difficulty with the first idea is that although extending the detector range may seem to help hold the signal within the detector, it also introduces more noise (only one of the correlator outputs contains the signal, all others contain noise only) in the loop, which is self-defeating. Therefore, the optimal detector characteristic must be chosen such that it minimizes the noise penalty associated with range extension.

The problem with the second idea, namely, reducing the loop noise bandwidth is that although it enhances the loop noise performance, it also increases the response times of the loop to vehicle dynamics. Therefore, if the excessive (severe) noise environment includes high dynamics, then the loop bandwidth has to be selected in some optimum way to trade off the noise performance with the loop dynamic performance. One way to reduce the loop dynamics tracking requirement is to integrate the GPS receiver with an inertial measurement unit (IMU). In the integrated-system configuration, the velocity measurements for the IMU can be used to provide the line-of-sight Doppler information to the tracking loop [6,11]. Furthermore, the GPS range—range-rate measurements can be used to estimate some of

---

[10] *Inertial measurement unit aiding the receiver and adaptive antenna ideas are not considered here.*

the inertial errors (e.g., gyro bias drift). In the following text, it is assumed that the GPS is integrated with the INS.

It has been shown [2,3] that the optimal selection of both the loop bandwidth and detector characteristics (i.e., number of correlators and their weights) can be formulated and solved as a problem in nonlinear filtering. Bowles [2] has provided a solution to the optimal PRN code-tracking problem for the GPS signal by invoking the Gaussian approximation, discussed in Section IV.C, on the conditional density. These results have been simulated and their performance evaluated for a tactical dynamic and electronic countermeasure (ECM) mission. The analytical results and the hardware implementation issues have been presented in [4,12]. In Section V.A.1 we shall formulate the GPS-PRN code-tracking problem mathematically and present a solution that employs the MMSE performance criterion.

1. *Mathematical Formulation of GPS Signal Tracking: MMSE*

The GPS signal, and L-band carrier signal biphase modulated by a PRN code, at satellite transmission time can be expressed as

$$S_T(t) = C(\omega t) \sin \omega_c t, \tag{54}$$

where ω is the code chipping rate ($\omega = 10^7$ MHz for the GPS P code); ω_c is the carrier frequency ($\omega_c = 1.575$ GHz at L_1); and $C(\cdot)$ is a ±1 pseudo-random-noise code.

The received signal (at a GPS receiver antenna) can be expressed as

$$S_R(t) = C(\omega t + \theta(t)) \sin[\omega_c t + \theta_c(t)] + n(t), \tag{55}$$

where $\theta(t)$ is the delay (transit time) between the GPS satellite and the receiver; θ_c is the phase shift in the carrier; and $n(t)$ is an additive noise representing receiver thermal noise and jamming noise, if any.

A GPS receiver would make range measurements by estimating the delay $\theta(t)$. However, in order to do this first the carrier has to be demodulated from $S_R(t)$. Assuming a perfect knowledge of carrier phase and frequency,[11] for the moment, the received signal can be multiplied by $\sin[\omega_c t + \theta_c(t)]$ (to remove the carrier). At this point, the code delay $\theta(t)$ can be measured by multiplying the carrier demodulated signal with a locally generated reference code $C(\omega t + \hat{\theta}(t))$ where $\hat{\theta}(t)$ is the code phase estimate yet to be determined. The resulting signal at this point is

$$z_1(t) = C(\omega t + \theta(t))C(\omega t + \hat{\theta}(t)) \sin^2[\omega_c t + \theta_c(t)]$$

$$+ n(t)C(\omega t + \hat{\theta}(t)) \sin[\omega_c t + \theta_c(t)]$$

$$= \frac{1}{2} C(\omega t + \theta(t))C(\omega t + \hat{\theta}(t))$$

$$\times [1 - \cos 2(\omega_c t + \theta_c(t)] + n_1(t), \qquad (56)$$

where $n_1(t)$ is the resulting noise and is white if $n(t)$ is white. The double-frequency term in Eq. (56), namely, $\cos 2(\cdot)$ can be removed by filtering, which leaves

$$z_1(t) \simeq \frac{1}{2} C(\omega t + \theta(t))C(\omega t + \hat{\theta}(t)) + n_1(t). \qquad (57)$$

The situation up to this point is described in Fig. 2. Note that $z_1(t)$ corresponds to the output of one correlator. The result of multiplication of the two shifted binary pseudo-random-noise codes, appearing in Eq. (57), when passed through the bandpass filter (shown in Fig. 2), is approximately equal

---

[11]*These quantities can be obtained by using other external measurements, e.g., an IMU, or can be measured using a carrier-tracking loop, e.g., a phase-lock loop. In the latter case, the received signal $S_R(t)$ is multiplied by a $\sin[\hat{\omega}_c t + \hat{\theta}_c(t)]$ and $\cos[\hat{\omega}_c t + \hat{\theta}_c(t)]$ to form the in-phase and quadrature phase (I, Q) measurements which are then low-pass filtered and squared to take care of any carrier phase error.*

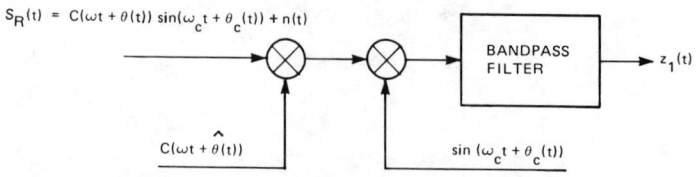

Fig. 2.   GPS signal demodulation.

to its autocorrelation function $R(\theta - \hat{\theta})$ defined as

$$R(\theta - \hat{\theta}) = \lim_{T \to \infty}(1/2T) \int_{-T}^{T} C(\omega t + \theta(t))C(\omega t + \hat{\theta}(t))dt$$

$$= \begin{cases} 1 - |\theta - \hat{\theta}|, & |\theta - \hat{\theta}| \le 1, \\ 0, & \text{otherwise,} \end{cases} \tag{58}$$

where T is the code chipping rate (in this case equal to 10^{-7} sec). Equation (57) can therefore be written approximately as

$$z_1(t) \simeq \frac{1}{2}R(\theta - \hat{\theta}) + n_1(t). \tag{59}$$

Equation (59) describes the output of one correlation when the reference code phase is centered at $\hat{\theta}(t)$ as shown in Fig. 2. In order to develop an error signal to track code phase $\theta(t)$, two correlator outputs, centered at some offset from the on-time estimate of true code phase $\hat{\theta}$ are needed as shown in Fig. 3. This loop configuration is known as the delay-lock loop (DLL).

The output of a standard (shifted by one code chip) delay-lock detector is

$$z_{DLL}(t) = \frac{1}{2}[R(\theta - \hat{\theta} + 1) - R(\theta - \hat{\theta} - 1)]$$

$$+ [n_{+1}(t) - n_{-1}(t)]. \tag{60}$$

As pointed out earlier, because of its limited tracking range the standard GPS will tend to lose the signal in a severe dynamic and noise environment. A generalized (or adaptively extended) DLL with adaptive bandwidth control feature will

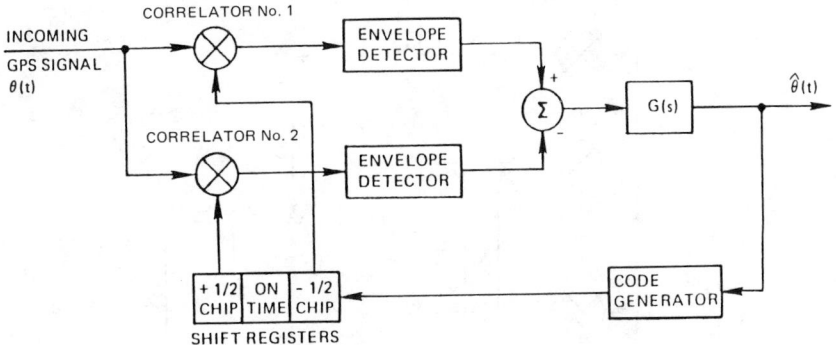

Fig. 3.   Conventional early-late delay lock loop.

alleviate these limitations.   A block diagram of a coherent
code-tracking loop using a generalized detector is shown in
Fig. 4.

The output of a generalized delay-lock detector employing
2N correlations, following Eq. (60), can be written as

$$z(t) = \sum_{i=1}^{N} \left\{ w_i \left[R\left(\theta - \hat{\theta} + \alpha_i\right) - R\left(\theta - \hat{\theta} - \alpha_i\right)\right]\right\}$$

$$+ N_f \eta(t), \tag{61}$$

where 2N is the total number of correlators, w_i is the weight
associated with the ith correlator, α_i is the offset of the
ith correlator from the on-time phase, and $N_f \eta(t)$ is resultant
postcorrelation noise; i.e.,

$$N_f \eta(t) = \sum_{i=1}^{N} w_i \eta_i(t), \qquad \eta_i(t) = (n_{\alpha_i}(t) - n_{-\alpha_i}(t)),$$

with its variance proportional to $\Sigma_{i-1}^{N} w_i^2$.   Let

$$h(\theta - \hat{\theta}) \equiv \sum_{i=1}^{N} \left\{ w_i \left[R\left(\theta - \hat{\theta} + \alpha_i\right) - R\left(\theta - \hat{\theta} - \alpha_i\right)\right]\right\}$$

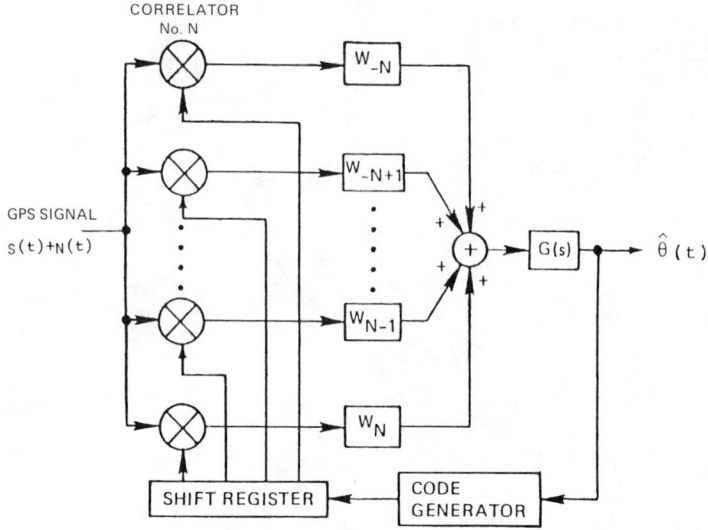

Fig. 4. Coherent code tracing loop using extended-range correlation.

be an arbitrary piecewise continuous function. Then, the measurements using extended range correlations for the code-tracking problem described earlier can be written as

$$z(t) = h(\theta - \hat{\theta}) + N_f \eta(t).$$

In order to formulate this problem as a problem in estimation theory, we need to develop a dynamic equation for the evolution in time of $\theta(t)$. To this end, it is assumed that the code phase dynamics can be adequately described by a finite-dimensional linear differential equation, $dx/dt = Fx + G\beta(t)$, where the first element of the state vector x is the code phase $\theta(t)$, the remaining elements of x are higher order derivatives of $\theta(t)$, and $\beta(t)$ is a zero mean white Gaussian process. Now we are ready to state the GPS code-tracking problem formally.

a. Statement of the problem

Select the detector function $h(\cdot)$ and tracking-loop parameters [parameters of the loop transfer function $G(\cdot)$, e.g., loop gains] to minimize the mean-square tracking error due to

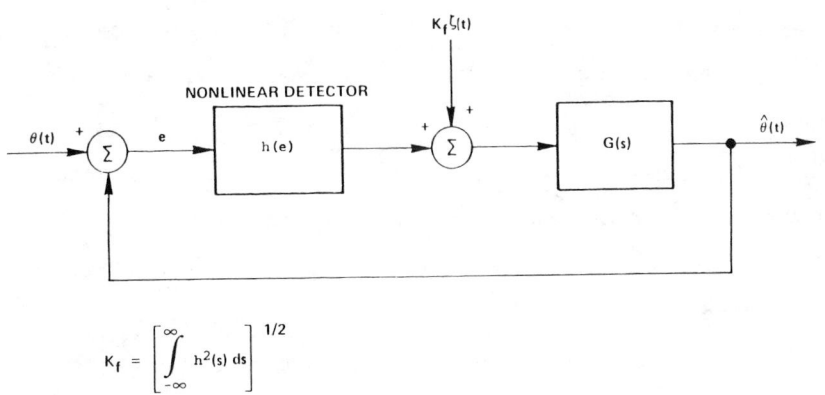

Fig. 5. Baseband model of an extended range tracking loop:
$\theta(t)$, *input code phase;* $\zeta(t)$, *white Gaussian measurement noise,*
$E[\zeta(t)\zeta(\tau)] = r(t)\delta(t - \tau)$; $\hat{\theta}(t)$, *estimate of code phase.*

noise and dynamics for the system (see Fig. 5) described by

Eqs. (62) and (63) subject to the constraint that the post-

correlation noise variance $N_f^2 r(t)$ depends on the detector shape

$h(\cdot)$. Note that $r(t)$ is the precorrelation noise variance.

Measurement equation: Scalar measurement,

$$z(t) = h(\theta - \hat{\theta}) + N_f \eta(t); \qquad Z_t = \{z(\tau): \ 0 \le \tau \le t\}. \quad (62)$$

Dynamics equation:

$$dx(t)/dt = Fx(t) + G\beta(t), \qquad \theta(t) = x_1(t), \qquad (63)$$

where $\beta(t)$ and $\eta(t)$ are zero mean white Gaussian noise pro-

cesses with

$$E\{\eta(t)\eta(\tau)\} = r(t)\delta(t - \tau),$$

$$E\{\beta(t)\beta(\tau) = q(t)\delta(t - \tau).$$

Given the measurements Z_t, the problem is to find the estimate

$\hat{\theta}(t)$ and the detector $h(\cdot)$ such that it minimizes the mean-

square error $E\{(x - \hat{x})(x - \hat{x})^T/Z_t\}$. It is well known [7,10]

that the MMSE estimate is the conditional mean, i.e., $\hat{\theta}(t) =$

$E\{\theta(t)/Z_t\}$. The equations for the conditional mean and vari-

ance for nonlinear dynamic and measurement equations have been presented in Eqs. (39) and (40). Note that in our case, the dynamics equation is linear, but the measurement equation is nonlinear.

Before applying nonlinear estimate equations to this situation it is worth pointing out that linearized extended Kalman filter has been shown to not work very well (the filter does not predict covariance correctly) for this problem [2]. The optimal nonlinear estimation equations for conditional mean and variance for the dynamics and measurement Eqs. (62 and (63) are now given as Eqs. (64) and (65) [2,4,12, Appendix A]:

$$d\hat{x}(t)/dt = F\hat{x}(t) + \left[(\widehat{xh} - \hat{x}\hat{h})/K_h^2 r \right][z(t) - \hat{h}(t)], \qquad (64)$$

$$dP(t)/dt = FP(t) + P(t)F^T(t) + qGG^T$$

$$- \left[(\widehat{xh} - \hat{x}\hat{h})(\widehat{xh} - \hat{x}\hat{h})^T/K_h^2 r \right]$$

$$+ \left[\overline{(x - \hat{x})(h - \hat{h})(x - \hat{x})}^T/K_h^2 r \right][z(t) - \hat{h}]. \qquad (65)$$

Equations (64) and (65) can be further simplified by explicitly evaluating the conditional expectations appearing on the right-hand side. Note that Eqs. (64) and (65) are optimal for any arbitrary L^2 piecewise linear function $h(\cdot)$. Invoking the property that $h(\cdot)$ is a generalized delay-lock detector, there-fore, it is an antisymmetric [h(x) = -h(-x)] function, and assuming that the conditional density function $p(x/Z_t)$ is sym-metric, results in $\hat{h}(\cdot) = 0$. In fact it can be shown that $[x(t)h(t) - \hat{x}(t)\hat{h}(t)]\hat{h}(t) = 0$. Thus, Eqs. (64) and (65) reduce to,

$$d\hat{x}(t)/dt = F\hat{x}(t) + \left(\widehat{xh}/K_h^2 r \right)z(t), \qquad (66)$$

$$dP(t)/dt = FP(t) + PF^T(t) + qGG^T - \left[(\widehat{xh})(\widehat{xh})^T/K_h^2 r \right]. \qquad (67)$$

Further simplification of Eqs. (66) and (67) is possible if we can evaluate $\widehat{xh}$. Invoking the Gaussian density approximation of Section IV.C, we have

$$\widehat{xh} = gP(t)H^T, \tag{68}$$

where

$$g = \left[P_{\theta\theta}(2\pi)^{1/2}P_{\theta\theta}\right]^{-1} \int_{-\infty}^{\infty} eh(e) \, \exp\left\{-e^2/2P_{\theta\theta}\right\};$$

$$H^T = [1, \, 0, \, \dots, \, 0]; \qquad e(t) = \theta(t) - \hat{\theta}(t);$$

$$P_{\theta\theta}(t) = P_{1,1}(t);$$

$$d\hat{x}(t)/dt = F\hat{x}(t) + g/K_h^2r \, P(t)H^Tz(t), \tag{69}$$

$$dP(t)/dt = FP(t) + P(t)F^T(t) + qGG^T$$
$$- \left(g/K_h^2r\right)P(t)H^THP(t). \tag{70}$$

Except for determining the optimal $h(\cdot)$, Eqs. (69) and (70) solve completely the GPS code-tracking problem.

The equations for the mean $\hat{x}(t)$ and the error covariance $P(t)$ are similar in appearance to Kalman filter equations, the major difference being the nonlinear term g, which appears in both equations.[12] The nonlinear term g involves the generalized detector $h(\cdot)$, yet unspecified. In order to numerically solve Eqs. (69) and (70), the detector characteristic needs to be specified.

---

[12]*It is also worth pointing out that in contrast to the covariance Eq. (48) in Section IV.C, the present Eqs. (67) or (70) do not depend on the observations $z(t)$. This simplification was possible because the measurement nonlinearity $h(\cdot)$ is assumed to be antisymmetric (generalization of a DLL) and the conditional density of $x(t)$ assumed to be a symmetric density function. This assumption imposes some structure on $h(\cdot)$, which is later relaxed [3].*

Summarizing the results so far here in Section V.A.1.a, we can state that because the approximation of the conditional density is Gaussian, the nonlinear estimation algorithm presented earlier is *optimal* for any antisymmetric, piecewise continuous, square integrable (L^2), detector function $h(\cdot)$.

b. A solution for the generalized delay-lock detector

A close examination of the estimation equations [Eqs. (69) and (70)] suggests that a reasonable criterion for selecting the detector is to choose $h(\cdot)$, which yields the largest negative first derivative of the covariance, namely, dP/dt. Since the function $h(\cdot)$ enters the covariance Eq. (70) only through g^2/K_h^2, it is sufficient to select an $h(\cdot)$ that maximizes this term (note the negative sign). The optimal detector shape is obtained by using the Schwartz inequality.

$$\frac{g^2}{K_h^2} = \frac{\left\{ \int_{-\infty}^{\infty} h(y) \, \exp\left[-\left(y^2/2P_{\theta\theta}\right)\right] dy \right\}^2}{P_{\theta\theta}^{5/2} 2\pi \int_{-\infty}^{\infty} h^2(y) \, dy}$$

$$\leq \frac{1}{2\pi P_{\theta\theta}^{5/2}} \frac{\int_{-\infty}^{\infty} h^2(y) \, dy \left\{ \int_{-\infty}^{\infty} y \, \exp\left[-\left(y^2/2P_{\theta\theta}\right)\right] dy \right\}^2}{\int_{-\infty}^{\infty} h^2(y) \, dy} \qquad \text{(Schwartz inequality)}$$

$$\leq \frac{1}{2\pi P_{\theta\theta}^{5/2}} \left\{ \int_{-\infty}^{\infty} y \, \exp\left[\frac{y^2}{2P_{\theta\theta}}\right] dy \right\}^2 . \qquad (71)$$

Equation (71) gives the upper bound for $g^2/K_h^{2'}$ the upper bound (equality) is reached when

$$h^*(y) = y \, \exp\left\{-y^2/2P_{\theta\theta}\right\} dt, \qquad (72)$$

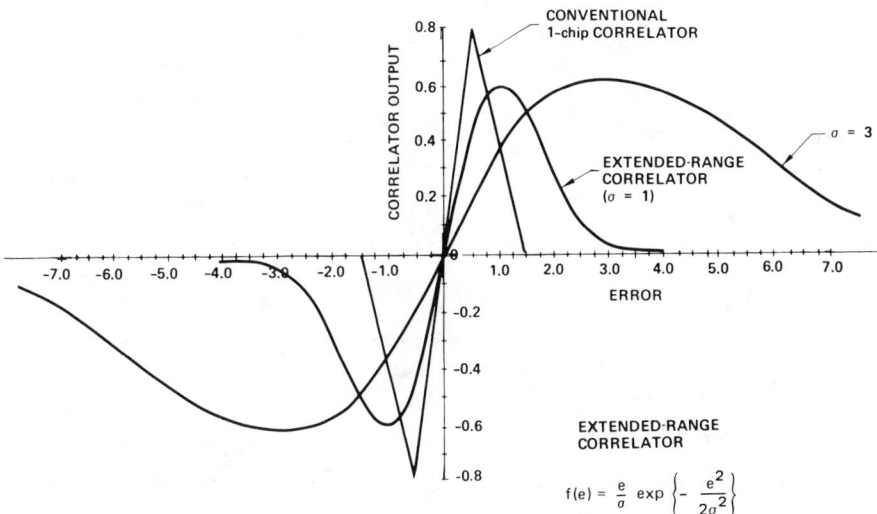

Fig. 6. Optimal generalized detector characteristic as a function of $\sigma = (P_{\theta\theta})^{1/2}$.

where $h^*(y)$ is the optimal detector characteristic which yields the largest negative variation in $P(t)$. This function is plotted in Fig. 6 for various values of $P_{\theta\theta}$.

Note some general properties of $h^*(\cdot)$. First of all, h^* is antisymmetric. It achieves its maxima at $y = (P_{\theta\theta})^{1/2}$. The latter property suggests that the optimal generalized delay-lock detector would assign higher weights to the correlators (outputs) placed near $(P_{\theta\theta})^{1/2}$ and lower weights elsewhere. This is a direct consequence of the Gaussian approximation. Note also that the generalized detector is *adaptive* (time-varying correlator weights) through its dependence on the co-variance $P_{\theta\theta}$, which in turn depends on the signal-to-noise power $r(t)$ and the dynamics $q(t)$. Substituting for $h(\cdot)$ from Eq. (72) into Eqs. (69) and (70), we have our final result:

$$d\hat{x}(t)/dt = F\hat{x}(t) + \left[r(2\pi P_{\theta\theta})^{1/2}\right]^{-1} P(t) H^T z(t), \tag{73}$$

$$dP(t)/dt = FP(t) + P(t)F^T + qGG^T - \left[4r\left(\pi P_{\theta\theta}^3\right)^{1/2}\right]^{-1}$$

$$\times\ P(t)HHP(t), \tag{74}$$

where $P_{\theta\theta}(t) = P_{1,1}(t)$.

This completes the derivation of the MMSE nonlinear estimator for the GPS pseudo-random-noise code-tracking problem.
In the process we have also specified a nonlinear measurement
system, namely, the generalized delay-lock detector, which is
optimal in the sense that it yields maximum reduction in covariance per time step. The nonlinear estimator presented here
has been analyzed and its performance evaluated in detail under
various environment conditions. These numerical results are
well documented in [12] and are therefore omitted here. It
suffices here to say that the results obtained and reported in
[12] are significantly superior to what has been obtained by
the standard GPS code-tracking approach [13,14].

2. *MMSE Tracking of GPS Signal:*
    Minimal Structural Assumptions
    on the Measurement System

The nonlinear estimator for the code phase-tracking problem
was derived in Section V.A.1 under the assumptions that conditional probability density of the state is Gaussian and that
the measurement nonlinearity is asymmetric (a generalization
of a DLL). These assumptions were responsible for the simplification and as such resulted in an implementable solution.
It is interesting to find out how these results may be affected
if the latter assumption, namely, the asymmetric measurement
nonlinearity is not invoked. Stating it differently, one would
like to know whether the generalized delay-lock detector is
indeed the optimal measurement system for the GPS code-tracking
problem.

Cartelli [3] solved this problem; however, he too had to invoke the Gaussian assumption on the conditional density to arrive at a closed-form solution. His formulation used the same dynamic equation [Eq. (63)] as described in Section V.A.1, but the measurement equation is different:

$$dx/dt = Fx(t) + g\beta(t), \quad z(t) = C(\omega t + \theta(t)) + n(t), \quad (75)$$

where the measurement $z(t)$ is the received PRN code itself, and $\theta(t)$ is the phase of the code delay from the transmitter to the receiver. Once again we are asked to find the MMSE of $x(t)$ conditioned on the measurements [given by Eq. (75)] up to time t, nemely, Z_t. Until this point, the previous measurement system is only a mathematical formulation, and the practicality of forming measurements using Eq. (75) has been set aside. Cartelli's [3] results show that the optimal nonlinear estimator equation provides the answer to the choice of an optimal measurement system. Specifically, he has shown that the generalized delay-lock detector is indeed the optimal detector (measurement system) for the PRN code-tracking problem.

Specifically, the conditional mean equation is [8]

$$d\hat{x}(t)/dt = F\hat{x}(t) + [\widehat{x(t)h} - \hat{x}(t)\hat{h}][z(t) - \hat{h}], \quad (76)$$

where $h(x, t) = C(\omega t + x_1)$. The term

$$[\widehat{x(t)h} - \hat{x}(t)\hat{h}] = \widehat{(x - \hat{x})h},$$

and it can be shown that for our choice of $h(\cdot)$

$$[\widehat{x(t)h} - \hat{x}(t)\hat{h}]\hat{h} = 0$$

under Gaussian assumption.[13] Therefore Eq. (76) becomes

$$d\hat{x}(t)/dt = F\hat{x}(t) + \{\widehat{[x(t) - \hat{x}(t)]h}\}z(t)$$

[Eq. (77) continues]

[13]*In fact, we only need the property that the density func-function is symmetric about $\hat{x}$.*

$$= F\hat{x}(t) + \int_{-\infty}^{\infty} [x(t) - \hat{x}(t)] C(\omega t + x_1) p(x/t) dxz(t),$$

$$= F\hat{x}(t) + \left[\int_{-\infty}^{\infty} y(t) C\left(\omega t + y_1 + \hat{x}_1\right) \right. $$
$$\left. \times p(y/t) dy \right] z(t), \tag{77}$$

where $y = x - \hat{x}$, and $y \approx N[0, P(t)]$. Equation (77) suggests
the feedback structure for the optimal nonlinear estimator,
i.e., the MMSE estimate of $x(t)$ requires multiplication (correlation) of the measurements $z(t)$ with the reference signal
$C(\omega t + x_1)$. If we substitute for $z(t)$ from Eq. (75) and define

$$R\left(y_1 + x_1 - \hat{x}_1\right) = C\left(\omega t + y_1 + \hat{x}_1\right) C(\omega t + x_1),$$

then Eq. (77) can be rewritten as

$$d\hat{x}(t)/dt = F\hat{x}(t) + \int_{-\infty}^{\infty} y(t) p(y/t) R\left(x_1 - \hat{x}_1 + y_1\right) dy$$

$$+ \int_{-\infty}^{\infty} y(t) p(y/t) C\left(\omega t + y_1 + \hat{x}_1\right) n(t) dy. \tag{78}$$

Equation (78) shows the structure of the optimal estimator.
The first integral in Eq. (78) points out that the conditional
mean $\hat{x}(t)$ requires all of the correlator's output $R(\cdot)$, weighting them by $y(t)p(y/t)$ and summing (integrating) their result.
Similarly, the precorrelation noise $n(t)$ is multiplied by the
same weight function, showing the noise effect of the generalized correlator on the mean estimator. This last term in Eq.
(78) is indeed the postcorrelation noise.

Equation (78) also points out clearly the feedback nature
of the estimator, the feedback being the code phase $\hat{x}_1(t)$.
Equation (78) is similar to Eq. (69) where an explicit feedback
structure, namely, generalized DLL, was assumed. We shall have
more to say about the comparison of the two results a bit later.

The covariance of $x(t)$, i.e., $P(t)$ is [3]

$$dP(t)/dt = [FP(t) + P(t)F^T(t) + qGG^T]$$

$$+ \int_{-\infty}^{\infty} [y^T y - P(t)]p(y/t)C\left(\omega t + \hat{x}_1 + y_1\right)dyz(t).$$

$$(79)$$

An important difference, between Eq. (79) and the previously derived Eq. (70) (Section V.A.1) is the data dependence $z(t)$ in Eq. (79). Furthermore, the correlator has a new weighting function $\{(y^T - P)p(y/t)\}$. This function is drawn in Fig. 7 and has the property that if $|y_1| > (P_{11})^{1/2}$, then it increases the covariance (positive contribution, to avoid loss of lock), whereas if $|y_1| < (P_{11})^{1/2}$, then it decreases the covariance to provide better tracking accuracy. Equations (78) and (79) specify the complete solution to the estimation problem discussed here. Cartelli [3] calls the above solution a data-dependent adaptive Gaussian (DDAG) detector solution. Equations (78) and (79) are tightly coupled through $z(t)$. Data

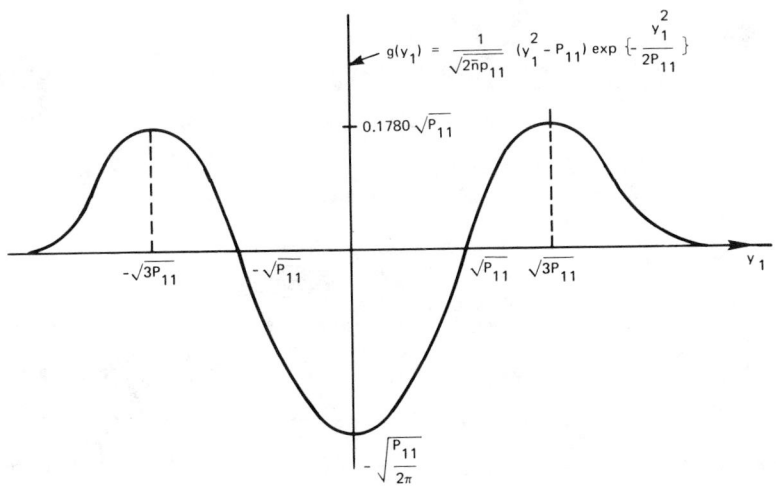

$$g(y_1) = \frac{1}{\sqrt{2\pi}P_{11}} (y_1^2 - P_{11}) \exp\left\{-\frac{y_1^2}{2P_{11}}\right\}$$

$0.1780 \sqrt{P_{11}}$

$-\sqrt{3P_{11}}$ $-\sqrt{P_{11}}$ $\sqrt{P_{11}}$ $\sqrt{3P_{11}}$ y_1

$-\sqrt{\dfrac{P_{11}}{2\pi}}$

Fig. 7. Nonlinear detector characteristic for the covariance equation.

dependence of the covariance equation makes $P(t)$ adaptive and offers a unique opportunity to account for any modeling errors and makes the solution more robust (stable). However, $P(t)$ is now stochastic through $z(t)$, and positive definiteness of the covariance matrix $P(t)$ cannot be guaranteed.

It is a relatively straightforward matter at this point to show that Cartelli's [3] mean Eq. (77) is identical to Bowles' [2] mean Eq. (69), assuming the same $P(t)$. Let us rewrite Bowles' [2] conditional mean Eq. (69):

$$d\hat{x}(t)/dt = F\hat{x}(t) + \left(g/K_h^2 r\right)P(t)H^T z_B(t).$$

Substituting for g and K_h^2, we have

$$\frac{d\hat{x}(t)}{dt} = F\hat{x}(t) + (2\pi P_{11})^{1/2} \begin{bmatrix} P_{11} \\ P_{12} \\ \vdots \\ P_{1n} \end{bmatrix} z_B, \qquad (80)$$

where z_B is the generalized detector output given by

$$z_B(t) = h\left(x_1 - \hat{x}_1\right) + n_0(t), \qquad (81)$$

and $h(x_1 - \hat{x}_1)$ given by Eq. (72) is the genralized detector shape, which is the convolution of the code correlation function $R(x_1 - \hat{x}_1)$ and the correlator weights $w(x_1 - \hat{x}_1)$, i.e.,

$$h\left(x_1 - \hat{x}_1\right) = \int_{-\infty}^{\infty} R\left(x_1 - \hat{x}_1 + s\right)w(s)\,ds, \qquad (82)$$

and $n_0(t)$ in Eq. (81) is the postcorrelation noise properly weighted by the correlator weights $w(\cdot)$. For implementation purposes, the correlator weights can be obtained from Eq. (82). It is sufficient here to say that if the correlator spacing is one chip, then the weights can be obtained by sampling the detector characteristics $h(\cdot)$.

Substituting for $z_B(t)$ in Eq. (80), we obtain

$$\frac{d\hat{x}(t)}{dt} = F\hat{x}(t) + (2\pi)^{1/2}\begin{bmatrix} P_{12}/P_{11} \\ \vdots \\ P_{1n}/P_{11} \end{bmatrix}$$

$$\times \left\{ \sum_{i=-N}^{N} (R(x - \hat{x} + i)w(i)) + n(t) \right\}. \tag{83}$$

Equation (83) is identical to Cartelli's mean equation [3, Eq. (6.27)].

As pointed out earlier the covariance equations are different. The prime differences are the data dependence in Eq. (79) and new correlator weights as shown in Fig. 7.

The results presented in Sections V.A.1 and V.A.2 have shown an application of nonlinear estimation theory to a practical PRN code-tracking problem that utilizes the MMSE criterion. An alternative approach, Widrow's [47] least mean-square (LMS) criterion, has also been applied successfully [1] where a new detector characteristic is obtained which is different from the DLL. This is the topic of Section V.B, next.

B. GPS SIGNAL TRACKING PROBLEM: LMS ALGORITHM

As a diversion from the Bayesian estimation approach where an a priori density function and/or a diffusion model of the process to be estimated is desired, the work described here in this Section V.B follows work of Widrow *et al.* [47,48] on adaptive filtering, which has come to be known as the least mean-square (LMS) technique or the adaptive noise-canceling technique.

The basic idea behind the adaptive noise-canceling technique is to develop a reference input (from sensors located at

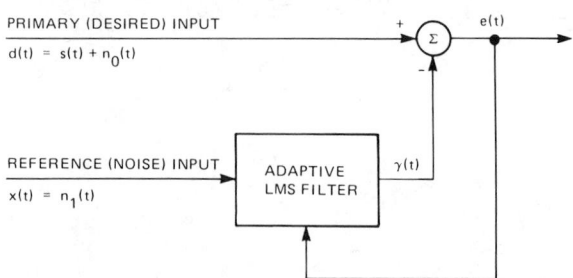

Fig. 8. Block diagram of adaptive noise canceling concept using LMS technique: $e(t) = d(t) - \gamma(t)$; $\gamma(t) = \int_0^\infty x(t - \tau) \times \omega(\tau)d\tau$, *where* $\omega(t)$ *is the LMS weight.*

points in the noise field) which when filtered and subtracted from the received signal cancels the noise and maximizes the signal-to-noise power at the output (Fig. 8). The success of the technique depends on the adaptive filter, which is optimized for the mean-square error (the difference between the reference and the received signal). The adaptive filter uses an LMS algorithm [47] which is merely a stochastic implementation of the method of steepest descent. The adaptive filter weights (assuming a transverse filter implementation) are computed recursively such that the "next" weight vector w_{k+1} is equal to the "present" weight vector w_k plus a change proportional to the negative of the mean-square-error gradient, i.e.,

$$w_{k+1} = w_k - \mu \nabla_k, \tag{84}$$

where ∇_k is the true gradient at the kth iteration (time), and μ is a constant that controls stability and rate of convergence. The gradient ∇_k is the instantaneous value calculated for a single error sample, i.e.,

$$\nabla_k = \left[\frac{\partial e_k^2}{\partial w_k} \right].$$

It has been shown that if x_k is uncorrelated, then the gradient estimate $\hat{V}_k$ used in the LMS algorithm is unbiased. Furthermore, starting with an arbitrary initial weight vector, the LMS algorithm will converge in the mean and will remain stable as long as parameter μ in Eq. (84) is positive and less than the reciprocal of the largest eigenvalue λ_{max} of the input (noise) correlation matrix R:

$$0 < \mu < 1/\lambda_{max}. \tag{85}$$

For stationary stochastic inputs, the performance of the LMS adaptive filter closely approximates that of fixed Wiener filters. However, when the input statistics are time varying, no general results on the convergence and/or performance of the LMS algorithm are available. The problems associated with the weight convergence are now described.

Let X_k be the reference input vector at time k, w_k the corresponding weight vector, and d_k the desired response. The mean-square error at time k is

$$E\left\{e_k^2\right\} = E\left[\left(d_k - w_k^T x_k\right)\right]^2 = E\left(d_k^2\right) - 2l^T W + w^T R W, \tag{86}$$

where $R = E\left\{x_k x_k^T\right\}$ is the autocorrelation matrix of the reference input and $l = E\{d_k x_k\}$ is the cross-correlation vector of the reference with the desired signal.

The gradient of the mean-square-error performance surface is $V = -2l + 2RW$. Setting the gradient to zero yields the optimal weight vector W^*:

$$w^* = R^{-1} l. \tag{87}$$

The MMSE can be written in terms of optimal weights W^* as

$$E\left\{e_k^2\right\}_{min} = E\left(d_k^2\right) - l^T w^*.$$

Now, the mean-square error [Eq. (86)] can be written in terms of the MMSE as

$$E\left\{e_k^2\right\} = E\left\{e_k^2\right\}_{min} + \left(W_k - W^*\right)^T R\left(W_k - W^*\right).$$ (88)

Equation (88) illustrates the effect of weight convergence on the mean-square error.

It is well known that for stationary processes the gradient estimation process introduces noise into the weight vector that is proportional to the speed of adaptation and the number of weights. For nonstationary processes, however, the gradient estimation process introduces noise inversely proportional to the speed of adaptation [48]. The latter result follows from the fact that for nonstationary processes the important thing is to track the variations in input statistics. These facts suggest that a compromise between fast and slow adaptation is required to achieve the best performance.

The LMS algorithm of Widrow *et al.* [47] is an implementation of the steepest descent using measured or estimated gradients to recursively compute the weights, i.e., $W_{k+1} = W_k - \mu\hat{\nabla}_k$, where $\hat{\nabla}_k$ is the estimate of the true gradient. In the LMS algorithm, $\hat{\nabla}_k$ is the single error sample gradient. From Eq. (86) we have $\hat{\nabla}_k = -2d_k X_k + X_k X_k^T W = -2e_k X_k$, and the recursion equation for the weights are

$$W_{k+1} = W_k + 2\mu e_k X_k.$$ (89)

Lamb [1] has successfully applied the LMS algorithm to obtain the extended-range correlator weights for the GPS code-tracking problem described earlier in Section V.A. A block diagram representation of LMS weight computation is shown in Fig. 9. The reference signal (using the terminology of this

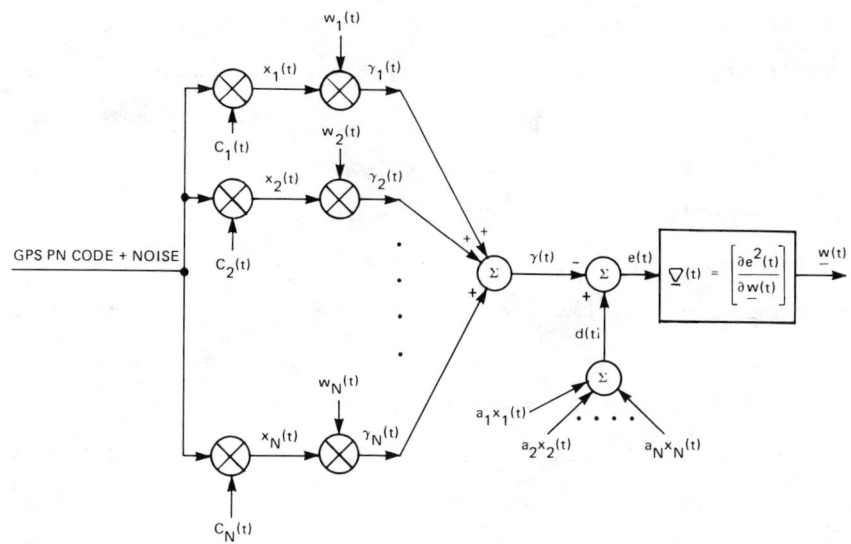

Fig. 9.   LMS weights for GPS extended-range correlator:
$x(t) = [x_1(t), x_2(t), ..., x_N(t)]; \underline{w}(t) = [w_1(t), w_2(t), ..., w_N(t)].$

section, Fig. 7) is the correlator output vector X, the weight

vector W is the adaptive filter, and the primary or training[14]

signal d(t) is the correlator output multiplied by a linear

fixed-weight vector a (such that a_i = i). Note that the linear

fixed weight a_i = i are the optimal correlator weights for the

noise-free situation.

The LMS algorithm for recursively computing the correlator

weight vector is given as [1]

$$W_{k+1} = W_k + \mu \tilde{e}_k X_k, \tag{90}$$

where $\tilde{e}_k = d_k - \gamma_k$ and $d_k = \Sigma_{i=1}^{N} a_{i_k} X_{i_k}$ and $\gamma_k = \Sigma_{i=1}^{N} W_{i_k} X_{i_k}$.

Lamb [1] derived the necessary conditions for the conver-

gence of the weights in terms of eigenvalues of the variance

matrix of the weights. The convergence of the weights were

[14]*The performance of the LMS adaptive filter depends on the
judicious choice of the training signal d(t).*

verified for stationary test input processes through a hardware
demonstration of the extended-range correlator with a CCD used
for weight storage and accumulation [1]. It is therefore correct to say that LMS algorithm performance has not yet been
tested completely for the GPS code-tracking problem.

Remarks. The potential for improved signal-acquisition
performance of the LMS algorithm can be explained from the fact
that the use of a large number of correlators does not contribute additional noise (after the weights have converged) because
at any time only a single correlator, corresponding to the signal, will have a nonzero weight. In contrast to the extended-
range correlator design of Section V.A, the above feature of
the LMS algorithm seems quite attractive. Another advantage
of the LMS technique is that it does not need a priori information on the signal and/or noise statistics. The disadvantages
of the LMS algorithm are (1) the performance depends heavily on
the choice of the "training" signal d(t); (2) the convergence
rate and stability of the algorithm can not be guaranteed for
nonstationary processes; and (3) the performance (e.g., mean
square) of the algorithm is difficult to evaluate in real time.

Contrasting these LMS features with the nonlinear filtering
approach, one can find analogous characteristics between the
two. It is felt that the selection of the desired (or training)
signal in the LMS approach is similar in complexity to the
optimal selection of the measurement subsystem h(t) in the
nonlinear estimation process. The scalar constant μ in Eq.
(84) which affects the convergence of the algorithm can be
modified to handle nonstationary processes using a diffusion
process model similar to the dynamic model for the states. It
is conjectured then, that in order for LMS weights to converge

under high dynamic reference processes, the constant μ should
be made a vector, its dimensionality probably equal to the dimension of the dynamic process.

This discussion suggests that both the LMS and nonlinear
estimation techniques have similar complexity. The LMS technique, however, seems to offer some potential advantage over
the nonlinear estimation technique for limited applications
dealing with stationary stochastic processes. The advantage
derives from the fact that fewer assumptions on the statistics
of the signal and noise processes are required in the LMS approach. Furthermore, recent developments in device technology
[e.g., CCD and surface acoustic wave (SAW)] have made the LMS
approach readily implementable. In the long run, this may in
fact turn out to be the driving reason for a wider acceptance
of the LMS technique.

C. COMBINED SIGNAL ACQUISITION, TRACKING, AND NAVIGATION PROBLEM

In Section V.B we presented some recent results on the application of suboptimal[15] nonlinear filtering to the GPS signal-
tracking problem. We showed how the filtering-theory formulation provides a solution to the signal tracking problem in the
GPS receiver. The adaptive solution was shown to be a function
of signal-to-noise ratio and dynamics. In addition, the formulation yielded the optimal detector characteristic which alleviates signal-loss-of-lock problems generally associated with
applications involving high-noise and high-dynamics environment
conditions. Furthermore, the additional correlators can also
be used to speed up signal reacquisition. It is worth pointing

---

[15] *Gaussian density approximation.*

out here that implicit to the signal tracking problem and the
resulting solution is the assumption that the signal is present
(with probability 1) within the detector range.

In Section V.C we relax this assumption and formulate the
combined signal detection and tracking problem. In fact, all
the tools required for tackling this problem have been previously discussed in Section II. We shall refer to the combined signal detection and tracking problem as a problem of
optimal signal tracking under uncertain observations.

*1. Signal Tracking
    under Uncertain Observations*

In many of the signal detection and tracking problems, e.g.,
radar target tracking, fault-failure detection, ranging and
communication systems (GPS and JTIDS[16]), it is desirable to
have the information on signal (fault) presence and its estimate simultaneously in order to improve system performance.
The classical solution to this problem is known as "detection-
directed estimation" which consists of a likelihood ratio detection followed by a MMSE estimation. The MMSE estimation
formulation assumes signal presence with probability 1. The
optimal solution to this problem was discussed in Section III,
and suboptimal implementable solutions were discussed in Section IV.

Applying the results of Section II to the GPS signal detection and tracking problem discussed in Section V.B, we develop
some intuitive and interesting results. In order to formulate
the GPS problem in the framework of signal tracking under uncertain observation, we write the dynamics and measurement

---

[16]*Joint tactical information distribution system.*

equations as

$$dx/dt = Fx(t) + G\beta(t), \tag{91}$$

$$z(t) = \gamma(t)h(x_1) + N_f\eta(t). \tag{92}$$

Note that the dynamics equation [Eq. (91)] is the same as before [Eq. (62)]. However, the measurement equation [Eq. (92)]
now includes a scalar binary variable $\gamma(t)$ which takes on values
0 (no signal) or 1 (signal present). All other terms have been
defined earlier. The problem once again is to obtain a MMSE
estimate of $x(t)$ given that the measurements are taken by Eq.
(92). Note that for the GPS problem under consideration, namely,
code and carrier phase tracking, $\gamma(t)$ is not a function of
time.[17] In other applications, e.g., communication systems
(blanking signal), $\gamma(t)$ will be a time-varying discrete parameter function and will be specified completely by its transition probability [49,50].

Rewriting Eq. (92) slightly differently, we have

$$z(t) = g(x_1, \gamma(t)) + v(t), \tag{93}$$

where $v(t) = N_f\eta(t)$ and $g(x_1, \gamma(t)) = \gamma(t)h(x_1)$. Recall that
if $\gamma(t)$ is specified, then the solution to this problem has
already been presented in Section V. Applying the results of
Section II to this problem [treating $\gamma(t)$ as an unknown, but
constant parameter], we can write the following integral equations for $\hat{x}(t/t)$ and $\hat{\gamma}(t/t)$:

$$\hat{x}(t/t) = \int \hat{x}(t/t, \gamma(t))p(\gamma(t)/t)d\gamma(t) \tag{94}$$

$$\hat{\gamma}(t/t) = \int \gamma(t)p(\gamma(t)/t)d\gamma(t). \tag{95}$$

---

[17]A $time$-$varying$ $\gamma(t)$ for GPS $could$ be $used$ to $model$ GPS
$navigation$ $data$ or $signal$ $amplitude.$

Note that Eqs. (94) and (95) are the same as Eqs. (19) and (20) with $\theta(t)$ replaced by $\gamma(t)$. The integral in Eqs. (94) and (95) is to be carried out over the γ space, which will be assumed in most applications to be a discrete parameter space. Therefore, for implementation purposes, the integral would be replaced by summation. Note also that $\hat{x}(t/t, \gamma(t))$ is the γ-conditional estimate of $x(t)$ and is given by Eq. (66) for this problem. The remaining term in Eq. (94) is the a posteriori probability density function $p(\gamma(t)/t)$, which is given by

$$p(\gamma(t)/t) = \frac{1 + \rho\Lambda(t/\gamma(t))}{1 + \rho\int\Lambda(t/\gamma(t))p(\gamma/t))d\gamma} p(\gamma(t)), \qquad (96)$$

where

$$\Lambda(\gamma(t)/t) = \exp\left\{\int_{t_0}^{t} \hat{g}_1(\sigma/\sigma, \gamma)R^{-1}z(\sigma)d\sigma\right.$$

$$\left. - \frac{1}{2}\int_{t_0}^{t} \|\hat{g}_1(\sigma/\sigma, \gamma)\|^2_{R^{-1}(\sigma)} d\sigma\right\}, \qquad (97)$$

$$\hat{g}_1(\sigma/\sigma, \gamma) = E[g(x(\sigma), \hat{x}(\sigma)/\sigma, \gamma = 1] = \hat{h}(x),$$

$$\rho = p(\gamma(t) = 1)/p(\gamma(t) = 0).$$

Equations (94)-(98) are easier to implement if $\gamma(t)$ takes on only discrete values and is time invariant, which is the case for the GPS signal-tracking problem. However, in some applications where $\gamma(t)$ is time varying, $\hat{x}(t/Z_t, \Gamma_t)$, where $\Gamma_t = [t_0 \le \gamma(t) \le t]$, has to be computed for each realization of $\gamma(t)$, which requires exponentially growing (with time) processing resources and is therefore difficult to implement. Various suboptimal techniques which attempt to control this growth have been suggested, and new ones are being developed. One particular technique [49,50] currently under investigation attempts to adaptively merge the a posteriori probability

densities of $p(\Gamma(t)/Z_t)$ while ensuring that the first two moments are invariant to mergin (see Section V.D.3). This approach is attractive because it not only provides a state estimate $\hat{x}(t/t)$, but it also provides a smoothed estimate of the
parameter $\gamma(t)$, which is needed in many applications.

Returning to the problem at hand, following the steps described in Section V.A.2 and after some algebra, we can write
a set of differential equations for the conditional mean and
covariance equations [51] for the states $x(t)$ and $\gamma(t)$:

$$d\hat{x}/dt = F\hat{x}(t) + \hat{\gamma}(t)(\widehat{xh} - \hat{x}\hat{h})R^{-1}(z - \hat{\gamma}\hat{h}), \tag{98}$$

$$d\hat{\gamma}/dt = \hat{\gamma}(1 - \hat{\gamma})\hat{h}R^{-1}(z - \hat{\gamma}\hat{h}), \tag{99}$$

$$dP_x(t)/dt = \left(FP_x + P_xF^T + Q\right) - \hat{\gamma}^2(\widehat{xh} - \hat{x}\hat{h})^2R^{-1}$$

$$+ \hat{\gamma}\left[\overline{(x - \hat{x})^2} - P_x\right]h(x)R^{-1}(z - \hat{\gamma}\hat{h}), \tag{100}$$

$$P_\gamma(t) = \hat{\gamma}(t)(1 - \hat{\gamma}(t)). \tag{101}$$

Note here that even though $\gamma(t)$ was assumed to be a discrete-
value parameter, the solution for its estimate is continuous.
For making decisions about signal presence, a likelihood ratio
test involving $\hat{\gamma}(t)$ is needed. Substituting for $h(x, t) =$
$C(\omega t + x_1(t))$, the PRN received code, Eqs. (98)-(101) can be
further simplified. These equations are presented here for
completeness for a first-order dynamical model [51]:

$$d\hat{x}_1(t)/dt = (\hat{\gamma}/r) \int_{-\infty}^{\infty} yC\left[\omega t - \left(\hat{x}_1 + y\right)\right]P_{x_1}(y)\,dy\ z(t), \tag{102}$$

$$dp_{x_1}/dt = Q + (\hat{\gamma}/r) \int_{-\infty}^{\infty} \left(y^2 - P_{x_1}\right)C\left[\omega t - \left(\hat{x}_1 + y\right)\right]$$

$$\times P_{x_1}(y)\,dy\ z(t), \tag{103}$$

$$d\hat{\gamma}/dt = \frac{1}{r} \hat{\gamma}(1 - \hat{\gamma}) \left\{ \int_{-\infty}^{\infty} C\left[\omega t - \left(\hat{x}_1 + y\right)\right] p_{x_1}(y) dy \; z(t) \right.$$

$$\left. - 4\hat{\gamma}\left[E\left(\frac{1}{\sqrt{2P_{x_1}}}\right) - \sqrt{\frac{P_{x_1}}{\pi}}\left(1 - \exp^{-\left(4P_{x_1}\right)^{-1}}\right)\right]\right\}, \qquad (104)$$

$$P_\gamma(t) = \hat{\gamma}(1 - \hat{\gamma}), \qquad\qquad\qquad\qquad\qquad\qquad (105)$$

where $E(x)$ = the error function. Except for the two terms in Eq. (104), these results [Eqs. (102)-(105)] were obtained without making any assumptions on the conditional density function. In fact, the conditional density $p_{x_1}(y)$ appears explicitly in Eqs. (102)-(104). If at this point, we invoke the Gaussian density approximation and in addition set $\gamma(t) = 1$ (signal present with probability 1), then Eqs. (102) and (103) reduce to Eqs. (78) and (79) of Section V.A.2.

2. *Combined Signal*
   Tracking and Navigation

Extending the approach of Section V.C.1 to include the navigation problem, we shall show that the GPS navigation and signal-tracking problems can be solved jointly by using the MMSE optimization criterion. Note that in this formulation, we are interested in minimizing the mean-square navigation error instead of the mean-square signal-tracking error.

Let the state vector x represent the navigation position, velocity, etc., in an orthogonal (e.g., earth-centered, earth-fixed, ECEF) coordinate frame. The time evolution of the navigation state can be described by the following linear differential equation:

$$dx(t)/dt = Ax(t) + Bw(t), \qquad\qquad\qquad\qquad (106)$$

where A and B are known matrices of appropriate dimensions, and $w(t)$ is a white-process noise assumed to have zero mean.

The measurement equation, as before, models the received PRN code, i.e.,[18]

$$z_i(t) = \gamma_i(t) C_i(\omega t + y_i(t)) + v(t),$$

$$z = 1, 2, \ldots, N, \tag{107}$$

where $y_i(t)$ is the phase of the ith satellite PRN code at time t [i.e., $y_i(t) = \theta_i(t)$], and $v(t)$ is the additive measurement noise. The code phase $y_i(t)$ (time delay) is related to the navigation state $x(t)$ through the following expression:

$$y_i(t) = \left\{ (x_1 - x_{s_i})^2 + (x_2 - y_{s_i})^2 + (x_3 - z_{s_i})^2 \right\}^{1/2}$$

$$+ x_4(t), \tag{108}$$

where x_{s_i}, y_{s_i}, and z_{s_i} are x, y, z coordinates of the ith satellite position; x_1, x_2, and x_3 are the x, y, z user position coordinates; and x_4 is the time bias between the GPS satellite and the user receiver clocks.

Equations (106)-(108) define the GPS MMSE navigation problem in all its glory. The task is to obtain an optimal estimate of the navigation state vector $x(t)$ given the measurement vector $\underline{z}(t) = [z_1(t), \ldots, z_4(t)]$, which is nothing but the PRN codes received from four different satellites, simultaneously, at the user antenna. The discrete parameter vector $\underline{\gamma}(t) = [\gamma_1(t), \ldots, \gamma_4(t)]$ models the fact that some of the received signals may in fact be only noise $v(t)$. This formulation, implicitly admits the case of different signal-to-noise power in each receiver channel.

The solution to the problem is difficult because of the nonlinear relationship between x and y as described by Eq. (108) and because of the structure of the receiver measurements

[18] *A minimum of four (N = 4) satellite measurements are required to solve for three position coordinates and time.*

[Eq. (108)]. In the previous sections we had invoked a Gaussian approximation on the conditional density function of $y(t)$ (i.e., the code phase error) to obtain closed-form expressions for the mean and variance. However, this approach does not work here because x and y cannot jointly be Gaussian [due to Eq. (108)] since this would imply that both x and y are Gaussian, a violation in light of Eq. (108).

Even though the above problem does not have a closed-form analytic solution,[19] the MMSE nonlinear estimation formulation provides some important clue to the coupling between the navigation and tracking problems. Weiss [51] has provided the following interpretation to this problem.

Let us rewrite equations for the mean of the navigation solution vector $\hat{x}$ (assume for simplicity that $\gamma_i = 1$ with probability 1 for all i):

$$d\hat{x}/dt = A\hat{x}(t) + \overline{(x - \hat{x})h^T(x)}(z - \hat{h}(x)),\qquad(109)$$

where

$$h(x) = \begin{bmatrix} c_1[t + y_1(t)] \\ \vdots \\ c_4[t + y_4(t)] \end{bmatrix}.$$

The ith element of the second term in Eq. (109) is

$$\sum_{j=1}^{4}\sum_{k=1}^{4}\left[\overline{\left(x_i - \hat{x}_i\right)h_k^T}(x)\,(z_j(t) - \hat{h}_j(x)\right]R_{kj}^{-1}$$

$$= \sum_{j=1}^{4}\sum_{k=1}^{4}\left[\overline{\left(x_i - \hat{x}_i\right)c_k(t} + y_k(t))z_j(t)\right]R_{kj}^{-1}$$

$$+ \sum_{j=1}^{4}\sum_{k=1}^{4}\left[\overline{\left(x_i - \hat{x}_i\right)c_k(t + y_k(t))}\,\overline{c_j(t + y_j(t))}\right]R_{kj}^{-1}.\quad(110)$$

[19] *A linearized extended Kalman filter solution to the problem has been investigated in [17] where the measurements $z_1(t)$ were assumed to be the I, Q phase measurements.*

Let $e_{x_i} = x_i - \hat{x}_i$ be the ith component of the navigation state error vector and $p(e_{x_i})$ be its probability density function. Following Section V.A.2, the first term of Eq. (110) can be written as

$$\sum_{j=1}^{4} \sum_{k=1}^{4} R_{kj}^{-1} \int e_{x_i}(t) p(e_{x_i}) C_k(t + y_k(t)) z_j(t) de_{x_i}. \qquad (111)$$

Equation (111) is similar in appearance to Eq. (77), which led to the extended-range correlator. The major difference here is that the weights of the extended-range correlator are not given by the statistics of the tracking error but the navigation error. Implementation of Eq. (111) would involve knowledge of the navigation estimation error statistics which, unfortunately, are part of the problem to be solved. Also, the transformation $y_k(x)$ is not one-to-one and so an implementation in terms of a derived pdf of is not directly computable.

Instead of pursuing this problem any further, we present the results for an important subset of this problem, namely, the multisatellite tracking problem. Recall that in Section V.A we were interested in solving the minimum mean-square tracking problem for one satellite PRN code. We know, however, at this point that four satellite PRN code measurements are needed to solve for the navigation position and the clock bias (time). We also know that each satellite line-of-sight measurement carries some information about each of the x, y, and z position components. In other words, there is a coupling (correlation) between the satellite channels which can be exploited to improve the tracking ability of other channels. This concept has sometimes been referred to as the "receiver interchannel aiding."

Weiss [51] has solved this problem of simultaneous tracking of four satellite signals and has obtained the expressions for the coupling terms between the channels. He followed the approach discussed in Section V.A.2 to obtain this result. Only the final results are presented here:

$$\frac{d\hat{\theta}_i}{dt} = \sum_{j=1}^{4} F_{ij}\hat{\theta}_j + \sum_{j=1}^{4} \frac{P_{\theta_i\theta_j}}{P_{\theta_j}} \int_{-\infty}^{\infty} \left(\frac{v \, \exp\left[-\left(v^2/2P_{\theta_i}\right)\right]}{(2\pi P_{\theta_j})^{1/2}} \right)$$

$$\times \, C_j[t - T(v + \theta_j)]z_j(t)\,dv, \qquad i = 1, 2, 3, 4. \qquad (112)$$

Note that the main result here suggests that the coupling between channels i and j is given by their correlation, namely, $P_{\theta_i\theta_j}/P_{\theta_j}$. This intuitive result is very appealing from an implementation point of view. Equations for the coupling coefficients are also given in [51].

D. *A GENERAL TARGET-TRACKING PROBLEM:*
    TIME-VARYING MULTIPLE-MODEL
    ESTIMATION APPROACH

Many important practical problems can be characterized by a multiple-model state space description. The signal-tracking problem of Section V.C.1, for example, considers two models for the measurement equation. A measurement of signal (a GPS PRN code) plus noise corresponded to a specific value of the model-specifying parameter $\gamma(t)$, namely, $\gamma = 1$. When $\gamma = 0$, a measurement of noise alone is considered. This problem is representative of a broader class of situations in which a set of possible models governing the time evolution of a state and/or observation process is hypothesized.

The equations for computing the a posteriori parameter probability density function in the system identification problem of Section II [26] can be applied to this class of problems only when the value of the model-specifying parameter $\gamma(t)$ is time invariant. In many problems of interest, however, it is quite natural for $\gamma(t)$ to vary with time. Weiss [50] has treated the discrete time version of the general time-varying multiple-model problem by supposing that the transitions be-tween models can be described by a Markov chain with known transition probabilities. This description has been considered by others [56,57] and, as mentioned previously, requires an exponentially growing processing resource for computation of the MMSE state estimate. In the likely event that such a re-source is not available, suboptimal schemes must be considered.

In this section we consider a class of suboptimal algorithms which mimic certain aspects of the optimal MMSE estimator. These algorithms are then applied to a general target-tracking problem.

1. Problem Definition

The general time-varying multiple-model problem is really one of state estimation for a hierarchal system model. We de-fine a status random process γ_k which takes on discrete values in a set S at each time instant k. The dynamics of this process evolve independently according to known one-step transition probabilities $p(\gamma_k/\gamma_{k-1})$. Furthermore, γ_k is a Markov process. That is,

$$\Pr\left(\gamma_k \mid \Gamma^{k-1}\right) = p(\gamma_k \mid \gamma_{k-1}),\tag{113}$$

where $\Gamma^k = \{\gamma_1, \gamma_2, \ldots, \gamma_k\}$. The status parameter is an indi-cator variable on a set of possible discrete time, linear, state space system models. For each value of γ_k, a complete description of the dynamics, noise, and observations of some random state vector process is known.

$$\underline{x}(k + 1) = \Phi(k; \gamma_{k+1}) \cdot \underline{x}(k) + \underline{w}(k + 1), \qquad (114)$$

$$\underline{y}(k) = H(k; \gamma_k) \cdot \underline{x}(k) + \underline{v}(k), \qquad (115)$$

where

$$\{w(k)\} \sim N(0, Q(\gamma_k)\delta_{kj}), \qquad \{v(k)\} \sim N(0, R(\gamma_k)\delta_{kj}),$$

$$Y^k = \{y(1), \ldots, y(k)\}.$$

The goal of the nonlinear filtering problem is to find the MMSE estimate of $x(k)$, given the observation sequence Y^k. To simplify description of the filtering algorithms, we consider only $S = \{0, 1\}$. The results, however, apply to any finite-dimensional parameter space S.

2. *Optimal Solution*

The optimal solution to the adaptive estimation (state estimation under parameter uncertainty) problem [Eqs. (113)-(115)] is obtained by applying what has been termed the "partitioning theorem" [31]. That is,

$$\underline{\hat{x}}(k|k) = \sum_{\Gamma^k} \Pr(\Gamma^k|Y^k)\underline{\hat{x}}(k|k, \Gamma^k), \qquad (116)$$

$$P(k|k) = \sum_{\Gamma^k} \Pr(\Gamma^k|Y^k)\{P(k|k, \Gamma^k)$$

$$+ [\hat{x}(k|k) - \hat{x}(k|k, \Gamma^k)][\cdot]^T\}. \qquad (117)$$

The optimal state estimate $\hat{x}(k|k)$ is just a weighted sum of conditional estimates $\hat{x}(k|k, \Gamma^k)$ with weights corresponding to the a posteriori probability of the status parameter sequence. Of course, other subsets of Γ^k could be used for conditioning. However, the entire sequence is chosen here because the solution then and only then admits a recursive formulation. From Eqs. (116) and (117), if we are given Γ^k, the MMSE estimate $\hat{x}(k|k, \Gamma^k)$ and its covariance, $P(k|k, \Gamma^k)$ are obtained from the

Kalman filter "matched" to the sequence Γ^k. The calculation of the probabilities associated with each Γ^k term is determined by Bayes' rule:

$$\Pr(\Gamma^k|Y^k) = \frac{p\left(y_k|\Gamma^k, Y^{k-1}\right)\Pr(\Gamma^k|Y^{k-1})}{p\left(y_k|Y^{k-1}\right)}, \tag{118}$$

The denominator of Eq. (118) is just a normalization constant so that the sum over all Γ^k of the probabilities $\Pr(\Gamma^k|Y^k)$ equals 1. Equation (118) is viewed as an update equation for the sequence probabilities. We also have

$$\Pr(\Gamma^k|Y^{k-1}) = p(\gamma_k|\gamma_{k-1})\Pr(\Gamma^{k-1}|Y^{k-1}). \tag{119}$$

This is due to the Markov assumption on the sequence of status events and the hierarchical system model. Equation (119) is viewed as a one-step prediction equation. The likelihood function in Eq. (118), after some thought, is given by

$$p\left(y_k|\Gamma^k, Y^{k-1}\right) = N_{y_k}\left(H(k; \gamma_k)\hat{x}(k|k - 1, \Gamma^k), H(k; \gamma_k)\right.$$

$$\left. \times P(k|k - 1, \Gamma^k)H^T(k; \gamma_k) + R(\gamma_k)\right) \tag{120}$$

where

$$N_x(m, \Lambda) \triangleq [2\pi\Lambda]^{-1/2} \exp\left[\frac{1}{2}(x - m)^T\Lambda^{-1}(x - m)\right].$$

3. Suboptimal Approximations

Because the overall estimation problem is nonlinear, a useful statistic for state estimation is the a posteriori probability density function (PDF) $p(\underline{x}(k)|Y^k)$. The suboptimal algorithms proposed here may all be thought of as methods of approximating this infinite-dimensional statistic with a finite number of parameters. Since each Γ^k conditioned model is linear, if the initial distribution of $x(0)$ is assumed to be Gaussian, then $p(\underline{x}(k)|Y^k)$ is a weighted sum of 2^k Gaussian densities.

That is,

$$p(\underline{x}(k) \,|\, Y^k) = \sum_{\Gamma^k} \Pr(\Gamma^k \,|\, Y^k) p(\underline{x}(k) \,|\, \Gamma^k, \, Y^k).\qquad(121)$$

Figure 10 provides a visualization of Eq. (21) for k = 3.   For
each "node" of the status sequence tree there is a separate
Gaussian density and corresponding a posteriori probability.
The extension of the tree to k = 4 can be separated into a predict cycle, which doubles the number of nodes in the tree, and
an update cycle.   The equations for both cycles are computed
recursively.

Figure 11 compares in block-diagram form the exact computation of the desired PDF (alternate predict and update operations) with a class of suboptimal, but finitely parameterized
approximate algorithms.   Up to some maximum window length N,
this class of suboptimal algorithms implements the equations
necessary to compute the actual PDF.   After this sequence of

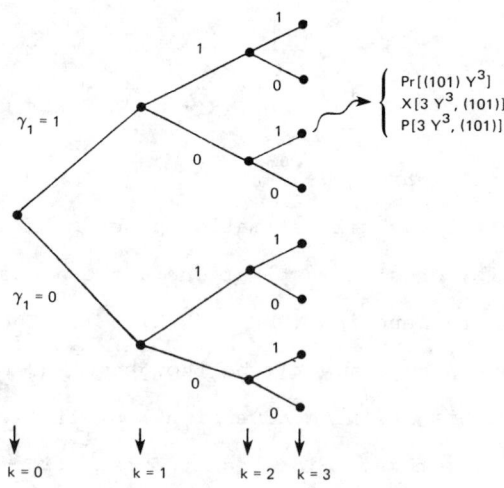

*Fig. 10.   Tree of possible status sequences for S = {0, 1}
and R = 3.*

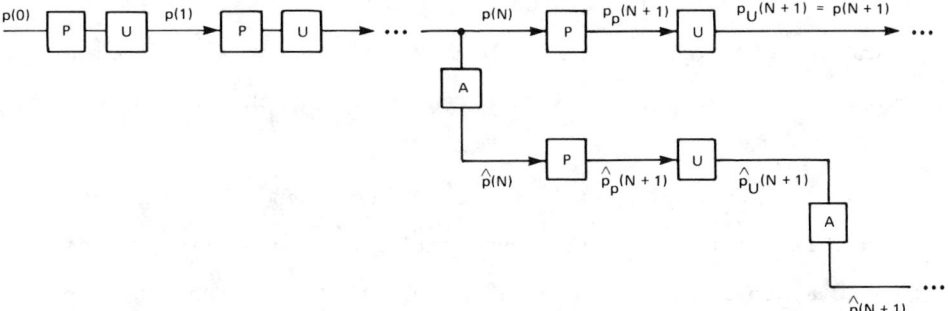

Fig. 11. Flow diagram for the exact and an approximate computation of the a posteriori probability density function: p(k), actual PDF at time k; p̂(k), approximate PDF at time k; P, predict operation; U, update operation.

predict and update cycles is performed, at k = N an approximation to the actual PDF is made. The purpose of the approximation is to reduce the number of parameters associated with the specification of the PDF so that a finite memory usage will be satisfied. Although many methods of such a parameter reduction could be conceived, we limit our consideration to algorithms which approximate the actual PDF by a fixed number of Gaussian terms. The predict and update cycles used for calculating the actual PDF are now applied to the approximate PDF. That is, it is assumed that each term in the approximation of Eq. (121) for k ≥ N corresponds to some node in the exact status sequence tree. The predict and update cycles operate on the parameters associated with these nodes instead of those associated with the exact sequence tree. In order to maintain a fixed number of parameters for all k, the approximation scheme is introduced after the predict and update cycles at every k ≥ N.

There are many types of approximation schemes which could be effective in approximating the desired PDF. In fact, the detection-directed estimation solution mentioned in Section V.C.1 can be viewed as an algorithm of the class with "window"

length of $N = 1$. In this scheme, a single Gaussian density
corresponding to a decision about γ_k is used as an approximation to the exact PDF. A more general algorithm is now considered.

Fixed merging approximation. It has been suggested [56]
that, in order to keep the computational burden fixed, a pos-
sible approximation to $p\left(\underline{x}_k \mid Y^k\right)$ would be to eliminate or ignore
the 2^{N-1} least likely nodes at time $k \geq N$. We could then per-
form a predict and update cycle on each of the remaining nodes
and drop the least likely nodes again. It is possible that
much of the information about $\underline{x}(k)$ could be lost when none of
the probabilities associated with the discarded nodes are small.
Instead of discarding possibly meaningful information, we pro-
pose the following:

Take two nodes at time N which have the same sequence
$\Gamma_2^N = \{\gamma_2, \ldots, \gamma_N\}$ *and replace the weighted sum of the two*
Gaussians these nodes represent in the distribution of x_k with
one Gaussian that has the same mean and covariance. If this
is done for each node, we will reduce the number of nodes by
$1/2$. The process describes the merging of predetermined components of the a posteriori PDF, and equations which define it
can be found in [50].

The approximation really represents a local Gaussian approximation in contrast to the global Gaussian approximation, described in Section V.A. Instead of approximating the PDF of
the entire state by a single Gaussian density, it is assumed
that a weighted sum of a fixed number of Gaussain terms is
sufficient. When the maximum processing time $N = 1$, the local
Gaussian approximation defined by the fixed merging algorithm

is equivalent to a global approximation. Note that "decision-directed estimation" can also be considered as a global Gaussian approximation, although it is very different from "fixed merging."

This algorithm has proved to be very successful in a number of applications. We now look at the application of this algorithm to a general target-tracking problem.

4. A General Target-Tracking Problem

The problem of estimating the location of a moving target based on a variety of measurements has long been one of interest. In Kalman filtering approaches, the dynamics equations must be specified when, in general, they are not precisely known. One [58] dynamics model has exponentially correlated accelerations (in a rectangular coordinate frame) as a driving term to the position coordinates. This model works well under certain conditions but has some drawbacks. When the target moves at a constant velocity, the "filter bandwidth" is larger than is necessary and results in a less precise estimate than the "correct" zero acceleration model. Although a zero-acceleration model works best for this kind of target motion, rms position estimates are severely degraded when other types of target motion occur.

Many target-motion scenarios are characterized by a series of distinct maneuvers. For example, a constant velocity trajectory might be followed by a constant angular velocity turn, which might then be followed by a fixed-acceleration trajectory, and so on. The Kalman filtering approach just described [58] is an attempt to make an MMSE estimation algorithm based on a single state space model more robust to modeling errors.

Clearly, an alternative approach is to hypothesize a set of
state space models which describe the dynamics of each maneuver
and a set of transition probabilities between maneuvers that
describe a specific scenario. This is the general time-varying
multiple-model formulation.

The suboptimal methods of Section V.D.3 were examined in a
two-dimensional target-tracking simulation with three target
motion models [50]. The set of target motions consisted of

1. a constant velocity motion;

2. a constant angular velocity turn in the clockwise
   direction; and

3. a constant angular velocity turn in the counterclock-
   wise direction.

Direct target-position measurements were assumed to be available. Note that all three of the target motions as well as
the observation equation are *linear*. The only nonlinearity in
this problem is that induced by the time-varying multiple-model
description.

The simulation results described in [50] indicate that the
fixed merging approximation works quite well in terms of rms
position error, even for fairly small "window" lengths. There
is, however, a distinct performance sensitivity to the assumed
transition probabilities. This suggests that for good performance of these algorithms (at least for small windows),
either the motion scenario (transition probabilities) should
be known or some means of adaptively estimating these probabilities should be implemented. Further work in this area should
address the questions of robustness and practical algorithm
design and implementation.

VI. VHSIC-VLSI IMPLEMENTATION

Application of the generalized filtering techniques to radio navigation (e.g., GPS) and communications problem discussed here suggests a processing architecture which consists of a generalized matched filter (correlator), followed by a signal (phase, frequency, amplitude, data, etc.) estimator. The latter could be viewed as a suboptimal nonlinear filter, e.g., a truncated second-order filter. The signal correlator can be implemented either digitally or by use of analog-device components, namely, CCDs or SAWs. The tradeoffs between the analog and digital implementation typically involve (1) the programmability of the device; (2) its dynamic range; (3) clock rate; and (4) size, power, weight, and cost. If a specific application can be worked at a lower clock rate (say, <20 MHz), then digital implementation seems to be the preferred choice based on programmability (flexibility) of the processor and its high-fidelity storage capability. Otherwise, CCDs, in particular GaAs CCDs, and SAWs seem to have the edge for implementing a generalized correlator (or convolver). The extended-range correlator developed in Section V.A for GPS code acquisition and tracking has been implemented in a CCD at the Massachusetts Institute of Technology Laboratory by using a pipe-organ architecture [52]. The clock rate for the device was demonstrated at 20 MHz. The key design features of this CCD correlator are an on-chip binary-analog charge multiplier; an on-chip charge integrator; and a 5-μm design.

In this section, we focus on the implementation of the nonlinear estimator, which has been discussed at great length in this chapter. This is a base-band function, and, in general,

its implementation is done digitally by using microprocessors
on a chip. Recent progress in microprocessor architecture,
namely, array processors and pipelined processors, together
with advances in integrated circuit (IC) design, has made it
feasible to consider real-time implementation of the nonlinear
filters presented in this chapter. There are two issues here.
The first one addresses the combined research work in the areas
of semiconductor materials, lithography, and IC design. This
brand of research has been popularly referred to as very large
scale integration (VLSI) technology, which implies certain
performance parameters, namely, on-chip integration (≥ 1000
gates), feature size (2-3 μm), and I/O pin counts. The counter-
part of VLSI for military applications is known as very high-
speed integrated circuits (VHSIC) technology. The published
[53] goal of VHSIC technology is the achievement of chip de-
signs with 0.5-μm feature sizes that can operate reliably in
military operational environments. VHSIC-VLSI technology permits development of large-memory, fast processors that are size,
power, and cost efficient and that are needed to implement
signal estimation-detection algorithms, developed in this chapter, for integrated navigation and communication systems.

Brute-force application of VHSIC processors to solve nonlinear partial differential equations arising in nonlinear
estimation have been proposed [40]. In this approach, the
Zakai partial differential equations for the nonlinear filtering
problem are discretized in space to obtain the system of ordinary differential equations. The resulting set of ordinary
differential equations are implemented by using VLSI processor
arrays in a pipeline arrangement to compute the requisite
matrix-vector multiplications. A similar approach can also be

used to implement the optimal Bayesian detector suggested by Cartelli [3] by discretizing in space the state a posteriori density function.

Even though the VHSIC processor speed (million operations per second [MOPS]) is helpful in exploring real-time implementation of nonlinear filters, a clever exploitation of the VHSIC processor architecture is needed to take full advantage of the VHSIC technology. Recent work [55] dealing with developing nonlinear filter algorithms which can be executed in parallel by using array-processor architecture seem to offer promise. Even though the state of the art [54] is currently focused on the fundamental Kalman filter equations, any understanding and experience gained in this area should be generally applicable to the nonlinear filtering equations presented in this chapter. The latter is beyond the scope of the work discussed here and should therefore be left for further maturation.

VII. SUMMARY

In this chapter, many of the results in nonlinear estimation theory have been discussed with applications to problems in communication-navigation systems. In particular, it has been shown that the detection-estimation-identification problems in communication-navigation systems can be formulated as a general problem in nonlinear estimation.

The optimal nonlinear estimation equations for the time evolution of the a posteriori PDF of the state vector were presented. The well-known Kushner and the Fokker-Planck equations were obtained as a special case of the general non-linear estimation equations. Using Kushner's equation for the a posteriori PDF, Jazwinski's equations for the conditional

mean and variance of the state vector were presented. It was
pointed out that even though Jazwinski's equations are in a
closed form, evaluation of the conditional expectations for
nonlinear terms on the right-hand side are difficult for an
arbitrary nonlinear function. Several suboptimal approximations
to these equations were discussed to ease their implementation.
One specific approximation, namely, the Gaussian density approximation, was detailed, and its application to the GPS signal detection, tracking, and navigation problem and a general
target-tracking problem were presented. These investigations
resulted in new results, namely, generalized extended-range
correlation and adaptive receiver control algorithms for improving the GPS antijam performance, and shed some new light on
the combined detection-tracking and navigation problem. The
role of an INS for improving GPS receiver performance was also
discussed. Extension of the signal-detection problem, i.e., a
fixed unknown parameter, to include time-varying unknown signal
parameters (modeled by a Markov chain) was discussed, and a
suboptimal implementable filter equation was presented. These
results are also applicable to problems in communication (e.g.,
intersymbol interference) as well as to target detection and
estimation. Digital implementation of the nonlinear estimators,
presented in this chapter, was discussed in view of the emerging microprocessor architectures (e.g., pipelined array processor). It is suggested that instead of attempts to brute-force
the application of these processors, future work should focus
on rederiving the estimators for specific processor architectures.

ACKNOWLEDGMENTS

This work was supported by the Charles Stark Draper Laboratory IR&D Projects No. 169 and 153.

The authors are grateful to Dr. Duncan B. Cox, Jr. for providing the opportunity and incentive to write this chapter.

REFERENCES

1. R. H. LAMB, "Adaptive Code Tracking," M.S. Thesis, Massachusetts Institute of Technology, Cambridge (Draper Lab. Rep. No. T-728, January 1981).

2. W. M. BOWLES, "Correlation Tracking," D.Sc. Dissertation, Massachusetts Institute of Technology, Cambridge (Draper Lab. Rep. No. T-706, June 1980).

3. J. A. CARTELLI, "Minimum Variance Tracking of Pseudo-Random Number Codes," M.S. Thesis, Massachusetts Institute of Technology, Cambridge (Draper Lab. Rep. No. T-749, June 1981).

4. T. N. UPADHYAY, R. S. ROTH, W. M. BOWLES, and C. W. HLAVATY, "Recent Results in GPS Receiver Code Tracking Technology," National Aerospace and Electronics Conference, Dayton, OH, 1980.

5. W. M. BOWLES, and T. N. UPADHYAY, "An Adaptive Data De-modulating Phase Lock Loop," National Aerospace and Electronics Conference, Dayton, OH, 1980.

6. D. B. COX, "Integration of GPS with Inertial Navigation Systems," AGARD No. 245, July 1979.

7. H. W. SORENSON, in "Advances in Control Systems," ed. C. T. Leondes, Vol. 3, p. 219, Academic Press, New York, 1966.

8. A. H. JAZWINSKI, "Stochastic Process and Filtering Theory," Academic Press, New York, 1970.

9. J. R. FISHER, in "Advances in Control Systems," ed. C. T. Leondes, Vol. 5, pp. 197-300, Academic Press, New York, 1968.

10. G. T. SCHMIDT, in "Control and Dynamic Systems," ed. C. T. Leondes, Vol. 12, pp. 63-98, Academic Press, New York, 1976.

11. B. A. KRIEGSMAN, D. B. COX, and W. M. STONESTREET, "An IMU Aided GPS Receiver," Draper Lab. Rep. No. P-490, Massachusetts Institute of Technology, Cambridge, 1977.

12. T. N. UPADHYAY, P. VAN BROEKHOVEN, J. R. DOWDLE, R. GENDRON, P. NICOLAIDES, T. SAVARINO, and B. KRIEGSMAN, "Advanced GPS/Inertial Integration Technology Program," Final Report AFWAL-TR-80-1148, Vol. 2, Draper Lab., Massachusets Institute of Technology, Cambridge, 1981.

13. R. W. CARROLL, and W. A. MICKELSON, "Velocity Aiding of Non-Coherent Receiver," National Aerospace and Electronics Conference, pp. 311-318, Dayton, OH, 1977.

14. E. H. MARTIN, "Aiding GPS Navigation Functions," National Aerospace and Electronics Conference, pp. 849-856, Dayton, OH, 1976.

15. R. L. BOGUSH, F. W. GUKILIANO, D. L. KNEPP, and A. H. MICHELET, *Proc. IEEE 69*, 787-795 (1981).

16. H. S. EILTS, *IEEE Trans. Aerosp. Electron. Syst. AES-16*, 800-810 (1980).

17. E. M. COPPS, G. J. GEIER, W. C. FIDLER, and P. A. GRUNDY, *J. Inst. Navigation 27*, (1980).

18. D. G. BOTHA, and F. W. SMEAD, *Proc. National Aerospace Electronics Conference 2*, 735-740 (1981)

19. C. R. WARD, and R. A. Reilly, *Proc. National Aerospace Electronics Conference 2*, 712-722 (1981).

20. P. C. CAMANA, S. K. OGI, and L. R. STINE, "Advanced Integrated CNI Architectures," *Proc. National Aerospace Electronics Conference 2*, 723-728 (1981).

21. J. C. DEYST, J. V. HARRISON, E. GAI, and K. C. DALY, "Fault Detection Identification and Reconfiguration for Spacecraft Systems," *J. Astronautical Sci. 29*, 113-126 (1981).

22. A. L. HOPKINS, T. B. SMITH, and J. H. LALA, *Proc. IEEE 66* (1978).

23. "An Integrated Fault-Tolerant Avionics System Concept for Advanced Aircraft," Draper Lab. Rep. No. R-1226, February 1979.

24. S. K. PARK, and D. G. LAINIOTIS, "An Unified Approach to Detection, Estimation, and System Identification," Technical Rep. AFOSR-TR-72-2013, Information Systems Res. Lab., Univ. of Texas, Austin, August 1972.

25. D. G. LAINIOTIS, *IEEE Trans. Inf. Theory IT-15* (1969).

26. D. G. LAINIOTIS, *Inf. Control 19*, 75-92 (1971).

27. D. MIDDLETON, and R. ESPOSITO, *IEEE Trans. Inf. Theory IT-14*, 434-444 (1968).

28. R. PRICE, *IRE Trans. Inf. Theory IT-2*, 125-135 (1956).

29. T. KAILATH, "Adaptive Matched Filters" in "Mathematical Optimization Techniques," ed. R. Bellman, pp. 109-140, Univ. of California Press, Berkeley, 1963.

30. T. KAILATH, *IEEE Trans. Inf. Theory IT-15*, 350-361 (1969).

31. D. G. LAINIOTIS, *IEEE Trans. Autom. Control AC-16*, 160-170 (1971).

32. R. E. KALMAN, and R. S. BUCY, *J. Basic Eng. 83*, 95-108 (1961).

33. H. J. KUSHNER, *J. Math. Anal. Appl. 8*, 332-344 (1964).

34. H. J. KUSHNER, "On the Differential Equations Satisfied by Conditional Probability Densities of Markov Processes," *J. SIAM Control Ser. A2*, Vol. 2, pp. 106-119 (1964).

35. R. S. BUCY, *IEEE Trans. Autom. Control AC-10*, p. 198 (1965).

36. T. KAILATH, *IEEE Trans. Autom. Control AC-13*, 646-654 (1968).

37. P. A. FROST, and T. KAILATH, *IEEE Trans. Autom. Control AC-16*, 217-226 (1971).

38. D. L. OSCONE, "Topics in Nonlinear Filtering," Ph.D. Dissertation, Massachusetts Institute of Technology, Cambridge, 1980 (LIDS-TH-1058).

39. D. L. OSCONE, J. S. BARAS, and S. I. MARKUS, "Filtering and Smoothing Equations for the Filtering Problem of Benes," *IEEE Conf. Decision Control, San Diego*, pp. 83-88 (1981).

40. J. S. BARAS, "Approximate Solution of Nonlinear Filtering Problems by Direct Implementation of the Zakai Equation," *IEEE Conf. Decision Control, San Diego*, pp. 309-310 (1981).

41. R. W. BROCKETT, "Asymptotically Optimal Solution," *IEEE Conf. Decision Control, San Diego*, pp. 76-79 (1981).

42. A. H. JAZWINSKI, *IEEE Trans. Autom. Control AC-11*, 765-766 (1966).

43. L. SCHWARTZ, and E. B. STEAR, *J. Math. Anal. Appl. 21*, 83-86 (1968).

44. H. J. KUSHNER, *IEEE Trans. Autom. Control AC-12*, 546-556 (1967).

45. J. L. DOOB, "Stochastic Processes," Wiley, New York, 1953.

46. D. L. ALSPACK, and H. W. SORENSON, *IEEE Trans. Autom. Control AC-17*, 437-447 (1972).

47.   B. WIDROW, J. H. GLOVER, J. M. McCOOL, J. KANMITZ, C. S.
      WILLIAMS, R. H. HEARN, J. R. ZEIDER, E. DOUG, and R. C.
      GOODLIN, *Proc. IEEE 63* (1975).

48.   B. WIDROW, J. M. McCOOL, M. G. LARIMORE, and C. R. JOHNSON,
      IEEE 64 (1976).

49.   J. WEISS, T. N. UPADHYAY, and R. R. TENNEY, *National
      Aerospace Electronics Conference Proc.*, Dayton, Ohio, May
      1983.

50.   J. L. WEISS, "Finite Methods for Estimation of Status
      Directed Methods," M.S. Thesis, Massachusetts Institute of
      Technology, Cambridge, May 1983.

51.   J. L. WEISS, "Integrated PRN Code Tracking for GPS,"
      Draper Lab. Internal Memo No. 15L-83-070, March 1983.

52.   B. E. BURKE, D. L. SMYTHE, D. J. SILVERSMITH, W. H.
      McGONAGLE, R. W. MOUNTAIN, and B. J. FELTON, "A 10-MHz CCD
      Time Integrating Correlator," *IEEE Int. Solid State Conf.*,
      1983.

53.   B. GROVES, "Getting VHSIC Into Real World Systems," *Def.
      Electron.*, January 1983.

54.   R. H. TRAVASSOS, and A. ANDREWS, "VLSI Implementation of
      Parallel Kalman Filters," *AIAA Proc.*, pp. 643-649 (1982).

55.   R. S. BUCY, and K. D. SENNE, *Computer Math. Appl. 6*, 317-
      338 (1980).

56.   J. TUGUAIT, *Automatica 18* (1982).

57.   A. G. JAFFER, and S. C. GUPTA, *IEEE Trans. Information
      Theory IT-17* (1971).

58.   R. A. SINGER, *IEEE Trans. Aerospace and Electron. Systems
      AES-6* (1970).

59.   D. MIDDLETON, "Introduction to Statistical Communication
      Theory," McGraw-Hill, New York, 1960.

INDEX